Pulsed Electromagnetic Fields for Clinical Applications

Pulsed Electromagnetic Fields for Clinical Applications

Edited by
Marko S. Markov, James T. Ryaby, and
Erik I. Waldorff

CRC Press
Taylor & Francis Group
Boca Raton London New York

CRC Press is an imprint of the
Taylor & Francis Group, an **informa** business

CRC Press
Taylor & Francis Group
6000 Broken Sound Parkway NW, Suite 300
Boca Raton, FL 33487-2742

First issued in paperback 2023

ISBN 13: 978-0-367-17971-7 (hbk)
ISBN 13: 978-1-03-265382-2 (pbk)
ISBN 13: 978-1-00-300195-9 (ebk)

DOI: 10.1201/9781003001959

Contents

Editors

Marko S. Markov received his BS, MS, and PhD from Sofia University, Bulgaria. He has been professor and chairman of the Department of Biophysics and Radiobiology at Sofia University for 22 years. He had been invited professor and lecturer in a number of European, Japanese, and American academic and industry entities. For years, his research has been on biological and clinical studies of the effects of various electromagnetic fields. He is a founder and member of the editorial board of the journal *Electromagnetic Biology and Medicine*. He is the author and/or co-author of more than 200 peer-reviewed papers. He is the author, editor, and/or co-editor of 26 books published by Plenum Press, Springer, Marcel Dekker, and CRC Press. His book *Electromagnetic Fields in Biology and Medicine* published by CRC Press is being translated into Mandarin in China.

James T. Ryaby, PhD, joined Orthofix as chief scientific officer in July 2013 after serving as a consultant. Before joining the company, he served as vice president of Research and Clinical Affairs for musculoskeletal applications at Mesoblast. He has also held the position of senior vice president of Research and Clinical Affairs and chief scientific officer at OrthoLogic Corp in Tempe, AZ, where he led the development of the CMF technology for spine fusion and nonunions and the TP508 peptide therapeutic development program in orthopedic and diabetic ulcer applications. Also, he has served as adjunct professor of Bioengineering at Arizona State University. He received his PhD degree in cellular and molecular biology from the Mt. Sinai School of Medicine, New York.

Erik I. Waldorff, PhD, is the principal scientist and research manager for the orthopedic medical device company, Orthofix, which he joined in 2012. Prior to this, he served as a clinical research scientist for the Neuromodulation Division at St. Jude Medical (now Abbott) where he performed preclinical studies in addition to postmarket clinical trials for spinal cord stimulation and deep brain stimulation patients. In addition, following his doctoral studies, he held a position as postdoctoral fellow at the Michigan Institute for Clinical and Health Research (MICHR) at the University of Michigan where he led orthopedic human and animal studies. He received his PhD and MSE degrees in Biomedical Engineering and MSE and BSE degrees in Aerospace Engineering from the University of Michigan (Ann Arbor, Michigan).

List of Contributors

Stephen J. Beebe
Frank Reidy Research Center for
    Bioelectrics
Old Dominion University
Norfolk, Virginia
sbeebe@odu.edu

Igor Belyaev
Department of Radiobiology
Cancer Research Institute, Biomedical
    Research Center, Slovak Academy of
    Sciences
Bratislava, Slovakia
igor.beliaev@savba.sk

Ryan Burke
Old Dominion University
Norfolk, Virginia

Matteo Cadossi
IGEA–Clinical Biophysics
Carpi, Italy

Ruggero Cadossi
IGEA–Clinical Biophysics
Carpi, Italy
r.cadossi@ikeamedical.com

Andrew F. Kuntz
McKay Orthopaedic Research
    Laboratory
University of Pennsylvania
Philadelphia, Pennsylvania

Leonardo Makinistian
Department of Physics and Instituto de
    Física Aplicada (INFAP)
Universidad Nacional de San
    Luis-CONICET
San Luis, Argentina

Marko S. Markov
Research International
Williamsville, New York
msmarkov@aol.com

Leo Massari
Department of Orthopaedic and
    Traumatology
Cona Sant'Anna Hospital, University
    of Ferrara
Ferrara, Italy

Mats-Olof Mattsson
SciProof International AB
Östersund, Sweden
and
Strömstad Academy
Strömstad, Sweden

Harvey Mayrovitz
Department of Medical Education
Nova Southeastern University
Dr. Kiran C. Patel College of Allopathic
    Medicine
Davie, Florida
mayrovit@nova.edu

Richard Nuccitelli
Chief Science Officer
Pulse Biosciences
Hayward, California
rich@bioelectromed.com

William Pawluk
Holistic and Preventive Medicine
Timonium, Maryland
bpaw@comcast.net

Arthur A. Pilla (Deceased)

James T. Ryaby
Orthofix Medical Inc.
Lewisville, Texas
jamesryaby@orthofix.com

Simona Salati
IGEA – Clinical Biophysiscs
Carpi, Italy

Stefania Setti
IGEA–Clinical Biophysics
Carpi, Italy

Myrtill Simkó
SciProof International AB
Östersund, Sweden
and
Strömstad Academy
Strömstad, Sweden
myrtill.simko@sciproof-international.se

P. Thomas Vernier
Old Dominion University
Norfolk, Virginia
pvernier@odu.edu

Erik I. Waldorff
Orthofix Medical Inc.
Lewisville, Texas
ErikWaldorff@Orthofix.com

Nianli Zhang
Orthofix Medical Inc.
Lewisville, Texas

1 Electric and Electromagnetic Stimulation
Electroporation

Marko S. Markov

CONTENTS

THE BEGINNING

Electromagnetic field (EMF) therapy has a long history and is still not widely popular among medical practitioners. James T. Ryaby, Erik Waldorf, and I had decided to integrate efforts of clinicians, engineers, and scientists in presenting the state of the art in international settings. This chapter will be a brief overview of the history and state of the art of EMF therapy and will include nearly the entire spectrum of frequency ranges, starting with static magnetic field and reaching terahertz signals.

I would like to start with a brief history. Very little known is the fact that contemporary magnetotherapy begun immediately after World War II by introducing both magnetic and EMFs, generated by various waveshapes of the supplied currents. It started in Japan and quickly moved to Europe, first in Romania and the former Soviet Union. During the period 1960–1985, nearly all European countries designed and manufactured their own magnetotherapeutic systems. Indeed, the first contemporary book on magnetotherapy was published in Bulgaria by Todorov (1982).

In the USA, the first approved signal was for treatment of nonunion fractures by employing a very specific biphasic low-frequency signal (Bassett et al., 1974, 1977). A decade later, the Food and drug administration (FDA) in the USA allowed the use of pulsed radio frequency (PRF) EMF for treatment of pain and edema in superficial soft tissues.

1

The worldwide physics, engineering, biology, and clinical applications of various modifications of EMFs for treatment of medical problems ranging from bone unification, to wound healing, to pain control has been the subject of several books (Polk and Postow, 1986; Rosch and Markov, 2004; Barnes and Greenebaum, 2007; Lin, 2011; Markov, 2015; Barnes and Greenebaum, 2019). In these and many papers in peer-reviewed journals was published a detailed explanation of the basic science and clinical evidence that time-varying magnetic fields can modulate molecular, cellular, and tissue function in a physiologically and clinically significant manner.

Therapy using EMF developed in various stages and involved any new discovery in the basic science, especially biophysics and bioelectrochemistry. I had the privilege to be involved in some stages in this development. In the early 1980s, I was in communication with members of the Bioelectrochemical Society, attended the meetings of this society in Europe, and organized two international conferences titled "Electromagnetic Fields and Biomembranes" in my "little" Bulgaria in the second half of the 1980s. In this society, electroporation was developed as a plausible method in biophysics and later suggested as a possible therapeutic method.

The fundamental questions for engineers, scientists, and clinicians is to identify the biochemical and biophysical conditions under which EMF signals could be recognized by cells in order to modulate cell and tissue functioning. It is also important for the scientific and medical communities to recognize that different EMFs applied to different tissues could cause different effects. What "different EMFs" means is at the same time difficult and easy to define. It is easy because a signal is new if it has at least one parameter different from the previous one. The difficulties arise because very few manufacturers, distributors, or authors of a published study exactly describe the parameters of the device.

NECESSARY STEPS

Especially since the beginning of this century and the development of the Internet, a large number of devices could be found in the market. In most cases, these devices are distributed without a proper engineering and physics description. Let me point out that the ideal situation in clinical use of this technology requires the following steps:

- medical diagnostics of the specific problem and desired EMF
- engineering of the appropriate devices
- physics/biophysics of dosimetry and selection of the device
- medical therapeutics.

The first part of the equation, medical diagnostics, should identify the exact target and the "dose" of EMF that the target needs to receive. Then, physicists and engineers should design the exposure system in such a way that the target tissue receives the required magnetic flux density. Any new device **must** be properly characterized by physical dosimetry. Once the appropriate signal is engineered, serious attention must be paid to the biophysical dosimetry capable of predicting the EMF signals that could be bioeffective and in monitoring device efficiency. This raises the question of using theoretical models and biophysical dosimetry in selection of the appropriate signals and in engineering and clinical application of new EMF therapeutic devices.

Unfortunately, in most cases, the chain starts with the question "what signal is available?" Then the clinician applies common sense and his/her experience and intuition in selecting the protocol of treatment. Dosimetry and analysis of the signal often remains out of consideration. For this reason, many laboratory and clinical studies cannot be replicated.

PHYSICAL DOSIMETRY AND BIOPHYSICAL DOSIMETRY

We introduce here two dosimetry terms, "physical dosimetry" and "biophysical dosimetry", because it appears that the issue of dosimetry is not properly interpreted in both the scientific and clinical communities. **Physical dosimetry** relates to characterization of the device with respect to engineering and physical parameters, including signal parameters such as frequency, signal shape, amplitude, repetition rate, rising time, etc. It should be pointed out that these are characteristics of the device and not the actual field distribution inside the target tissue.

Biophysical dosimetry characterizes the "dose" received by the target tissue. I suggest that the only valuable information for a clinician is that which explains the field values at the target site. This goal is easy to achieve because the magnetic properties of air and biological tissue, especially for low-frequency fields, are almost the same.

Another important issue is the fact that any physical body attenuates the electric field component and allows only the magnetic component to penetrate deep into the body. Therefore, biophysical dosimetry is mainly dosimetry of the magnetic fields. It should also be taken into account that any time-varying magnetic field induces an electric field, which in some cases might be of importance in analyzation and interpretation of the observed effects. As it was pointed out earlier (Markov, 2002, 2009), the induced electric field generates "back" magnetic fields, and this fact should be taken into account in any measurements, especially in measuring high-frequency signals.

The pulsed electromagnetic field (PEMF) signals in clinical use have a variety of designs, which in most cases are selected without any motivation for the choice of the particular waveform, field amplitude, or other physical parameters.

It is reasonable that the first and widely used waveshape is the sine wave with a frequency of 60 Hz in North America and 50 Hz in the rest of the world. From symmetrical sine waves, engineers moved to an asymmetrical waveform by means of rectification. These types of signals basically flip-flop the negative part of the sine wave into positive, thereby creating a pulsating sine wave. Textbooks usually show the rectified signal as a set of ideal semi-sine waves. However, due to the impedance of any particular design, such an ideal waveshape is impossible to achieve. As a result, the ideal form is distorted, and a short DC-type component appears between two consecutive semi-sine waves.

In addition to sine-wave-type signals, a set of therapeutic devices which utilize unipolar or bipolar rectangular signals are available in the market. Probably for those signals, the most important is to know that, due to the electrical characteristics (mostly the impedance) of the unit, these signals could never be rectangular. There should be a short delay both in raising the signal up and in its decay to zero; thereby the rectangular signal is basically trapezoid. The rise time of such a signal could be of extreme importance because the large value of dB/dt could induce significant electric current into the target tissue. Some authors consider that neither frequency, nor the amplitude is so important for the biological response but the dB/dt rate is the factor responsible for observed beneficial effects. Kotnik and Miklavcic (2006) suggested that the rectangular signals should be replaced by more realistic trapezoid signals. It appears easy to say, but it is nearly impossible technologically to generate trapezoid signals.

As the signal choice in any particular device, in most cases, is not a function of medical and biophysical criteria, especially with free Internet offering, one may find large variation in signals. Unfortunately, manufacturers and distributors have one only goal – to make money, in most cases at free press variety of claims could be found.

Which are the signals that could be the most effective and at what conditions? Are certain signal parameters better than others? It should be pointed out that many EMF signals used in research and as therapeutic modalities have been chosen in some arbitrary manner. Very few studies assessed the biological and clinical effectiveness of different signals by comparing the physical/biophysical dosimetry and biological/clinical outcomes.

It was nearly half a century since the concept of "biological windows" was introduced. Interestingly enough, three groups located in different countries and unknown to one another published almost similar papers in which it was stated that during evolution, Mother Nature created preferable levels of recognition of the signals from exogenous magnetic fields. The "biological windows" could be identified by amplitudes and frequencies and their combinations (Adey, 1993; Markov, 1994). The research in this direction requires assessment of the response in a range of amplitudes and frequencies. It has been shown that at least three amplitude windows exist: at 5–10 Gauss, 150–200 Gauss, and 450–500 Gauss (Markov, 2005). Using cell-free myosin phosphorylation to study a variety of signals, Markov's group has shown that the biological response depends strongly on the parameters of applied signal, confirming the validity of the last two "windows".

The first clinical signal approved by FDA for treatment of nonunions or delayed fractures (Bassett et al., 1974, 1977) exploited the pulse-burst approach. Having a repetition rate of 15 burst/s, the clinical use of this asymmetrical signal (with a long positive and very short negative component) has been very successful in healing nonunion bones for more than 30 years. The engineers assume that the cell/tissue would ignore the short negative part of the pulse and would respond only to the envelope of the burst which has a duration of 5 μs, enough to induce sufficient amplitude in the kHz frequency range. A series of modalities generate signals that consist of single narrow pulses separated by long "signal-off" intervals. This approach allows modification of not only the amplitude of the signal but also the duty cycle (time on/time off).

Let me point to another problem with time-varying EMF. We performed the Fourier analysis for a rectified semi-sine wave signal implemented in the therapeutic system THERAMAG™· The spectrum shows that, at basic frequency of 120 pulses per second (pps), the first harmonic (240.3 pps) has an amplitude equal to 20% of the amplitude of the basic frequency. The third harmonic (360.4 pps) has an amplitude equal to 5% of the basic amplitude. In addition, the first 20 harmonics, up to 2,643 pps, were identified by the Fourier analysis (Figure 1.1).

The importance of the impedance of the waveshape and harmonics in physics and biology remains to be determined. As for many other signals, the biophysical analysis of the rectified signal implemented in this type of therapeutic unit is practically absent. More research is needed to further clarify the importance of impedance of the system in generating and delivering appropriate signals (and their frequency components for treatment of the desired target). What does all this mean? It means that, at least for this specific rectified signal, in addition to the basic frequency, the target receives a set of signals with higher frequencies and amplitudes which might be of importance for certain processes. It is possible that different generating systems, having different impedances, would generate signals with different spectrum compositions (in respect of harmonics' amplitudes). The impedance of the system could change not only the harmonics' spectrum, but also the general view of the signal.

In another chapter in this book, we discuss the parameters of EMF that need to be described and evaluated in any scientific or clinical study. There are two groups of parameters – one related to physical characteristics of signal (described by physical dosimetry) and another more related to the target (described by the biophysical dosimetry).

Now comes the specific absorption rate (SAR).

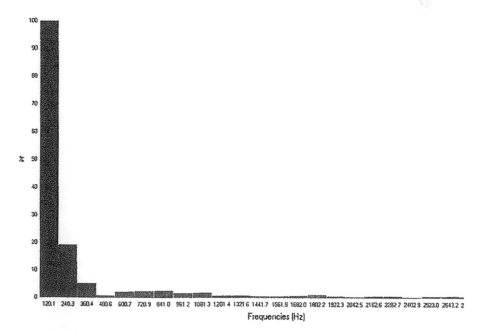

FIGURE 1.1   Spectrum of harmonics at the rectified 60 Hz signal.

SPECIFIC ABSORPTION RATE

Let us start with the definition. SAR basically means the specific absorption of a given signal into a given substance or material. In the sense of technicality, the term was introduced when EMF energy was used for dielectric heating of a homogeneous object.

The aim of SAR is to estimate the heating of a physical body exposed to EMF energy. It assumes that penetration of this energy into a body with homogeneous electrical properties is not dependent on geometrical size of the body.

In reality, the penetration of EMF energy flow is a function of the body size and its electric structure, frequency and polarization of the field, and reflection and refraction of the energy, not to mention energy dispersion due to radiation and thermoregulation *in vivo*.

Thus, SAR estimation and its spatial distribution within an exposed body by performing external measurement of the electric field may be acceptable only in a limited number of cases. Nevertheless, SAR remains the only available physical method to estimate the extent of absorption of the EMF energy within biological tissues.

However, I strongly refuse to accept SAR as a criteria for delivered energy. Unfortunately, until now, many scientific and clinical studies have reported that the target tissue **receives** EMF at the SAR. Every scientist and every clinician **must remember that the SAR relates to the target, not to the source of EMF. In other words, the term is related to the specific absorption of the EMF by a specific tissue.**

THERMAL VERSUS NONTHERMAL BIOLOGICAL EFFECTS

Obviously, SAR is associated with temperature effects and measurements. Despite so many years of existence of ICNIRP (International Commission on Non-Ionizing Radiation Protection), there is no definition for the threshold level that determines which effect is thermal.

In the 1990s, I was involved in estimating the extent to which the therapeutic device SofPulse could elevate the temperature at the target. After checking the literature, we created a 2 L saline phantom and found that the prescribed 30 min session with this signal did not elevate temperature more than 0.8°C. Our report stated that, during the therapeutic session, the temperature did not increase more than 1°C, and a plausible clinical effect was observed (Markov, 2001). Since that time, I have not seen any "standard" for this level. **Should we consider 1°C as a criteria for thermal effect?**

There is a principal difference between the biophysical and engineering approaches in determining EMF-initiated bioeffects in biological mechanisms. While biophysical approach is based on experimentally obtained data on biological responses to the applied EMF, the engineering approach strongly relies on SAR values and computations, not measurements. In most cases, SAR is used to determine the hazardous threshold.

It is clear that classical physics and equilibrium thermodynamics have an easy model to look for the potential mechanism of EMF interactions within biological bodies: heat. Unfortunately, it does not work in biology. It is well known that the structure of any living tissue is far from homogeneity, and for that reason, we need to apply the principles of nonequilibrium thermodynamics.

Let me recall some facts from history. In 1953, the late Herman Schwan proposed the value of $100\,W/m^2$ ($10\,mW/cm^2$) as a safe limit for human exposure to microwave energy, based on calculations (Foster, 2005). Even if one accepts the accuracy of calculation, there is no biological proof for this value. Since then, many efforts have been taken and millions of dollars for funding have been spent to prove this value, all without success.

Becker (1990) pointed out that based solely on calculations, the magic figure of $10\,mW/cm^2$ was adopted by the air force as the standard for safe exposure. Subsequently the thermal effects concept dominated policy decisions for the complete exclusion of nonthermal effects. While the $10\,mW/cm^2$ standard was limited to microwave frequencies, the thermal effects concept was extended to all other parts of the electromagnetic spectrum. This view led to "the policy of denying any nonthermal effects from any electromagnetic usage, whether military or civilian".

Probably, it is time to say few words in regard to what health effect is. It is important since for many years until now, ICNIRP has been distributing statements "that only thermal effects can be biologically important". From here, one should conclude that since, according to ICNIRP, EMF cannot produce heat in target, they cannot have biological and health importance.

Here we are dealing with categories such as biological effect, health effect, and health hazard. WHO policy is that **not every biological effect is a health effect**. Obviously, by saying "health effect" WHO is considering the adverse effects in the sense of diseases, pathologies, and injuries. However, there is at least one reason to say that such a statement is incorrect. In clinical medicine, any chemical or physical method prescribed by physician causes a biological effect that results in health improvement. Therefore, the correct statement should be as follows: **Not every biological effect initiated by EMF is a health hazard**. For that reason, the word "hazard" should be kept for irreversible effects caused by short or prolonged exposure to EMF.

I wouldn't speak about "thermal noise" if this term is not directly related to the understanding of biological mechanisms of action. In the last two decades of the 20th century, basic science was very much involved in discussion of the problems of thermal noise and "kT". Based upon the basic principle of classical thermodynamics, many physicists and physical chemists affirmed that, because of "kT", no effects of static and low-frequency EMF cannot be observed. I strongly disagree with such statement. In brief, life on this planet originated and has evolved in the presence of natural magnetic fields. Contemporary conditions of life also include a number of EMFs that are different in amplitudes and frequencies.

It would be plausible to remind the reader that the entire biology and physiology of living creature(s) is based upon three types of transfer:

- energy
- matter
- information.

While the first two processes might be described in terms of classical (equilibrium) thermodynamics, information transfer obviously needs another approach, and this may be found in nonequilibrium thermodynamics. As the late Ross Adey (2004) wrote in his last paper, "Current equilibrium thermodynamics models fail to explain an impressive spectrum of observed bioeffects at non-thermal exposure levels. Much of this signaling within and between cells may be mediated by free radicals of the oxygen and nitrogen species". Cell signaling, signal transduction cascade and conformational changes are events and processes that may be explained only by using nonequilibrium thermodynamics.

For unicellular organisms, the cellular membrane is both the detector and effector of physical and chemical signals. As a sensor, it detects alterations in conditions in the environment and further provides pathways for signal transduction. As effector, the membrane may also transmit a variety of electrical (and chemical) signals to the neighboring cells with an invitation to "whisper together" as suggested by Ross Adey (2004).

An estimate of the electrostatic interactions involving different biological structures is assumed to result primarily from electronic polarization, reorientation of dipolar groups, and changes in the concentrations of charged species in the vicinity of charges and dipoles. These effects could be well characterized for interactions in isotropic, homogeneous media. However, biological structures represent complex inhomogeneous systems in which it is difficult to predict ionic and dielectric properties. In these cases, factors such as the shape and composition of the surface and presence/absence of charged or dipolar groups appear to be especially important. It is clear now that in order to induce EMF bioeffects, the applied signal should not only satisfy the dielectric properties of the target, but also induce sufficient voltage to be detectable above thermal noise (Markov and Pilla, 1995). Such an approach relies on conformational changes and transfer of information (Markov, 2004a,b).

There is a whole series of biologically important modifications occurring under weak, static, or alternating EMF action that could be explained only from the view point of nonthermal mechanisms. The spectrum includes changes at various levels: alterations in membrane structure and function, changes in a number of subcellular structures as proteins and nucleic acids, protein phosphorylation, cell proliferation, free radical formation, adenosine triphosphate (ATP) synthesis, etc. The wide range of reported beneficial effects of using electric current or EMF/EF/MF therapy worldwide shows that more than 3 million patients received relief from their medical problems: from bone unification, pain relief, and wound healing to relatively new applications for victims of multiple sclerosis and Alzheimer's.

I will conclude this section with the question, what is the criteria for "thermal mechanism" or by how much should the temperature of the tissue be increased for the thermal mechanism to be considered? There are no standards and criteria for the thermal threshold, neither for methods to evaluate the actual rise of temperature in the target tissues. Once again, dosimetry has to be performed based on science principles and requirements, not with assumed calculations and criteria. We have to collect actual data that would allow 3D distribution of the EMF and especially the values of the fields at the **target site**.

THE HAZARDS OF HIGH-FREQUENCY COMMUNICATION

The problems of dosimetry gained importance with the increasing distribution of satellite and cellular communications, especially with latest 5G mobile phone system which has dramatically changed the electromagnetic environment. The entire biosphere and every organism living on this planet is exposed to continuous action of complex and unknown (by sources, amplitudes, frequencies) EMFs. Wi-Fi technologies include not only mobile phones but also all means of emitters and distributors of Wi-Fi signals, mainly antennas, base stations, and satellites.

Generally speaking, we do not know how or to what extent the Wi-Fi radiation alters the physiology of normal, healthy organisms. The situation becomes more complex when we are asking about its influence on children, aging adults, or sick individuals.

Based upon my long experience in radiobiology and bioelectromagnetics, I believe that it will be fair to say that biological effects of EMFs are nonthermal and have to be discussed from the low energy considerations, such as conformational changes in biological structures, effects on signal-transduction cascade, as well as in their manifestations on the surface of important biomolecules. More likely, the effects of EMF have informational character, and because of that they need certain time to manifest and initiate further changes in biochemistry and physiology (Markov, 2012; Markov and Grigoriev, 2013).

The major standards and guidelines established by the engineering community, Institute of Electrical and Electronics Engineers (IEEE) in 2005, and ICNIRP in 2009 provide the approach and terminology which are not accepted by physics and biological communities but nevertheless remain the guiding rules (mainly for the industry).

It has been pointed out elsewhere (Markov, 2006) that when the engineering committees state that "Nonthermal RF biological effects have not been established", basically they are guiding science and society in the wrong direction. Declining the possibility of nonthermal effects is not reasonable, but more important is that they mixed the terms "effect" and "hazard".

The epidemiological teams claim that "there is no consistent evidence for the occurrence of the modification". They also state that "there is no conclusive and consistent evidence that nonionizing radiation emitted by cell phone is associated with cancer risk" (Boice and Tarone, 2011). It is remarkable that this paper was published after International Agency of Research on Cancer (IARC) defined radiofrequency fields (RFs) as a "possible cancerogenic for humans".

At the WHO meeting on harmonization of standards close to two decades ago, I made a statement that **neglecting the hazard of high-frequency EMF for children is a crime against humanity** (Markov, 2001) and am ready to repeat it.

Unfortunately, mobile communication technology, especially the newest 5G, is developing with high speed without **ANY** evaluation of the potential hazard of this new modality. Even at the recent bioelectromagnetic society meetings, the biological side of this technology had been neglected, even in the light of the WHO position on the potential hazard of microwaves.

REFERENCES

Adey WR (1993) Electromagnetic technology and the future of bioelectromagnetics. In Blank M. (ed) *Electricity and Magnetism in Biology and Medicine.* New York, Plenum Press, pp. 101–108.

Adey WR (2004) Potential therapeutic application of nonthermal electromagnetic fields: ensemble organization of cells in tissue as a factor in biological tissue sensing. In Rosch PJ, Markov MS (eds) *Bioelectromagnetic Medicine.* New York, Marcel Dekker, pp. 1–15.

Barnes F, Greenebaum B (2007) *Handbook of Biological Effects of Electromagnetic Fields,* Third edition. CRC Press, Boca Raton, FL.

Barnes F, Greenebaum B (2019) *Handbook of Biological Effects of Electromagnetic Fields,* Fourth edition. CRC Press, Boca Raton, FL.

Bassett CAL Pawluk RJ, Pilla AA (1974) Acceleration of fracture repair by electromagnetic fields. *Ann NY Acad Sci* 238:242–262. doi:10.1111/j.1749-66.

Bassett CAL, Pilla AA, Pawluk R (1977) A non-surgical salvage of surgically-resistant pseudoarthroses and non-unions by pulsing electromagnetic fields. *Clin Orthop Relat Res* 124:117–131.

Becker R (1990) *Cross Current.* Jeremy Tarcher Inc., New York, p. 324.

Boice J, Tarone RE (2011) Cell phone, cancer and children. *J Natl Cancer Inst* 103(16):1211–1213.

Foster K (2005) Bioelectromagnetics pioneer Herman Schwan passed away at age 90. *Bioelectromagn Newsletter* 2:1–2.

Kotnik T, Miklavcic D (2006) Theoretical analysis of voltage inducement on organic molecules. In Kostarakis P (ed) *Proceedings of Forth International Workshop Biological Effects of Electromagnetic Fields,* Crete 16–20 October 2006, ISBN 960-233-172-0, pp. 217–226.

Lin J (2011) *Electromagnetic Fields in Biological System.* CRC Press, Boca Raton, FL.

Markov M (1994) Biological effects of extremely low frequency electromagnetic fields. In Ueno S (ed) *Biomagnetic Stimulation,* Plenum Press, New York, pp. 91–103.

Markov MS (2001) Magnetic and electromagnetic field dosimetry: necessary step in harmonization of standards. Proceedings of WHO Meeting, Varna, April 2001.

Markov MS (2002) How to go to magnetic field therapy? In Kostarakis P (ed) *Proceedings of Second International Workshop of Biological Effects of Electromagnetic Fields,* Rhodes, Greece, 7–11 October 2002, ISBN 960-86733-3-X, pp. 5–15.

Markov MS (2004a) Magnetic and electromagnetic field therapy: basic principles of application for pain relief. In Rosch PJ and Markov MS (eds) Bioelectromagnetic Medicine, Marcel Dekker, New York, pp. 251–264.

Markov MS (2004b) Myosin light chain phosphorylation modification depending on magnetic fields I. Theoretical. *Electromagn. Biol. Med.* 23:55–74.

Markov MS (2005) Biological windows: a tribute to Ross Adey. *Environmentalist* 25(2–3):67–74.

Markov MS (2006). Thermal vs. nonthermal mechanisms of interactions between electromagnetic fields and biological systems. In Ayrapetyan SN, Markov MS (eds) *Bioelectromagnetics: Current Concepts.* Springer, Dordrecht, The Netherlands. pp. 1–16.

Markov MS (2009) What need to be known about therapy with magnetic fields. *Environmentalist* 29(2):169–176.

Markov MS (2012) Cellular phone hazard for children. *Environmentalist* 32(2):201–209.

Markov MS (2015) Benefit and hazard of electromagnetic fields. In Markov M (ed) *Electromagnetic Fields in Biology and Medicine.* CRC Press, Boca Raton, FL, pp. 15–29.

Markov M, Grigoriev Y (2013) Wi-Fi technology – an uncontrolled global experiment on the health of mankind. *Electromagn Biol Med* 32(2):200–208.

Markov MS, Pilla AA (1995) Electromagnetic field stimulation of soft tissue: Pulsed radio-frequency treatment of post-operative pain and edema. Wounds 7(4):143–151.

Polk C, Postow E (eds) (1986) *CRC Handbook of Biological Effects of Electromagnetic Fields*. CRC Press, Boca Raton, FL.

Rocsh PJ, Markov MS (eds) (2004) Bioelectromagnetic Medicine, Marcel Dekker, New York, 851p.

Todorov N (1982) *Magnetotherapy*. Meditzina i Physcultura Publishing House, Sofia, p. 106.

2 Pulsed Electromagnetic Fields

From Signaling to Healing

Arthur A. Pilla*

CONTENTS

INTRODUCTION

Pulsed electromagnetic fields (PEMFs), from extremely low frequency (ELF) to pulsed radio frequency (PRF), have been successfully employed as adjunctive therapy for the treatment of delayed, nonunion, and fresh fractures and fresh and chronic wounds. Recent increased understanding of the mechanism of action of PEMF signals has permitted technologic advances allowing the development of PRF devices that are portable and disposable; can be incorporated into dressings, supports, and casts; and can be used over clothing. This broadens the use of nonpharmacological, noninvasive PEMF therapy to the treatment of acute and chronic pain and inflammation.

The vast majority of PEMF signals employed clinically do not directly cause a physiologically significant temperature rise in the target cell/tissue area. Furthermore, many experimental and clinical studies show there are preferential responses to the waveform configuration that cannot be explained on the basis of power or energy transfer alone. It was proposed more than 40 years ago that PEMF signals could be configured to be bioeffective by matching the amplitude spectrum of the *in-situ*

---

* Deceased. This chapter is reprinted from a previous CRC publication in recognition of his pioneering contribution for clinical application of EMF in the USA.

electric field to the kinetics of ion binding (Pilla, 1972, 1974a,b, 1976). Models have been developed that show that a nonthermal PEMF could be configured to modulate electrochemical processes at the electrified interfaces of macromolecules by assessing its detectability versus background voltage fluctuations in a specific ion-binding pathway (Pilla et al., 1994, 1999; Pilla, 2006). This has allowed the a priori configuration of nonthermal pulse-modulated radio-frequency (RF) signals that have been demonstrated to modulate calmodulin (CaM) dependent nitric oxide (NO) signaling in many biological systems (Pilla et al., 2011; Pilla, 2012, 2013).

Given the aforementioned points, PEMF therapy is rapidly becoming a standard part of surgical care, and new, more significant, clinical applications for osteoarthritis, brain and cardiac ischemia, and traumatic brain injury are in the pipeline. This study reviews recent evidence that suggests that CaM-dependent NO signaling, which modulates the biochemical cascades of living cells and tissues employed in response to external challenges, is an essential PEMF transduction pathway. Cellular, animal, and clinical results are presented that provide support for this mechanism of action of PEMF and may provide a unifying explanation for the reported effects of a variety of PEMF signal configurations on tissue repair, angiogenesis, pain, and inflammation.

PEMF AS A FIRST MESSENGER

The electrochemical information transfer (ECM) model (Pilla, 1974b, 2006) proposed that weak nonthermal PEMF could be configured to modulate voltage-dependent electrochemical processes at electrified aqueous cell and molecular interfaces. Use of the ECM model for PEMF signal configuration required knowledge of the initial kinetics of ion binding in the target pathway from which physiologically meaningful equivalent electric circuits could be developed. The model provided a means for configuring nonthermal EMF signals to couple to electrochemical kinetics to modulate target processes such as ion binding and transmembrane transport. Application of the ECM model led to the a priori configuration of bone growth stimulator (BGS) signals that are now part of the standard armamentarium of orthopedic practice worldwide for the treatment of recalcitrant bone fractures (Basset et al., 1977; Fontanesi et al., 1983; Aaron et al., 2004). Signal-to-thermal-noise ratio (SNR) analysis was later added to the ECM model to enable more precise signal configurations that could be effective in specific ion-binding pathways (Pilla et al., 1994).

The proposed PEMF signal transduction pathway begins with voltage-dependent Ca^{2+} binding to CaM, which is driven by increases in cytosolic Ca^{2+} concentrations in response to chemical and/or physical challenges (including PEMF itself) at the cellular level. Ca/CaM binding has been well characterized, with a binding time constant reported to be in the range of 1–10 ms (Blumenthal and Stull, 1982), whereas the dissociation of Ca^{2+} from CaM requires the better part of a second (Daff, 2003). Thus, Ca/CaM binding is kinetically asymmetrical; that is, the rate of binding exceeds the rate of dissociation by several orders of magnitude ($k_{on} \gg k_{off}$), driving the reaction in the forward direction according to the concentration and voltage dependence of Ca^{2+} binding. The asymmetry in Ca/CaM binding kinetics provides an opportunity to configure any PEMF waveform to induce an electric field that can produce a net increase in the population of bound Ca^{2+} (activated CaM) (Pilla et al., 1999; Pilla, 2006).

However, this is only possible if pulse duration or carrier period is significantly shorter than bound Ca^{2+} lifetime (Pilla et al., 2011). Thus, Ca^{2+} binds to CaM when the voltage at the binding site increases; however, Ca^{2+} does not immediately dissociate from CaM when the voltage decreases as the waveform decays or changes polarity, because Ca^{2+} bound in the initial phase of the waveform is sequestered for the better part of a second during which activated CaM activates its target enzyme. Thus, the Ca/CaM signaling pathway can exhibit rectifier-like properties for any PEMF signal because ion-binding kinetics are asymmetrical, not because there is a nonlinearity in electrical response to RF signals (Kowalczuk et al., 2010). What follows will demonstrate how a nonthermal PEMF signal can be configured to optimally modulate Ca^{2+} binding to CaM using the ECM model by assessing its electrical detectability in the binding pathway.

PEMF Signal Configuration

Initial configuration of a PEMF signal that can modulate CaM-dependent enzyme activity starts with an analysis of the kinetic equations describing the two-step Ca/CaM/enzyme binding process using the ECM model (Pilla et al., 1999; Pilla, 2006). Thus,

$$Ca^{2+} + CaM \underset{k_{off}}{\overset{k_{on}}{\rightleftarrows}} CaM^* + Enz \underset{k'_{off}}{\overset{k'_{on}}{\rightleftarrows}} Enz^*, \qquad (2.1)$$

where CaM* is the activated CaM (Ca^{2+} bound) and Enz* is the activated enzyme target (CaM* bound). In all cases, the forward reaction of Ca^{2+} binding is significantly faster than Ca^{2+} dissociation ($k_{on} \gg k_{off}$). It is also important to note that CaM*/enzyme binding is significantly slower than Ca/CaM binding ($k_{on} \gg k'_{on}$). Under these conditions, Ca^{2+} binding per se can be treated as a linear system for the small changes in voltage at the binding site produced by nonthermal PEMF. Thus, the change in bound Ca^{2+} concentration with time, $\Delta Ca(s)$, may be written as

$$\Delta Ca(s) = \frac{k_{on}}{sCa^\circ} \left[-\Delta Ca(s) + \kappa E(s) + \Delta Enz^*(s) \right], \qquad (2.2)$$

where s is the real-valued Laplace transform frequency, E(s) the induced voltage from PEMF, κ the voltage dependence of Ca^{2+} binding, and Ca° the initial concentration of bound Ca^{2+} (homeostasis, too small to activate CaM). Note that the Laplace transform is positive-valued at all frequencies and inherently accounts for the kinetic asymmetry of Ca/CaM binding (Pilla, 2006).

The net increase in Ca^{2+} bound to CaM from PEMF is proportional to the binding current, $I_A(s)$, which, in turn, is proportional to $\Delta Ca(s)$ from Equation 2.2. From this, as shown in detail elsewhere (Pilla, 2006), binding impedance, $Z_A(s)$, may be written as

$$Z_A(s) = \frac{E(s)}{I_A(s)} = \frac{1}{q_e \kappa} \left[\frac{\dfrac{1 + sCa^\circ}{k_{on}}}{sCa^\circ} \right]. \qquad (2.3)$$

Equation 2.3 shows that the kinetics of Ca^{2+} binding to CaM may be represented by a series R_A–C_A electrical equivalent circuit, where R_A is the equivalent resistance of binding, inversely proportional to k_{on}, and C_A is the equivalent capacitance of binding, proportional to bound Ca^{2+}. The time constant for Ca^{2+} binding is, thus, $\tau_A = R_A C_A$, which is proportional, as expected, to $1/k_{on}$. To evaluate SNR for the Ca/CaM target, the quantity of interest is the effective voltage, $E_b(s)$, induced across C_A, evaluated for simple ion binding (Equation 2.3) using standard circuit analysis techniques (Cheng, 1959):

$$E_b(s) = \frac{\left(\dfrac{1}{sC_A}\right)E(s)}{\left(R_A^2 + \left(\dfrac{1}{sC_A}\right)^2\right)^{1/2}}. \tag{2.4}$$

Equation 2.4 clearly shows that $E_b(s)$ is dependent upon $E(s)$ for any waveform, for example, rectangular, sinusoidal, arbitrary, or chaotic. In every case, only that portion of the applied electric field that appears at the Ca/CaM binding pathway, $E_b(s)$, can increase surface charge (bound Ca^{2+}). Thus, as long as it is nonthermal, the total applied energy in $E(s)$ is not the dose metric. Rather, the frequency spectrum of $E_b(s)$, for any $E(s)$, is taken as a measure of biological reactivity. The calculation of SNR has been described in detail elsewhere (Pilla et al., 1994; Pilla, 2006). Briefly, thermal voltage noise in the binding pathway may be evaluated via (DeFelice, 1981)

$$\mathrm{RMS}_{noise} = \left[4kT \int_{\omega_1}^{\omega_2} \mathrm{Re}\left[Z_A(\omega)\right] d\omega\right]^{1/2}, \tag{2.5}$$

where RMS_{noise} is the root mean square of the thermal voltage noise spectral density across C_A; Re is the real part of the total binding impedance, Z_A; $\omega = 2\pi f$; and the limits of integration (ω_1, ω_2) are determined by the band pass of binding, typically 10^{-2}–10^7 rad/s. SNR is evaluated using

$$\mathrm{SNR} = \frac{E_b(s)}{\mathrm{RMS}_{noise}}. \tag{2.6}$$

The ECM model with SNR analysis can be applied to any ion-binding process, provided it is kinetically asymmetrical with $k_{on} \gg k_{off}$, as well as to any nonthermal weak PEMF signal configuration (Pilla, 2006; Pilla et al., 2011). The model has proven to be very useful for the a posteriori analysis of many different PEMF signals particularly for studies in which the response was either marginal or did not exist.

ECM Model Verification

An example of the use of Equation 2.6 to compare the expected efficacy of PRF signals with different pulse modulations, assuming Ca/CaM binding as the transduction pathway, is shown in Figure 2.1a. As may be seen, the rate of Ca^{2+} binding to CaM is expected to be increased identically using either a 2 ms burst of 27.12 MHz repeating at 1 burst/s with 5 µT peak amplitude (signal A) or a 0.065 ms burst of 27.12 MHz repeating at 1 burst/s with 200 µT peak amplitude (signal B). However, the 0.065 ms burst

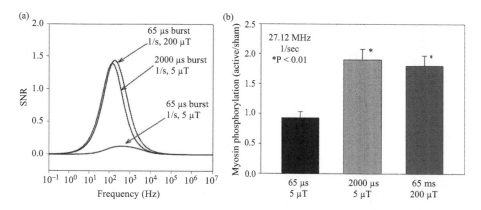

FIGURE 2.1 ECM model applied to configure RF PEMF. (a) SNR analysis in Ca/CaM binding pathway shows 2000 µs burst at 5 µT and 65 µs burst at 200 µT should be equally effective versus 65 µs at 5 µT that was predicted ineffective. (b) Five-minute PEMF exposure on MLCK cell-free assay verified ECM prediction. (Adapted from Pilla, A.A., Mechanisms and therapeutic applications of time varying and static magnetic fields, in *Biological and Medical Aspects of Electromagnetic Fields*, F. Barnes and B. Greenebaum, eds., CRC Press, Boca Raton FL, 2006, p. 351.)

signal at 5 µT peak amplitude (signal C) is not expected to be effective. This was tested on CaM-dependent myosin light chain kinase (MLCK) in a cell-free enzyme assay for myosin light chain (MLC) phosphorylation (Markov et al., 1994; Pilla, 2006). MLC is a contractile protein of physiological importance in muscle and blood and lymph vessel tone. As may be seen in Figure 2.1b, MLC phosphorylation was increased twofold versus control after a 5 min exposure for signals A and B, whereas there was no significant difference versus shams for signal C, just as predicted by the ECM model.

The cell-free MLC assay was also employed to assess the effect of burst duration of the 27.12 MHz RF carrier, keeping burst repetition and amplitude constant (Pilla, 2006). These results are shown in Figure 2.2a wherein the increase in bound Ca^{2+} appears to reach saturation as burst duration approaches 2–3 ms. Further verification of the ECM model was obtained from a study of the effect of PRF on wound healing in the rat (Strauch et al., 2007). The study was designed as a prospective, placebo-controlled, blinded trial in which rats were treated for 30 min twice daily with PRF signals with identical carrier frequency (21.12 MHz), burst repetition rate (2 bursts/s), and amplitude (5 µT), but with progressively increasing burst duration. The results are shown in Figure 2.2b, wherein it may be seen that, as for the MLCK assay, peak effect occurred at a burst duration of between 2 and 3 ms. However, instead of a saturation effect, wound strength was lower as burst duration increased. Thus, in a complete tissue, there appears to be a window within which maximum acceleration of healing occurs. The possible reasons for this behavior in living cells and tissue and its implications for the clinical applications of this PRF signal will be considered in detail in the following.

These and other studies suggested an RF signal configuration consisting of a 27.12 MHz carrier pulse modulated with a 2 ms burst repeating at 2 bursts/s, having

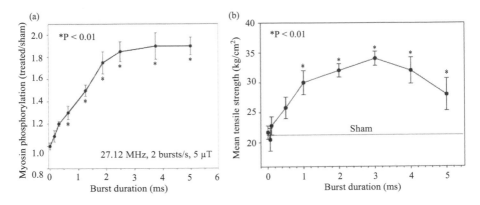

FIGURE 2.2 Pulse modulation of RF PEMF impacts outcome. (a) MLC phosphorylation is increased nearly twofold as burst duration of 27.12 MHz RF PEMF signal increases. Maximum effect occurs at 2–3 ms beyond which saturation occurs. (Adapted from Pilla, A.A., Mechanisms and therapeutic applications of time varying and static magnetic fields, in *Biological and Medical Aspects of Electromagnetic Fields*, F. Barnes and B. Greenebaum, eds., CRC Press, Boca Raton FL, 2006, p. 351.) (b) Wound strength in a rat model similarly increases to 2–3 ms; however, in contrast to the cell-free enzyme assay, tensile strength does not saturate but peaks as burst duration increases. (Adapted from Strauch, B. et al., *Plast. Reconstr. Surg.*, 120, 425, 2007.)

an amplitude of 4 ± 1 μT. Unless otherwise indicated, this signal configuration was employed in the studies reviewed here. Single-exposure times varied between 15 and 30 min, except where indicated.

NO SIGNALING

PEMF signals with a vast range of waveform parameters have been reported to reduce pain and inflammation (Ross and Harrison, 2013b) and enhance healing (Pilla, 2006). Using the ECM model, a common unifying mechanism has been proposed, which involves Ca^{2+}-dependent NO signaling, to quantify the relation between signal parameters and bioeffect (Pilla et al., 2011). Intracellular calcium ions play an important role in the signal transduction pathways a cell utilizes to respond to external challenges, for example, cell growth and division, apoptosis, metabolism, synaptic transmission, and gene expression (Bootman et al., 2001; Mellstrom et al., 2008). Regulation of cytosolic Ca^{2+} concentration is orchestrated by an elaborate system of pumps, channels, and binding proteins found both in the plasma membrane and on intracellular organelles such as the endoplasmic reticulum (Harzheim et al., 2010). High-affinity proteins (e.g., CaM, troponin) mediate the multiple physiological responses regulated by changes in intracellular Ca^{2+} concentrations produced by challenges that cause free Ca^{2+} to increase above its normal and tightly regulated value of approximately 100 nM in mammalian cells (Konieczny et al., 2012). In terms of PEMF mechanism, CaM is of particular interest because it is the first responder to changes in cytosolic Ca^{2+} and because of the many roles it plays in cell signaling and gene regulation pathways once activated by bound Ca^{2+} (Faas et al., 2011). In an immediate response to stress or injury, activated CaM binds to its primary enzyme target, constitutive nitric oxide synthase (cNOS, neuronal [nNOS], and/or endothelial [eNOS]), which, in turn, binds to and catalyzes L-arginine resulting in the release of the signaling molecule NO. As a gaseous free radical with an *in-situ* half-life of about 5 s (Ignarro et al., 1993), NO can diffuse through membranes and organelles and act on molecular targets at distances up to about 200 μm (Tsoukias, 2008). Low transient concentrations of NO activate its primary enzyme target, soluble guanylyl cyclase (sGC), which catalyzes the synthesis of cyclic guanosine monophosphate (cGMP) (Cho et al., 1992). The CaM/NO/cGMP signaling pathway is a rapid response cascade that can modulate peripheral and cardiac blood flow in response to normal physiologic demands as well as to inflammation and ischemia (Bredt and Snyder, 1990). This same pathway also modulates the release of cytokines, such as interleukin-1beta (IL-1β), which is proinflammatory (Ren and Torres, 2009), and growth factors, such as basic fibroblast growth factor (FGF-2) and vascular endothelial growth factor (VEGF), which are important for angiogenesis, a necessary component of tissue repair (Werner and Grose, 2003).

Following a challenge such as a bone fracture, surgical incision, or other musculoskeletal injury, repair commences with an inflammatory stage during which the proinflammatory cytokines, such as IL-1β, are released from macrophages and neutrophils that rapidly migrate to the injury site. IL-1β upregulates inducible nitric oxide synthase (iNOS), which is not Ca^{2+} dependent and therefore not modulated directly by PEMF. Large sustained amounts of NO that are produced by iNOS in the wound bed (Lee et al., 2001) are proinflammatory and can lead to increased cyclooxygenase-2 (COX-2) and prostaglandins (PGEs). These processes are a natural and necessary component of healing but are often unnecessarily prolonged, which can lead to increased pain and delayed or abnormal healing (Broughton et al., 2006). The natural anti-inflammatory regulation produced by CaM/NO/cGMP signaling attenuates IL-1β levels and downregulates iNOS (Palmi and Meini, 2002).

PEMF MODULATES NO SIGNALING

PEMF that produces sufficient SNR in the Ca/CaM binding pathway causes a burst of additional NO to be released from a cell, which has already been challenged. In addition to causing cytosolic Ca^{2+} to increase, inflammatory challenges, injury, and temperature all cause proinflammatory cytokines to be released from macrophages and neutrophils, the first cellular responders. This upregulates iNOS, resulting in the release of large sustained amounts of NO, which is proinflammatory. A recent study specifically examined the effect of PRF on CaM-dependent NO signaling in real time. Dopaminergic cells (MN9D) in phosphate buffer were challenged acutely with a nontoxic concentration of lipopolysaccharide (LPS), which causes an immediate increase in cytosolic Ca^{2+}. As reviewed earlier, any free cytosolic Ca^{2+} above approximately 100 nM instantly activates CaM, which, in turn, instantly activates cNOS. The result is immediate NO production. PRF was applied during LPS challenge. The results (Figure 2.3a) showed that PRF approximately tripled NO release

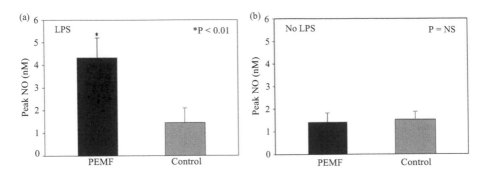

FIGURE 2.3   Effect of PEMF on neuronal cells challenged acutely with LPS (a) or phosphate buffer (b). NO was measured electrochemically with a NO-selective membrane electrode in real time during challenge. PEMF augmented CaM-dependent NO by nearly threefold when cells were challenged with LPS. In contrast, there was no PEMF effect on NO release for cells challenged with phosphate buffer. (Adapted from Pilla, A.A., *Biochem. Biophys. Res. Commun.*, 426, 330, 2012.)

from challenged MN9D cells within seconds, as measured electrochemically in real time using a NO-selective membrane electrode specially designed for use in cell cultures (Pilla, 2012). It was also verified in this study that cells challenged with phosphate buffer alone did not respond to this PRF signal, as shown in Figure 2.3b. This indicates that PEMF can act as an anti-inflammatory only when intracellular Ca^{2+} increases and provides one explanation for the lack of adverse effects reported in all clinical studies that employ this PRF signal.

Other studies have confirmed that this PRF signal can augment CaM-dependent NO and cGMP release from human umbilical vein endothelial cell (HUVEC) and fibroblast cultures (Pilla et al., 2011; Pilla, 2013). These studies additionally showed that a PRF effect could be obtained only if the cells were challenged (e.g., temperature shock) sufficiently to cause increases in cytosolic Ca^{2+} large enough to satisfy the Ca/CaM dependence of cNOS (Bredt and Snyder, 1990). Direct measurement of cytosolic Ca^{2+} binding to CaM under physiological conditions in living cells or tissue has not yet been successfully performed under PEMF exposure, primarily because of the submicromolar concentrations of Ca^{2+} involved. However, the CaM antagonists N-(6-aminohexyl)-5-chloro-1-naphthalenesulfonamide hydrochloride (W-7) and trifluoperazine (TFP) were able to block the PEMF effect on additional NO release, supporting CaM activation, which only occurs if Ca^{2+} binds to CaM, as a principle PEMF target for the modulation of tissue repair. A typical example of the effect of PRF on NO release in heat shock–challenged fibroblast cultures is summarized in Figure 2.4a, which shows a single 15 min PRF exposure produced a nearly twofold increase in NO, which was blocked with W-7. Another example can be seen in Figure 2.4b that shows that a single 15 min PRF exposure of heat shock–challenged HUVEC cultures produced a threefold increase in cGMP, which was blocked by TFP. Note that the CaM antagonists employed in these studies had, as expected, no effect on NO produced by CaM-independent iNOS in control cultures subjected to the identical heat shock challenge.

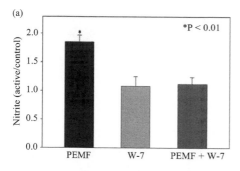
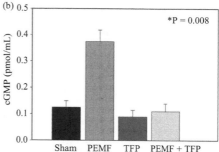

FIGURE 2.4 Effect of PEMF on CaM/NO/cGMP signaling in fibroblast and HUVEC cultures. (a) A single 15 min exposure of RF PEMF in fibroblast cultures increased NO by twofold, which could be blocked by the CaM antagonist W-7. (Adapted from Pilla, A.A. et al., *Biochem. Biophys. Acta.*, 1810, 1236, 2011.) (b) Similar PEMF exposure in HUVEC cultures increased cGMP by threefold, which, in this case, was blocked by the CaM antagonist TFP. These results are consistent with CaM/NO/cGMP as the transduction pathway for PEMF bioeffects. (Adapted from Pilla, A.A., *Electromagn. Biol. Med.*, 32, 123, 2013.)

PEMF IS ANTI-INFLAMMATORY

Thus far, this review has shown the considerable evidence that a PRF signal can be configured to have an immediate effect on CaM/NO/cGMP signaling. This, in turn, has an immediate effect on inflammation because transient bursts of cGMP have been found to inhibit proinflammatory nuclear factor-kappaB (NF-κB), which downregulates inflammatory cytokines such as IL-1β (Lawrence, 2009). This sequence is schematized in Figure 2.5, which summarizes the pathways from the initial PEMF effect on CaM activation to the modulation of cytokines and growth factors.

To illustrate, RF PEMF has been reported to modulate inflammatory cytokines in fibroblasts and keratinocytes (Moffett et al., 2012). Cell cultures were placed at room temperature (heat shock challenge) and exposed to PRF for 15 min. Cytokine expression was assayed 2 h after PEMF exposure. The results in Figure 2.6 show that PRF downregulated the proinflammatory cytokine IL-1β and upregulated the anti-inflammatory cytokines IL-5, IL-6, and IL-10, consistent with an anti-inflammatory PEMF effect via modulation of CaM/NO/cGMP signaling.

PEMF has been reported to downregulate iNOS at the mRNA and protein levels in monocytes (Reale et al., 2006) and proinflammatory cytokines in human keratinocytes (Vianale et al., 2008). Weak electric fields partially reversed the decrease in the production of extracellular matrix caused by exogenous IL-1β in full-thickness articular cartilage explants from osteoarthritic adult human knee joints (Brighton et al., 2008). PRF reduced IL-1β in cerebrospinal fluid 6 h after posttraumatic brain injury in a rat model (Rasouli et al., 2012). PEMF downregulated IL-1β and upregulated IL-10 in a mouse cerebral ischemia model (Pena-Philippides et al., 2014b) and upregulated IL-10 within 7 days in a chronic inflammation model in the mouse (Pena-Philippides et al., 2014a).

In the clinical setting, PRF has been shown to enhance the management of postoperative pain and inflammation. Several double-blind, placebo-controlled,

PEMF: Enhances CaM activation (millisec, real-time);

$$Ca^{2+} + CaM \rightleftharpoons Ca^{2+}CaM \; (\textit{kinetic asymmetry})$$

Increases enzyme activity (millisec, real-time);

$$Ca^{2+}CaM + cNOS \longrightarrow NO \; (signaling)$$

Modulates inflammation and repair (real-time)

$$NO \longrightarrow cGMP \longrightarrow Cytokines, Growth\,Factors$$

PEMF: Accelerates PDE inhibition of cGMP (real-time)

$$Ca^{2+}CaM + PDE + cGMP \longrightarrow GMP$$

Competes with PEMF-enhanced cGMP release

FIGURE 2.5 Schematic representation of the proposed PEMF mechanism of action. Induced electric field from the PEMF signal enhances CaM activation, which enhances cNOS activation. This, in turn, enhances CaM-dependent NO release, which enhances cGMP release. This reduces inflammatory cytokines and increases anti-inflammatory cytokines. As inflammation is reduced, CaM/NO/cGMP signaling modulates growth factor release, allowing tissue repair to be accelerated. Also shown is PEMF acceleration of PDE regulation of cGMP, which places limits on PEMF dosing.

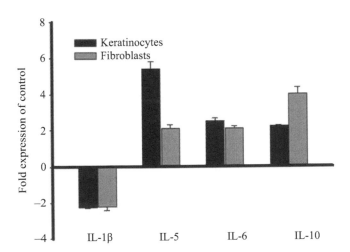

FIGURE 2.6 Effect of RF PEMF on inflammatory cytokine panel in fibroblasts and keratinocytes. A single 15 min exposure of cultures placed at room temperature (heat shock) produced severalfold changes in inflammatory cytokines. PEMF downregulated the proinflammatory cytokine IL-1β and upregulated the anti-inflammatory cytokines IL-5, IL-6, and IL-10. These results are consistent with CaM/NO/cGMP signaling as the PEMF transduction pathway. (Adapted from Moffett, J. et al., *J. Pain. Res.*, 12, 347, 2012.)

randomized studies, including breast reduction (Rohde et al., 2010), breast augmentation (Hedén and Pilla, 2008; Rawe et al., 2012), and autologous flap breast reconstruction (Rohde et al., 2012), have reported that nonthermal PRF fields significantly accelerate postoperative pain and inflammation reduction and, concomitantly, reduce postoperative narcotic requirements. Two of these studies examined the effect of PRF on levels of IL-1β in wound exudates, as well as wound exudate volume in the first 24 h postoperatively (Rohde et al., 2010, 2012). Figure 2.7a shows that IL-1β in wound exudates of active patients in both studies was 50% of that for sham patients at 6 h post operation. Figure 2.7b shows wound exudate volume at 12 h post operation was approximately twofold higher in sham versus active patients in both studies. Reductions in IL-1β and wound exudate volume are consistent with a PEMF effect on inflammation via CaM/NO/cGMP signaling.

An important result of PEMF modulation of CaM/NO/cGMP signaling is the modulation of nociception (Cury et al., 2011), perhaps since NO signaling modulates the sensitivity of opioid receptors (Chen et al., 2010). Clinically, NO enhances the actions of narcotics for postoperative analgesia (Lauretti et al., 1999), which may play a role in the reported effects of RF PEMF signals on reduction of postoperative narcotic usage (Rohde et al., 2010, 2012, 2014; Rawe et al., 2012). CaM-dependent NO also reduces IL-1β, which in turn reduces phosphodiesterase (PDE) and COX-2. Thus, addition of an NO donor to NSAIDs and aspirin enhances analgesia (Velazquez et al., 2005; Borhade et al., 2012). Modulation of the endogenous opioid pathway by physical modalities has previously been described for ELF magnetic fields (Kavaliers and Ossenkopp, 1991; Prato et al., 1995) and transcutaneous

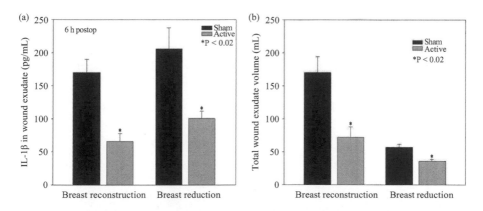

FIGURE 2.7 Effect of RF PEMF on inflammation in two double-blinded, randomized, and placebo-controlled clinical studies. (a) PEMF reduced IL-1β in wound exudates, which was approximately twofold higher in sham versus active patients at 6 h post operation in both breast reconstruction and breast reduction studies. (b) Total wound exudate volume was approximately twofold higher in sham versus active patients at 12 h post operation in both studies. These results are consistent with a PEMF effect on inflammation via CaM/NO/cGMP signaling. (Adapted from Rohde, C. et al., *Plast. Reconstr. Surg.*, 125, 1620, 2010; Rohde, C. et al., *Plast. Reconstr. Surg.*, 130, 91, 2012.)

electrical nerve stimulation (Sluka et al., 1999). It has recently been reported that PRF signals increase endogenous opioid precursors in human epidermal keratinocytes and dermal fibroblasts both at the mRNA and protein levels (Moffett et al., 2012). The same PRF signal was also used in a rat pain behavior model, wherein pain reduction was, in part, related to PRF modulation of the release of β-endorphin (Moffett et al., 2010, 2011). Thus, PEMF can reduce inflammation and potentially enhance the action of narcotics, NSAIDS, and aspirin, all of which reduce patient morbidity.

Clinical Test of ECM Model

In the clinical arena, the ECM/SNR model can be used a posteriori to analyze the effectiveness of different configurations of PRF signals assuming the transduction pathway is CaM-dependent NO signaling. This was performed for two independent double-blind, placebo-controlled, and randomized studies that assessed the effect of PRF on postoperative pain relief and narcotic requirements in breast augmentation patients (Hedén and Pilla, 2008; Rawe et al., 2012). One study used PRF signal A consisting of a 2 ms burst of 27.12 MHz repeating at 2 bursts/s and inducing a peak electric field of 5 ± 2 V/m (predicted effective by ECM). The second study used PRF signal B consisting of a 0.1 ms burst of 27.12 MHz repeating at 1,000 bursts/s and inducing a peak electric field of 0.4 ± 0.1 V/m. Both studies reported accelerated postoperative pain reduction. However, sham patients in the first study (signal A) had 2.7-fold more pain at POD2 compared to active patients. In contrast, sham patients in the second study (signal B) had only 1.3-fold more pain than active patients at POD2. Clearly, pain reduction in patients treated with signal A was substantially faster than that in patients treated with signal B. Relative effectiveness of both signals is shown in Figure 2.8, wherein it may be seen that dosimetry for signal A, in terms of SNR at Ca/CaM binding sites, is significantly larger than that for signal B. Clearly, signal A would be expected to produce a burst of NO significantly higher than that produced by signal B that could lead to a larger anti-inflammatory effect.

PEMF CAN MODULATE ANGIOGENESIS

As tissue responds to injury, PEMF can rapidly modulate the relaxation of the smooth muscles controlling blood and lymph vessel tone through the CaM/NO/cGMP cascade (McKay et al., 2007). PEMF also enhances growth factor release through the same cascade in endothelial cells to modulate angiogenesis. PEMF modulation of eNOS activity may, therefore, be a useful strategy to augment angiogenesis for tissue repair and possibly other conditions that require vascular plasticity, such as ischemia (Cooke, 2003). An early study showed that PEMF augmented the creation of tubular, vessel-like structures from endothelial cells in culture in the presence of growth factors (Yen-Patton et al., 1998). Another study confirmed a sevenfold increase in endothelial cell tubule formation *in vitro* (Tepper et al., 2004). Quantification of angiogenic proteins demonstrated a fivefold increase in FGF-2, suggesting that PEMF modulates angiogenesis by increasing FGF-2 production. This same study also reported PEMF increased vascular in-growth more than twofold when applied

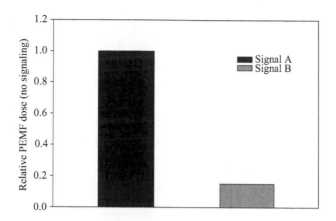

FIGURE 2.8 Effect of pulse modulation parameters of RF PEMF on clinical efficacy for postoperative pain reduction. Relative PEMF dose is measured assuming the transduction mechanism is the modulation of CaM/NO/cGMP signaling. A posteriori analysis of SNR for each signal reveals that signal B is expected to enhance CaM activation by 20% versus 100% for signal A. Clinical data confirm sham patients in signal A study had 2.7-fold more pain at POD2 compared to active patients, compared to only 1.3-fold more pain in the signal B study. (Adapted from Hedén, P. and Pilla, A.A., *Aesthetic. Plast. Surg.*, 32, 660, 2008; Rawe, I.M. et al., *Aesthetic Plast. Surg.*, 36, 458, 2012.)

to an implanted Matrigel plug in mice, with a concomitant increase in FGF-2, similar to that observed *in vitro*. PEMF significantly increased neovascularization and wound repair in normal mice, and particularly in diabetic mice, through an endogenous increase in FGF-2, which could be eliminated by using an FGF-2 inhibitor (Callaghan et al., 2008). Similarly, a PRF signal of the type used clinically for wound repair was reported to significantly accelerate vascular sprouting from an arterial loop transferred from the hind limb to the groin in a rat model (Roland et al., 2000). This study was extended to examine free flap survival on the newly produced vascular bed (Weber et al., 2005). Results showed 95% survival of PRF-treated flaps compared to 11% survival in the sham-treated flaps, suggesting a significant clinical application for PRF signals in reconstructive surgery. Another study (Delle Monache et al., 2008) reported that PEMF increased the degree of endothelial cell proliferation and tubule formation and accelerated the process of wound repair, suggesting a mechanism based upon a PEMF effect on VEGF receptors. In the clinical setting, PRF has been reported to enhance fresh and chronic wound repair (Kloth et al., 1999; Kloth and Pilla, 2010; Strauch et al., 2007; Guo et al., 2012).

A recent study evaluated the effect of PRF on cardiac angiogenesis in a reproducible thermal myocardial injury model. The injury was created in the region of the distal aspect of the left anterior descending artery at the base of the heart in a blinded rat model (Patel et al., 2006; Strauch et al., 2006, 2009; Pilla, 2013). PRF exposure was 30 min twice daily for 3, 7, 14, or 21 days. Sham animals were identically exposed but received no PRF signal. A separate group of animals treated for 7 days received L-nitroso-arginine methyl ester (L-NAME), a general NOS inhibitor, in their drinking water. Upon sacrifice, myocardial tissue specimens were stained

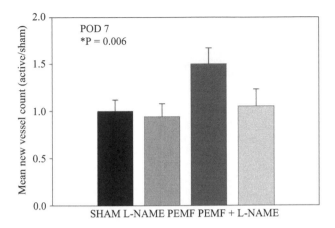

FIGURE 2.9 Effect of RF PEMF on angiogenesis in a thermal myocardial injury model. The results show that PEMF significantly increased new vessel growth and that L-NAME, a general NOS inhibitor administered in drinking water, blocked the PEMF effect. These results suggest PEMF was acting through the CaM/NO/cGMP pathway since blockage occurred far upstream of the production of growth factors necessary for angiogenesis. (Adapted from Pilla, A.A., *Electromagn. Biol. Med.*, 32, 123, 2013.)

with CD-31, and the number of new blood vessels was counted on histological sections at the interface between normal and necrotic muscle at each time point by three independent blinded histologists. The results showed mean new vessel count was not significantly increased by PEMF at day 3 but was significantly increased at day 7 (+50%, P = 0.006), day 14 (+67%, P = 0.004), and day 21 (+99%, P < 0.001). The results shown in Figure 2.9 for day 7 indicate L-NAME completely blocked the PEMF effect on angiogenesis, suggesting the transduction pathway for the PRF effect on angiogenesis in this study involved CaM-dependent NO signaling.

PEMF, NO SIGNALING, AND PDE

It will be recalled that PRF effects in living systems appear to occur within a window defined by a combination of pulse modulation parameters and amplitude of induced electric field (see Figure 2.2b). It will also be recalled that activated CaM can activate a number of enzymes, including MLCK, cNOS, CaM-dependent protein kinase II (CaMPKII), and some PDE isoforms. It is the latter that appears to play a large role in the effectiveness of PEMF to enhance NO signaling. This is because the dynamics of Ca^{2+}-dependent NO release are tightly regulated through a negative feedback mechanism that involves PDE isoenzymes (Mo et al., 2004; Batchelor et al., 2010). These studies show the rate of Ca^{2+}-dependent NO production in challenged tissue will be modulated not only by PEMF but also by PDE activity, which, itself, is also Ca^{2+} dependent. In essence, the augmentation of cGMP and therefore the modulation of cytokines and growth factors that follows from the augmentation of NO by PEMF (see Figure 2.5) can be curtailed by PDE that converts cGMP to GMP, which is inactive. Thus, PEMF can enhance both the activity of cNOS, which increases NO release, and the activity of PDE, which decreases NO release in tissue by inhibiting cGMP. Thus, if PEMF increases NO too rapidly, PDE isoenzyme activity may also be enhanced sufficiently to reduce or block the effects of PEMF on NO signaling.

Thus, the predictions of the ECM model for signal configuration now need to take into account CaM/NO/cGMP/PDE signaling dynamics. The PRF signal considered here produces an instantaneous burst of NO during each signal burst, in real time (Pilla, 2012). The amount of increased NO is directly dependent upon induced electric field amplitude, burst duration, repetition rate, and total exposure time. Each of these parameters will modify the accumulated increase of NO in tissues produced by PEMF. As NO increases, so does cGMP, and this dynamic is regulated by a CaM-dependent PDE isoform. It has been suggested that the extreme sensitivity of NO receptors is due to the low affinity of PDE to cGMP, meaning NO activates sGC faster than PDE inhibits cGMP (Batchelor et al., 2010). However, PEMF also enhances PDE activation, suggesting that the natural CaM/NO/cGMP dynamic can be disrupted if the PEMF signal enhances NO too rapidly. As PEMF-enhanced NO production increases, PDE activation also increases, and the dynamic can favor cGMP inhibition (see Figure 2.5). This may be the explanation for the existence of peak responses to PRF in cellular and animal studies (Pilla, 2006; Strauch et al., 2007; Pilla et al., 2011). It may also explain recent clinical results in which, for an identical PRF configuration, a treatment regimen, which produced NO at fourfold the rate of a known effective signal configuration for postoperative pain relief, was ineffective (Taylor et al., 2014).

This modification of the ECM model may be tested in other studies. Two recent publications show that PEMF effects depend upon signal configuration. In the first, a PEMF effect on breast cancer cell apoptosis was significant when the same waveform, applied for the same exposure time, was repeated at 20 Hz but not at 50 Hz (Crocetti et al., 2013). The second study showed that PEMF significantly reduced the expression of inflammatory markers, tumor necrosis factor (TNF), and NF-κB in challenged macrophages when the same waveform, applied for the same exposure time, was repeated at 5 Hz, but not at 15 or 30 Hz (Ross and Harrison, 2013a). In both studies, Ca^{2+}-dependent NO signaling modulates the expressions of these inflammatory markers (Ha et al., 2003; Yurdagul et al., 2013), suggesting the effect of increased repetition rate is consistent with increased production of NO at a rate high enough for negative feedback from PDE isoforms to predominate, thus blocking the PEMF effect on Ca^{2+}-dependent NO signaling.

DISCUSSION AND CONCLUSIONS

There is now sufficient evidence to support the proposal that CaM-dependent NO signaling is a primary PEMF transduction pathway. There is abundant recent data showing PEMF effects on NO, cGMP, and inflammatory cytokines at the cellular, animal, and clinical levels, as well as evidence for enhanced wound repair and tissue healing. Many studies have suggested NO plays a role in biological responses to weak nonthermal EMF without proposing a transduction mechanism. Perhaps the first was a report that a burst-type RF PEMF increased NO and cGMP in a rat cerebellum cell-free supernatant (Miura et al., 1993). This study determined that

L-NAME and EDTA could inhibit the effect, suggesting CaM-dependent NO release was involved. Interestingly, when a continuous frequency waveform was applied, there were no effects on NO, confirming that bioeffects require matching amplitude spectrum of the signal to the kinetics of the proposed ion-binding target (Pilla, 2006; Pilla et al., 2011). Several studies have reported that NO mediates the effect of PEMF on osteoblast proliferation and differentiation (Diniz et al., 2002; Cheng et al., 2011) and chondrocyte proliferation (Fitzsimmons et al., 2008).

In this review, emphasis was placed on the evidence that a PRF signal could be configured a priori to modulate CaM/NO/cGMP signaling. However, the predictions of the ECM model apply to any PEMF signal. For example, a study on the effect of PEMF on bone repair in a rat fibular osteotomy model compared the original BGS signal (Pilla, 2006) with a newer pulse-burst waveform having much shorter pulse duration (Midura et al., 2005). This study showed the original BGS waveform accelerated bone repair by producing a twofold increase in callus volume and a twofold increase in the rate of hard callus formation at 2 weeks, followed by a twofold increase in callus stiffness at 5 weeks. In contrast, the new pulse burst waveform had no significant effect on any of these parameters in this model. Application of the ECM model, assuming a CaM/NO/cGMP signaling target, provides an explanation for the differing PEMF effects. Although both PEMF signals were repetitive asymmetrical pulse bursts, they had large differences in the amplitude of their respective frequency components in the Ca/CaM binding pathway. The new pulse burst induced an electric field consisting of approximately a 1 ms burst of relatively short, asymmetrical pulses (4/12 μs) having a peak amplitude of 0.05 V/m, repeating at approximately 1 Hz. In contrast, the induced electric field from the original BGS signal consisted of a 5 ms burst of significantly longer asymmetrical pulses (200/20 μs) having significantly higher peak amplitude (2 V/m) and repetition rate (15 Hz). Comparison of SNR in the Ca/CaM binding pathway for these EMF signals shows that the SNR frequency spectrum of SNR for the new pulse-burst signal does not effectively couple to the Ca/CaM binding pathway. Thus, application of the ECM model, assuming Ca/CaM/cGMP as the transduction pathway, would have predicted this signal ineffective for bone repair, as was indeed reported. It is of interest to note that peak SNR in the Ca/CaM binding pathway for the original BGS signal is approximately fivefold lower than that for the PRF signal considered in this review (Pilla et al., 2011). This large difference in dosimetry may provide one explanation for the generally observed slow response to BGS signals in clinical applications.

It is also possible to use the ECM model to predict whether nonthermal bioeffects may be produced from the RF signals emitted by cellular phones using the ECM model. Indeed, the asymmetrical kinetics of Ca/CaM binding may also be a potential signaling target for a GSM waveform. Evaluation of SNR in the Ca/CaM target pathway indicates that SNR for a single 577 μs pulse of a 1,800 MHz GSM signal at 20 V/m falls within the same frequency range and has a peak value similar to that for a single 2,000 μs pulse of a 27.12 MHz RF signal at 12 V/m, typical measured field strengths within the biological target for these signals (Pilla and Muehsam, 2010). Thus, GSM signals may have nonthermal bioeffects, including therapeutic effects, as suggested by recent reports showing that long-term exposure to a GSM signal protects against and reverses cognitive impairment in a mouse model of Alzheimer's disease (Arendash et al., 2010, 2012; Dragicevic et al., 2011). Also, it has been suggested that inflammation, mediated by both IL-1β and iNOS, will enhance the deposition of β-amyloid (Chiarini et al., 2006). According to the transduction mechanism proposed here, a GSM signal would be expected to downregulate both factors, which may prevent or reverse the effects of Alzheimer's disease and any other neurodegenerative disease with an inflammatory component. It is to be noted that the mechanism proposed here is based upon kinetic asymmetry in ion-binding kinetics and does not depend upon nonlinearity in the electrical response of the target to RF signals (Kowalczuk et al., 2010).

While the results presented here support an electric field effect on CaM signaling via the asymmetrical voltage-dependent kinetics of Ca^{2+} binding to CaM (k_{on}), it is important to emphasize that the slow dissociation kinetics of Ca/CaM (k_{off}) can be responsive to weak DC and combined DC/AC magnetic (B) fields. Indeed, it has been reported that weak B fields can affect NO signaling (Palmi et al., 2002; Reale et al., 2006). Of the many models proposed to explain the bioeffects of weak magnetic fields, those involving modulation of the bound trajectory of a charged ion by classic Lorentz force (Chiabrera et al., 1993; Edmonds, 1993; Muehsam and Pilla, 2009a,b) are most relevant to asymmetrical ion-binding kinetics, suggesting that weak B-field effects on the trajectory of the ion within the binding site itself could affect reactivity. One interpretation of this is that weak DC and certain combinations of weak AC/DC magnetic fields could enhance or inhibit the exit of the target ion from the binding site, thereby accelerating or inhibiting the overall reaction rate by manipulating dissociation kinetics (k_{off}), even in the presence of thermal noise. This has been tested with success for CaM-dependent myosin phosphorylation (Muehsam and Pilla, 2009b) suggesting that analysis of B-field effects on bound ion trajectories could be used to explain and predict weak magnetic field effects on CaM-dependent NO signaling via modulation of Ca/CaM dissociation kinetics.

Taken together, the results reviewed here strongly support that PEMF signals can be configured to instantaneously modulate NO release in injured tissue. At the cellular level, PEMF-mediated NO signaling could be the common transduction mechanism in studies that report up- and downregulation of anti- and proinflammatory genes (Brighton et al., 2008; Moffett et al., 2012; Pena-Philippides et al., 2014a,b) and modulation of adenosine pathways (De Mattei et al., 2009; Vincenzi et al., 2013; Adravanti et al., 2014). At the clinical level, there is strong support that PEMF-mediated signaling is anti-inflammatory and can enhance angiogenesis. NO release via cNOS is dynamic (Batchelor et al., 2010) and closely linked to the negative feedback provided by PDE inhibition of cGMP (Mo et al., 2004). This causes ensuing dynamic consequences in the rates of up- or downregulation of growth factors and cytokines. For example, iNOS activity in the inflammatory stage of healing can be rapidly downregulated by inhibition of NF-kB in a negative feedback mechanism (Chang et al., 2004).

The proposed transduction mechanism is consistent with the hypothesis that a nonthermal PEMF signal can be configured a priori to act as a first messenger in CaM-dependent signaling pathways that include NO and cyclic nucleotides relevant to tissue growth, repair, and maintenance. Of the multitude of intracellular Ca^{2+} buffers, CaM is an important early responder because CaM-dependent cytokine and growth factor release orchestrates rapid cellular and tissue response to physical and mechanical challenges. Ca^{2+} binding to CaM is voltage dependent, and its kinetic asymmetry ($k_{on} \gg k_{off}$) yields rectifier-like properties, allowing the use of an electrochemical model to configure effective EMF signals to couple efficiently with this pathway. The ECM model is potentially powerful enough to unify the observations of many groups and may offer a means to explain both the wide range of reported bioeffects as well as the many equivocal reports from PEMF studies. The predictions of the proposed model open a host of significant possibilities for configuration of nonthermal EMF signals for clinical and wellness applications that can reach far beyond fracture repair and wound healing. Active studies are underway to assess the utility of the known anti-inflammatory activity of PRF for the treatment of traumatic brain injury, neurodegenerative diseases, cognitive disorders, degenerative joint disease, neural regeneration, and cardiac and cerebral ischemia.

REFERENCES

Aaron RK, Ciombor DM, Simon BJ. Treatment of nonunions with electric and electromagnetic fields. *Clin Orthop.* 2004;419:21–29.

Adravanti P, Nicoletti S, Setti S, Ampollini A, de Girolamo L. Effect of pulsed electromagnetic field therapy in patients undergoing total knee arthroplasty: A randomised controlled trial. *Int Orthop.* 2014;38(2):397–403.

Arendash GW, Mori T, Dorsey M, Gonzalez R, Tajiri N, Borlongan C. Electromagnetic treatment to old Alzheimer's mice reverses β-amyloid deposition, modifies cerebral blood flow, and provides selected cognitive benefit. *PLoS One.* 2012;7(4):e35751.

Arendash GW, Sanchez-Ramos J, Mori T et al. Electromagnetic field treatment protects against and reverses cognitive impairment in Alzheimer's disease mice. *J Alzheimers Dis.* 2010;19:191–210.

Basset CAL, Pilla AA, Pawluk RA Non-surgical salvage of surgically-resistant pseudoarthroses and nonunions by pulsing electromagnetic fields. *Clin Orthop.* 1977;124:117–131.

Batchelor AM, Bartus K, Reynell C, Constantinou S, Halvey EJ, Held KF, Dostmann WR, Vernon J, Garthwaite J. Exquisite sensitivity to subsecond, picomolar nitric oxide transients conferred on cells by guanylyl cyclase-coupled receptors. *Proc Natl Acad Sci USA* 2010;107:22060–22065.

Blumenthal DK, Stull JT. Effects of pH, ionic strength, and temperature on activation by calmodulin and catalytic activity of myosin light chain kinase. *Biochemistry* 1982;21:2386–2391.

Bootman MD, Lipp P, Berridge MJ. The organisation and functions of local Ca(2+) signals. *J Cell Sci.* 2001;114:2213–2222.

Borhade N, Pathan AR, Halder S et al. NO-NSAIDs. Part 3: Nitric oxide-releasing prodrugs of non-steroidal anti-inflammatory drugs. *Chem Pharm Bull (Tokyo)* 2012;60:465–481.

Bredt DS, Snyder SH. Isolation of nitric oxide synthetase, a calmodulin-requiring enzyme. *Proc Natl Acad Sci USA.* 1990;87:682–685.

Brighton CT, Wang W, Clark CC. The effect of electrical fields on gene and protein expression in human osteoarthritic cartilage explants. *J Bone Joint Surg Am.* 2008;90:833–848.

Broughton G 2nd, Janis JE, Attinger CE. Wound healing: An overview. *Plast Reconstr Surg.* 2006; 117(7 Suppl):1e-S–32e-S.

Callaghan MJ, Chang EI, Seiser N, Aarabi S, Ghali S, Kinnucan ER, Simon BJ, Gurtner GC. Pulsed electromagnetic fields accelerate normal and diabetic wound healing by increasing endogenous FGF-2 release. *Plast Reconstr Surg.* 2008;121:130–141.

Chang K, Lee SJ, Cheong I, Billiar TR, Chung HT, Han JA, Kwon YG, Ha KS, Kim YM. Nitric oxide suppresses inducible nitric oxide synthase expression by inhibiting posttranslational modification of IkappaB. *Exp Mol Med.* 2004;36:311–324.

Chen Y, Boettger MK, Reif A et al. Nitric oxide synthase modulates CFA-induced thermal hyperalgesia through cytokine regulation in mice. *Mol Pain.* 2010;6:13–17.

Cheng DK. *Analysis of Linear Systems.* Addison-Wesley, London, U.K., 1959.

Cheng G, Zhai Y, Chen K, Zhou J, Han G, Zhu R, Ming L, Song P, Wang J. Sinusoidal electromagnetic field stimulates rat osteoblast differentiation and maturation via activation of NO-cGMP-PKG pathway. *Nitric Oxide.* 2011;25:316–325.

Chiabrera A, Bianco B, Moggia E. Effect of lifetimes on ligand binding modeled by the density operator. *Bioelectrochem Bioenerg.* 1993;30:35–42.

Chiarini A, Dal Pra I, Whitfield JF, Armato U. The killing of neurons by beta-amyloid peptides, prions, and pro-inflammatory cytokines. *Ital J Anat Embryol.* 2006;111:221–246.

Cho HJ, Xie QW, Calaycay J, Mumford RA, Swiderek KM, Lee TD, Nathan C. Calmodulin is a subunit of nitric oxide synthase from macrophages. *J Exp Med.* 1992;176:599–604.

Cooke JP. NO and angiogenesis. *Atheroscler Suppl.* 2003;4:53–60.

Crocetti S, Beyer C, Schade G, Egli M, Fröhlich J, Franco-Obregón A. Low intensity and frequency pulsed electromagnetic fields selectively impair breast cancer cell viability. *PLoS One.* 2013;8(9):e72944.

Cury Y, Picolo G, Gutierrez VP et al. Pain and analgesia. The dual effect of nitric oxide in the nociceptive system. *Nitric Oxide.* 2011;25:243–254.

Daff S. Calmodulin-dependent regulation of mammalian nitric oxide synthase. *Biochem Soc Trans.* 2003;31:502–505.

DeFelice LJ. *Introduction to Membrane Noise.* Plenum, New York, 1981, pp. 243–245.

Delle Monache S, Alessandro R, Iorio R, Gualtieri G, Colonna R. Extremely low frequency electromagnetic fields (ELF-EMFs) induce in vitro angiogenesis process in human endothelial cells. *Bioelectromagnetics.* 2008;29:640–648.

De Mattei M, Varani K, Masieri FF, Pellati A, Ongaro A, Fini M, Cadossi R, Vincenzi F, Borea PA, Caruso A. Adenosine analogs and electromagnetic fields inhibit prostaglandin E2 release in bovine synovial fibroblasts. *Osteoarthritis Cart.* 2009;17:252–262.

Diniz P, Soejima K, Ito G. Nitric oxide mediates the effects of pulsed electromagnetic field stimulation on the osteoblast proliferation and differentiation. *Nitric Oxide.* 2002;7:18–23.

Dragicevic N, Bradshaw PC, Mamcarz M, Lin X, Wang L, Cao C, Arendash GW. Long-term electromagnetic field treatment enhances brain mitochondrial function of both Alzheimer's transgenic mice and normal mice: A mechanism for electromagnetic field-induced cognitive benefit? *Neuroscience* 2011;185:135–149.

Edmonds DT. Larmor Precession as a mechanism for the detection of static and alternating magnetic fields. *Bioelectrochem Bioenerg.* 1993;30:3–12.

Faas GC, Raghavachari S, Lisman JE, Mody I. Calmodulin as a direct detector of Ca^{2+} signals. *Nat Neurosci.* 2011;14:301–304.

Fitzsimmons RJ, Gordon SL, Kronberg J, Ganey T, Pilla AA. A pulsing electric field (PEF) increases human chondrocyte proliferation through a transduction pathway involving nitric oxide signaling. *J Orthop Res.* 2008;26:854–859.

Fontanesi G, Giancecchi F, Rotini R, Cadossi R. Treatment of delayed union and pseudarthrosis by low frequency pulsing electromagnetic stimulation. Study of 35 cases. *Ital J Orthop Traumatol.* 1983;9:305–318.

Guo L, Kubat NJ, Nelson TR, Isenberg RA. Meta-analysis of clinical efficacy of pulsed radio frequency energy treatment. *Ann Surg.* 2012;255:457–467.

Ha KS, Kim KM, Kwon YG et al. Nitric oxide prevents 6-hydroxydopamine-induced apoptosis in PC12 cells through cGMP-dependent PI3 kinase/Akt activation. *FASEB J.* 2003;17:1036–1047.

Hedén P, Pilla AA. Effects of pulsed electromagnetic fields on postoperative pain: A double-blind randomized pilot study in breast augmentation patients. *Aesthetic Plast Surg.* 2008;32:660–666.

Ignarro LJ, Fukuto JM, Griscavage JM, Rogers NE, Byrns RE. Oxidation of nitric oxide in aqueous solution to nitrite but not nitrate: Comparison with enzymatically formed nitric oxide from L-arginine. *Proc Natl Acad Sci USA.* 1993;90:8103–8107.

Kavaliers M, Ossenkopp KP. Opioid systems and magnetic field effects in the land snail, Cepaea nemoralis. *Bio Bull.* 1991;180:301–309.

Kloth LC, Berman JE, Sutton CH, Jeutter DC, Pilla AA, Epner ME. Effect of pulsed radio frequency stimulation on wound healing: A double-blind pilot clinical study. In *Electricity and Magnetism in Biology and Medicine*, Bersani F (ed.) Plenum, New York, 1999, pp. 875–878.

Kloth LC, Pilla AA. Electromagnetic stimulation for wound repair. In *Wound Healing: Evidence Based Management*, 4th edn., McCulloch JM, Kloth LC (eds.) Davis, Philadelphia, PA, 2010, pp. 514–544.

Konieczny V, Keebler MV, Taylor CW. Spatial organization of intracellular Ca^{2+} signals. *Semin Cell Dev Biol.* 2012;23:172–180.

Kowalczuk C, Yarwood G, Blackwell R et al. Absence of nonlinear responses in cells and tissues exposed to RF energy at mobile phone frequencies using a doubly resonant cavity. *Bioelectromagnetics.* 2010;31:556–565.

Lauretti GR, de Oliveira R, Reis MP et al. Transdermal nitroglycerine enhances spinal sufentanil postoperative analgesia following orthopedic surgery. *Anesthesiology.* 1999;90(3):734–739.

Lawrence T. The nuclear factor NF-kappaB pathway in inflammation. *Cold Spring Harb Perspect Biol.* 2009;1(6):a001651.

Lee RH, Efron D, Tantry U, Barbul A. Nitric oxide in the healing wound: A time-course study. *J Surg Res.* 2001;101:104–108.

Markov MS, Muehsam DJ, Pilla AA. Modulation of cell-free myosin phosphorylation with pulsed radio frequency electromagnetic fields. In *Charge and Field Effects in Biosystems 4*, Allen MJ, Cleary SF, Sowers AE (eds.) World Scientific, Hackensack, NJ, 1994, pp. 274–288.

McKay JC, Prato FS, Thomas AW. A literature review: The effects of magnetic field exposure on blood flow and blood vessels in the microvasculature. *Bioelectromagnetics* 2007;28:81–98.

Mellstrom B, Savignac M, Gomez-Villafuertes R et al. Ca^{2+}-operated transcriptional networks: Molecular mechanisms and in vivo models. *Physiol Rev.* 2008;88:421–449.

Midura RJ, Ibiwoye MO, Powell KA, Sakai Y, Doehring T, Grabiner MD, Patterson TE, Zborowski M, Wolfman A. Pulsed electromagnetic field treatments enhance the healing of fibular osteotomies. *J Orthop Res.* 2005;23:1035–1046.

Miura M, Takayama K, Okada J. Increase in nitric oxide and cyclic GMP of rat cerebellum by radio frequency burst-type electromagnetic field radiation. *J Physiol.* 1993;461:513–524.

Mo E, Amin H, Bianco IH, Garthwaite J. Kinetics of a cellular nitric oxide/cGMP/ phosphodiesterase-5 pathway. *J Biol Chem.* 2004;279:26149–26158.

Moffett J, Fray LM, Kubat NJ. Activation of endogenous opioid gene expression in human keratinocytes and fibroblasts by pulsed radiofrequency energy fields. *J Pain Res.* 2012;12:347–357.

Moffett J, Griffin NE, Ritz MC et al. Pulsed radio frequency energy field treatment of cells in culture results in increased expression of genes involved in the inflammation phase of lower extremity diabetic wound healing. *J Diabetic Foot Complications.* 2010;2:57–64.

Moffett J, Kubat NJ, Griffin NE et al. Pulsed radio frequency energy field treatment of cells in culture results in increased expression of genes involved in angiogenesis and tissue remodeling during wound healing. *J Diabetic Foot Complications.* 2011;3:30–39.

Muehsam DJ, Pilla AA. A Lorentz model for weak magnetic field bioeffects: Part I— Thermal noise is an essential component of AC/DC effects on bound ion trajectory. *Bioelectromagnetics.* 2009a; 30:462–475.

Muehsam DJ, Pilla AA. A Lorentz model for weak magnetic field bioeffects: Part II— Secondary transduction mechanisms and measures of reactivity. *Bioelectromagnetics.* 2009b;30:476–488.

Palmi M, Meini A. Role of the nitric oxide/cyclic GMP/Ca^{2+} signaling pathway in the pyrogenic effect of interleukin-1beta. *Mol Neurobiol.* 2002;25:133–147.

Patel MK, Factor SM, Wang J, Jana S, Strauch B. Limited myocardial muscle necrosis model allowing for evaluation of angiogenic treatment modalities. *J Reconstr Microsurg.* 2006;22:611–615.

Pena-Philippides JC, Hagberg S, Nemoto E, Roitbak T. Effect of pulsed electromagnetic field (PEMF) on LPS induced chronic inflammation in mice. In *Pulsed Electromagnetic Fields*, Markov MS (ed.) CRC Press, Boca Raton, FL, 2015, pp. 165–172 (this volume).

Pena-Philippides JC, Yang Y, Bragina O, Hagberg S, Nemoto E, Roitbak T. Effect of pulsed electromagnetic field (PEMF) on infarct size and inflammation after cerebral ischemia in mice. *Transl Stroke Res.* 2014b;5:491–500.

Pilla AA. Electrochemical information and energy transfer in vivo. *Proceedings 7th IECEC*, Washington, DC, 1972, pp. 761–764.

Pilla AA. Electrochemical information transfer at living cell membranes. *Ann NY Acad Sci.* 1974a;238:149–170.

Pilla AA. Mechanisms of electrochemical phenomena in tissue growth and repair. *Bioelectrochem Bioenerg.* 1974b;1:227–243.

Pilla AA. On the possibility of an electrochemical trigger for biological growth and repair processes. *Bioelectrochem Bioenerg.* 1976;3:370–373.

Pilla AA. Mechanisms and therapeutic applications of time varying and static magnetic fields. In *Biological and Medical Aspects of Electromagnetic Fields*, Barnes F, Greenebaum B (eds.) CRC Press, Boca Raton, FL, 2006, pp. 351–411.

Pilla AA. Electromagnetic fields instantaneously modulate nitric oxide signaling in challenged biological systems. *Biochem Biophys Res Commun.* 2012;426:330–333.

Pilla AA. Nonthermal electromagnetic fields: From first messenger to therapeutic applications. *Electromagn Biol Med.* 2013;32:123–136.

Pilla AA, Fitzsimmons R., Muehsam DJ, Rohde C, Wu JK, Casper D. Electromagnetic fields as first messenger in biological signaling: Application to calmodulin-dependent signaling in tissue repair. *Biochem Biophys Acta.* 2011;1810:1236–1245.

Pilla AA, Muehsam DJ. Non-thermal bioeffects from radio frequency signals: Evidence from basic and clinical studies, and a proposed mechanism. *Proceedings, 32nd Annual Meeting, Bioelectromagnetics Society*, Frederick, MD, June 2010.

Pilla AA, Muehsam DJ, Markov MS, Sisken BF. EMF signals and ion/ligand binding kinetics: Prediction of bioeffective waveform parameters. *Bioelectrochem Bioenerg.* 1999;48:27–34.

Pilla AA, Nasser PR, Kaufman JJ. Gap junction impedance, tissue dielectrics and thermal noise limits for electromagnetic field bioeffects. *Bioelectrochem Bioenerg.* 1994;35:63–69.

Prato FS, Carson JJ, Ossenkopp KP et al. Possible mechanisms by which extremely low frequency magnetic fields affect opioid function. *FASEB J.* 1995;9:807–814.

Rasouli J, Lekhraj R, White NM, Flamm ES, Pilla AA, Strauch B, Casper D. Attenuation of interleukin-1beta by pulsed electromagnetic fields after traumatic brain injury. *Neurosci Lett.* 2012;519:4–8.

Rawe IM, Lowenstein A, Barcelo CR, Genecov DG. Control of postoperative pain with a wearable continuously operating pulsed radiofrequency energy device: A preliminary study. *Aesthetic Plast Surg.* 2012;36:458–463.

Reale M, De Lutiis MA, Patruno A, Speranza L, Felaco M, Grilli A, Macrì MA, Comani S, Conti P, Di Luzio S. Modulation of MCP-1 and iNOS by 50-Hz sinusoidal electromagnetic field. *Nitric Oxide.* 2006;15:50–57.

Ren K, Torres R. Role of interleukin-1beta during pain and inflammation. *Brain Res Rev.* 2009;60:57–64.

Rohde C, Chiang A, Adipoju O, Casper D, Pilla AA. Effects of pulsed electromagnetic fields on IL-1β and post operative pain: A double-blind, placebo-controlled pilot study in breast reduction patients. *Plast Reconstr Surg.* 2010;125:1620–1629.

Rohde C., Hardy K, Asherman J, Taylor E, Pilla AA. PEMF therapy rapidly reduces postoperative pain in TRAM flap patients. *Plast Reconstr Surg.* 2012;130(5S-1):91–92.

Rohde C, Taylor E, Pilla A. Pulsed electromagnetic fields accelerate reduction of post-operative pain and inflammation: Application to plastic and reconstructive surgical procedures. 2014, this volume.

Roland D, Ferder MS, Kothuru R, Faierman T, Strauch B. Effects of pulsed magnetic energy on a microsurgically transferred vessel. *Plast Reconstr Surg.* 2000;105:1371–1374.

Ross CL, Harrison BS. Effect of pulsed electromagnetic field on inflammatory pathway markers in RAW 264.7 murine macrophages. *J Inflamm Res.* 2013a;6:45–51.

Ross CL, Harrison BS. The use of magnetic field for the reduction of inflammation: A review of the history and therapeutic results. *Altern Ther Health Med.* 2013b;19(2):47–54.

Sluka KA, Deacon M, Stibal A et al. Spinal blockade of opioid receptors prevents the analgesia produced by TENS in arthritic rats. *J Pharmacol Exp Ther.* 1999;289:840–846.

Strauch B, Herman C, Dabb R, Ignarro LJ, Pilla AA. Evidence-based use of pulsed electromagnetic field therapy in clinical plastic surgery. *Aesthet Surg J.* 2009;29:135–143.

Strauch B, Patel MK, Navarro A, Berdischevsky M, Pilla AA. Pulsed magnetic fields accelerate wound repair in a cutaneous wound model in the rat. *Plast Reconstr Surg.* 2007;120:425–430.

Strauch B, Patel MK, Rosen D, Casper D, Pilla AA. Pulsed magnetic fields increase angiogenesis in a rat myocardial ischemia model. *Proceedings, 28th Annual Meeting*, Bioelectromagnetics Society, Frederick, MD, June 2006.

Taylor E, Hardy K, Alonso A, Pilla A, Rohde C. Pulsed electromagnetic field (PEMF) dosing regimen impacts pain control in breast reduction patients. *J Surg Res.* 2014 [Epub ahead of print].

Tepper OM, Callaghan MJ, Chang EI et al. Electromagnetic fields increase in vitro and in vivo angiogenesis through endothelial release of FGF-2. *FASEB J.* 2004;18:1231–1233.

Tsoukias NM. Nitric oxide bioavailability in the microcirculation: Insights from mathematical models. *Microcirculation.* 2008;15:813–834.

Velazquez C, Praveen Rao PN, Knaus EE. Novel nonsteroidal antiinflammatory drugs possessing a nitric oxide donor diazen-1-ium-1,2-diolate moiety: Design, synthesis, biological evaluation, and nitric oxide release studies. *J Med Chem.* 2005;48:4061–4067.

Vianale G, Reale M, Amerio P, Stefanachi M, Di Luzio S, Muraro R. Extremely low frequency electromagnetic field enhances human keratinocyte cell growth and decreases proinflammatory chemokine production. *Br J Dermatol.* 2008;158:1189–1196.

Vincenzi F, Targa M, Corciulo C, Gessi S, Merighi S, Setti S, Cadossi R, Goldring MB, Borea PA, Varani K. Pulsed electromagnetic fields increased the anti-inflammatory effect of A_2A and A_3 adenosine receptors in human T/C-28a2 chondrocytes and hFOB 1.19 osteoblasts. *PLoS One.* 2013;8(5):e65561.

Weber RV, Navarro A, Wu JK, Yu HL, Strauch B. Pulsed magnetic fields applied to a transferred arterial loop support the rat groin composite flap. *Plast Reconstr Surg.* 2005;114:1185–1189.

Werner S, Grose R. Regulation of wound healing by growth factors and cytokines. *Physiol Rev.* 2003;83:835–870.

Yen-Patton GP, Patton WF, Beer DM et al. Endothelial cell response to pulsed electromagnetic fields: Stimulation of growth rate and angiogenesis in vitro. *J Cell Physiol.* 1998;134:37–39.

Yurdagul A Jr, Chen J, Funk SD, Albert P, Kevil CG, Orr AW. Altered nitric oxide production mediates matrix specific PAK2 and NF-κB activation by flow. *Mol Biol Cell.* 2013;24:398–408.

3 Biophysical Stimulation of Bone Growth in Fractures

Ruggero Cadossi, Stefania Setti,
Matteo Cadossi, and Leo Massari

CONTENTS

INTRODUCTION

The effect of physical stimuli on bone tissue is well established; bone cells respond to both mechanical and electrical forces. The cell membrane plays the fundamental role in recognizing and transferring the physical stimulus to the various metabolic pathways in the cell. Brighton identified the intracellular events that are activated when biophysical stimuli are applied *in vitro* to cells (Brighton et al. 2001; Varani et al. 2017) (Figure 3.1).

The effects of mechanical forces on bone tissue are linked to the mineralized component of the bone or to the vital one. When a load is applied to the bone, the mineralized component, in particular the asymmetric crystals of hydroxyapatite, and

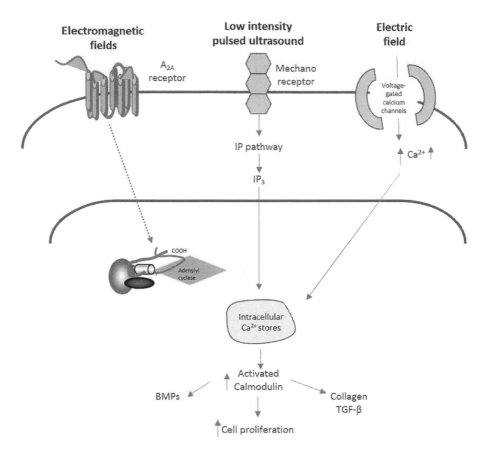

FIGURE 3.1 Schematic representation of the biophysical stimuli targets on the cell surface and corresponding metabolic pathways within the cell.

the collagen matrix become electrically charged. In a bone under compression, negative charges accumulate, while a bone under tension is positively charged (Black 1987; Guzelsu 1993). The electrical response of a bone to mechanical load is described as piezoelectric effect. Furthermore, the electrical properties of the bone are modified by the electrokinetic phenomenon occurring when ion flow occurs within Haversian and endocanalicular spaces (Green and Kleeman 1991; Otter et al. 1992; Pollack 1984). The mechanical load is thus transformed into an electrical signal that can be perceived by bone cells. Thus electronegative areas will undergo bone deposition, and electropositive areas will undergo bone resorption, finally leading to bone remodeling. This phenomenon is responsible for bone adaptation to mechanical load and for bone remodeling that occurs in the last phase of fracture healing to optimize the mechanical competence of healed bone. Furthermore, bone cells possess surface mechanical receptors that can be activated directly by mechanical stimuli (Govey et al. 2014).

Electrical charges can be observed in the vital bone as a consequence of cell activity; the surface stationary bioelectric potential and stationary electric (ionic) current can be measured *ex vivo*. Friedenberg and Brighton described how the typical distribution of the bioelectric potentials is immediately altered by bone fracture (Friedenberg and Brighton 1966; Friedenberg et al. 1971, 1973). Specifically the area of fracture becomes electronegative compared to the surrounding tissue.

The above observations promoted a series of researches to investigate if biophysical stimuli (electromagnetic, electrical, mechanical) could be used in clinical practice to enhance fracture healing. Different procedures to apply biophysical stimulation to fractured bone have been investigated: pulsed electromagnetic fields (PEMFs), capacitively coupled electric fields (CCEFs), low-intensity pulsed ultrasound (LIPU) (Figure 3.2). Biophysical stimulation was approved by Food and Drug Administration (FDA) 40 years ago.

The principles of pharmacological research have been adopted to identify, characterize, and optimize the biophysical stimuli parameters: amplitude, frequency, waveform, and exposure length. Ultimately, dose response effects have been described.

As pharmacodynamics for drug development, Physical dynamics is the basis of biophysical stimulation of bone healing. Physical dynamics proves the relationship between the biological effect and the specific physical parameters used for treatment (Figure 3.3).

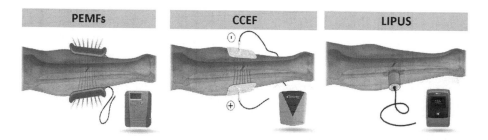

FIGURE 3.2 Technologies used to apply physical stimuli to bone fracture: inductive (PEMF), capacitive (CCEF), and mechanical (LIPU).

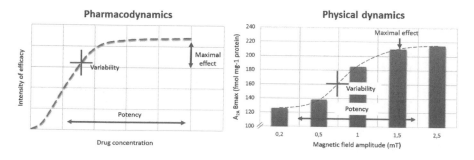

FIGURE 3.3 Physical dynamics shows that biological system response depends on signal's physical parameters.

PRECLINICAL RESEARCH

The above observations stimulated research activities to investigate the possibility of active intervention with physical stimuli in bone cell activity *in vitro* (Sollazzo et al. 1996, 2011; Bianco and Chiabrera 1992; Vincenzi et al. 2013; Guerkov et al. 2001; Lin and Lin 2011; Aaron et al. 1989) and in reparative osteogenesis *in vivo* (Canè et al. 1993, 1997).

IN VITRO STUDIES

Many researches have focused on the effect of biophysical stimulation on cell proliferation and synthesis of extracellular bone matrix (Aaron and Ciombor 1996). Physical stimuli increase synthesis of bone matrix and growth factors and favor the proliferation and differentiation of the osteoblast-like primary cells (Aaron et al. 2004; Ciombor and Aaron 2005). Fassina et al. investigated the effect of PEMF on SAOS-2 human osteoblast proliferation and on calcified matrix production over a polyurethane porous scaffold. The authors showed a higher cell proliferation and a greater expression of decorin, fibronectin, osteocalcin, osteopontin, transforming growth factor-β (TGF-β1), type I collagen, and type III collagen in PEMF-stimulated culture compared to controls (Fassina et al. 2006). It has been shown that bone cell proliferation is related to the length of exposure to PEMF; no effect was observed for exposures shorter than 12 h (Sollazzo et al. 1996) (Figure 3.4). Recently, Tschon et al. have demonstrated the effect of PEMF and platelet-rich plasma (PRP) in preventing osteoclastogenesis in an *in vitro* model of osteolysis (Tschon et al. 2018). The authors concluded that biophysical stimulation seems to be the treatment of choice for counteracting osteoclastogenesis rather than PRP treatment; PEMF and PRP showed a synergic effect in reducing tumor necrosis factor-α (TNF-α) levels.

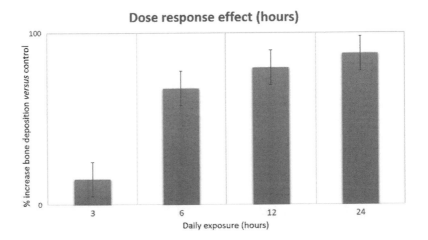

FIGURE 3.4 Effect of PEMF exposure length on proliferation of human osteosarcoma cell lines (MG63).

Hartig et al. demonstrated that the exposure to CCEF of osteoblast-like primary cells increases the synthesis of bone matrix and favors their proliferation and differentiation (Hartig et al. 2000).

Ryaby et al. reported that LIPU increased calcium incorporation in bone-cell cultures, reflecting a change in cell metabolism (Ryaby et al. 1992). This increase in second messenger activity was paralleled by the modulation of adenylate cyclase activity and TGF-β1 synthesis in osteoblastic cells.

The effect of biophysical stimuli on bone morphogenetic proteins (BMPs) has been repeatedly demonstrated using different cells and exposure conditions (Table 3.1). Ultimately, the different biophysical stimulation technologies result in increased endogenous production of growth factors belonging to the TGF-β family, leading to increased bone formation. Furthermore, the study of additional effects when cells are exposed to both growth factors and physical stimuli is a fertile area for investigation. Based on the effect of physical stimuli on BMP synthesis and releases, PEMF stimulation has been proposed as an exogenous technology to increase endogenous synthesis of BMPs able to promote osteogenesis for nonunion treatment.

In Vivo Studies

Several *in vivo* studies demonstrate that biophysical stimulation can shorten the time to healing of experimental fractures in rats (Fontanesi et al. 1986). Moreover, PEMFs alone or PEMFs enriched with PRP are able to limit bone loss by reducing the number of bone resorbing cells and fibrous tissue interposition between implants and bone tissue in rats (Figure 3.5) (Veronesi et al. 2018). PEMF enhances the apposition of bone by increasing bone to implant contact and bone volume, appearing to be a promising treatment for accelerating bone ingrowth in both trabecular and cortical bone of replacement models (Fini et al. 2006).

PEMFs' effect on osteoblast activity has been also investigated. Holes were drilled in both metacarpal bones of horses, half of which were stimulated. Tetracycline labeling was used to monitor mineral apposition rate in newly formed trabeculae. The results showed that the mineral apposition rate was double in PEMF-exposed bone defects (Canè et al. 1991). The effect of PEMFs was dependent on daily exposure and disappeared when treatment time was <6 h/day. Midura exposed fibular osteotomies in rats to two PEMFs employing different pulse parameters but the same exposure time: 3 h/day (Midura et al. 2005). The beneficial effect of the stimulation was observed in only one of the two PEMFs settings. The authors concluded that there is a specific relationship between waveform characteristics and biological outcomes.

Using CCEF (Brighton et al. 1985) in experimental fractures of rabbit fibula, a significant increase in fracture stiffness was observed. The effect was frequency dependent and peaked at 60 kHz. In 1994, Rijal et al. created a nonunion experimental model and showed an 18% increase in density (p < 0.05) in the CCEF group compared to the controls (Rijal et al. 1994).

Wang et al. reported a 67% increase in stiffness of bilateral closed femoral fractures in rats treated with LIPU compared to controls (p < 0.02) (Wang et al. 1994).

TABLE 3.1

Effect of Different Biophysical Stimulation Techniques on Bone Cell Cultures

Author	Biophysical stimuli	In vitro models	Results
Bodamyali et al. (1998)	PEMF	Rat calvarial osteoblasts	↑Proliferation ↑BMP-2,4 mRNA
Aaron et al. (1999)	PEMF	Endochondral ossification in vivo	↑Differentiation ↑TGF-β1mRNA protein
Lohmann et al. (2000)	PEMF	MG63 Osteoblasts	↑Differentiation ↑TGF-β1
Guerkov et al. (2001)	PEMF	Human nonunion cells	↑TGF-β1
Fassina et al. (2006)	PEMF	SAOS-2 Osteoblasts	↑Proliferation ↑TGF-β1
Jansen et al. (2010)	PEMF	hBMSCs	↑TGF-β1 ↑BMP-2mRNA ↑Differentation
Esposito et al. (2012)	PEMF	hBMSCs	↑Proliferation ↑Differentation
Ceccarelli et al. (2013)	PEMF	hBMSCs	↑Proliferation ↑ECM deposition
Lim et al. (2013)	PEMF	Human Alveolar BMSCs	↑Proliferation ↑Differentation
Zhou et al. (2014)	PEMF	Rat calvarial osteoblasts	↑Proliferation
Zhuang et al. (1997)	CCEF	Osteoblastic cells (MC3T3-E1)	↑Proliferation ↑TGF-β1mRNA
Hartig et al. (2000)	CCEF	Osteoblast from peropsteum explants	↑Proliferation ↑Differentiation
Wang et al. (2006)	CCEF	Osteoblastic cells (MC3T3-E1)	↑BMP-2,4,5,6,7mRNA
Bisceglia et al. (2011)	CCEF	Osteoblast-like cell lines (SAOS-2)	↑Proliferation
Clark et al. (2014)	CCEF	Human calvarial osteoblasts	↑BMP-2,4 mRNA ↑TGF-β1, β2, β3mRNA ↑FGF-2
Hauser et al. (2009)	LIPUS	Osteoblast-like cell lines (SAOS-2)	↑Proliferation
Fassina et al. (2010)	LIPUS	SAOS-2 human osteoblasts	↑Proliferation ↑ECM deposition
Xue et al. (2013)	LIPUS	Alveolar bone in vivo	↑BMP-2 mRNA
Carina et al. (2017)	LIPUS	Human Mesenchymal Stem Cells	↑Proliferation ↑MgHA/coll Hybrid composit scaffold ↑VEGF gene expression

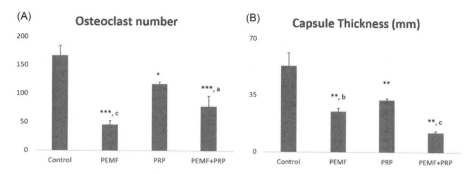

FIGURE 3.5 Histograms of osteoclast number (A) and capsule thickness (B) values in an *in vitro* model of osteolysis. *, p < 0.05; **, p < 0.005; ***, p < 0.0005: Vs no treatment. a, p < 0.05, b, p < 0.05, c, p < 0.005: Vs PRP.

CLINICAL EXPERIENCE AND INDICATION FOR USE

In USA and Europe, research on the use of physical energy as a noninvasive and safe therapy strategy to accelerate bone repair was going on throughout the past century (Yuan et al. 2018). Every year, tens of thousands of patients undergo treatment all over the world. An inquiry to medical hospitals in USA found that 72% of interviewed were offering biophysical stimulation to patients with fractures not yet healed at 3 months from trauma (Huang et al. 2004).

TECHNOLOGIES

Inductive Systems: PEMF

Regarding PEMF, the enhancement of osteoblast activity may occur by means of both the time-varying component of the magnetic field and the electrical component, i.e., the induced electric field. These are signals with a complex waveform, whose predominant spectral content ranges between a few tenths to a few ten thousandths of hertz. Thermal effects are excluded, and the presence of synthesis devices, plates, or nails, as well as of infection, does not contraindicate the treatment.

The first application of PEMF for fracture healing was reported in Germany by Kraus and Lechner (1972). In 1977, Bassett and coworkers reported their initial results on the treatment of nonunions (Bassett et al. 1977). Following these reports, extensive clinical experience has been gathered in Italy, Belgium, UK, and the Netherlands.

Capacitive Systems: CCEF

The CCEF method entails the use of electrodes placed in contact with the skin by means of conductive gel. The voltage applied ranges between 1 and 10 V at frequencies from 20 to 200 kHz. Optimal values lie, however, between 50 and 100 kHz. The electric field within the tissue ranges from 1 to 100 mV/cm. The density of the electric current produced in the tissue varies between 0.5 and 50 µA/cm²

(Brighton et al. 1985). Originally developed by Brighton in the USA, the technology has been further developed in Italy and applied with good results to patients with failed union and/or chronic pain in vertebral fractures (Impagliazzo et al. 2006; Rossini et al. 2010).

Mechanical Stimulation: LIPU

The LIPU method is based on the assumption that the mineral component of bones, in response to mechanical vibration, converts it into an electrical signal that enhances osteogenesis. To enhance fracture healing, optimal dosage has been identified at 30 mW/cm^2 (Wang et al. 1994; Jingushi et al. 1998; Rubin et al. 2001). Ultrasound is used at a frequency of 1.5 MHz and delivered in pulse bursts of 200 μs at 1 kHz. Exposure length does not exceed 20 min/day. The employment of ultrasound dates back to 1950s and was initiated by Corradi and Cozzolino (1952), who reported the positive effect of applying ultrasound on the fracture site in order to enhance healing.

CLINICAL EXPERIENCES WITH BIOPHYSICAL STIMULATION

Stimulation of Reparative Osteogenesis in Congenital Pseudoarthrosis

Congenital pseudoarthrosis is a rare disease; it occurs most commonly in the tibia of infants with stigmata of neurofibromatosis, particularly café-au-lait spots, or a family history of neurofibromatosis. An extensive review of the treatment of congenital pseudoathrosis with PEMFs was prepared by Sharrard (1985). The experience is limited to the use of the inductive systems, and the only clinical series have been reported in UK (Sutcliffe and Goldberg 1982) and Italy (Dal Monte et al. 1986). The healing rate reaches 80% when PEMF stimulation is associated with nailing and is able to limit dysmetry of limbs (Poli et al. 1985). Authors underline the importance of the orthopedic procedure to be associated with the electrical stimulation (Bassett and Schinkascani 1991; Kort et al. 1982; Lavine et al. 1977) to maintain alignment and to protect against the risk of refracture.

Stimulation of Reparative Osteogenesis in Nonunion

The FDA suggested that any fracture failing to heal in 6 months after trauma is to be considered a nonunion.

In international literature, there is abundant clinical evidence of the effectiveness of biophysical stimulation in nonunions with healing rates ranging from 73% to 85% (Bassett 1994; Fontanesi et al. 1986; Bassett et al. 1977; Schmidt-Rohlfing et al. 2011; Assiotis et al. 2012). In 1985, Hinsenkamp et al. (1985) reported the results of a European multicenter study, with success rate above 70%. The same positive outcome was obtained in France by Sedel et al. (1981). In Italy, Marcer et al. (1984) reported the results of a series of 147 patients treated with external fixation and PEMFs, with a 73% overall healing rate, the humerus being the least successful site. In 1983, an 88.5% success in a group of 31 patients was reported by Fontanesi et al. (1983). Dal Monte et al. (1986) reported a success rate of 84% in a clinical series of 248 patients, with average time to healing 4.3 months. The presence of infection did not influence the outcome of the treatments. In Spain, a multicenter study, including

1,710 patients suffering from nonunion, reported positive results with an average treatment time of 4.8 months. Hernandez Vaquero et al. (1999), in a retrospective study on the effect of PEMF on nonunions, reported a success rate of 74%; among the factors influencing the results were the age of the patient (p = 0.048), the fracture site (p < 0.001), the type of nonunion (p = 0.02), and the presence of infection (p = 0.01). In the Netherlands, using high-frequency electromagnetic fields, Fontijne reported a positive experience with 85% success (Fontijne and Konings 1998).

Sharrard in 1990 demonstrated the efficacy of PEMF stimulation in a double-blind study involving patients suffering from delayed union (Sharrard 1990).

Traina reported a higher success rate with PEMF stimulation compared to surgery (87.8% versus 69%) and a shorter time to heal (Traina et al. 1991) in nonunions. In another study in patients suffering from tibial nonunions treated with endomedullary nailing alone or in association with PEMF, healing rate increased from 83% to 91%. The healing time decreased from 4.9 to 3.3 months after surgery when PEMF was used (Cebrian et al. 2010).

Scott and King (1994), in a prospective double-blind study of patients suffering from nonunion, reported a successful healing rate of 60% in the CCEF active group, while none of the patients healed in the placebo group (p = 0.004). In agreement with these findings, Impagliazzo et al. (2006) reported a success rate of 84% in patients suffering from nonunion and stimulated with CCEF.

In a prospective study, Romanò et al. included 49 patients affected by septic nonunions treated with LIPU and antibiotics therapy (Romanò et al. 2009). Three patients decided to discontinue the treatment, and the healing was achieved in 39 patients (85.1%). The authors conclude that LIPU is a conservative treatment that may avoid the need for additional complex surgery. Mechanical stimulation by means of ultrasound has recorded a fairly ample experience in Europe, particularly in Germany and Italy with success rate of over 75% in the treatment of nonunions (Mayr et al. 1999).

Stimulation of Reparative Osteogenesis in Recent Fractures

Biophysical stimulation has been shown to be capable of accelerating healing of recent fractures; the stimulation succeeded in shortening the average time of healing. Early biophysical stimulation has been proposed in those cases where the site, type of exposure, morphology of the fracture, or conditions of the patient foreshadow difficulties in the repair process (Hinsenkamp et al. 1978; Luna et al. 1999).

Faldini et al. reported the effects of PEMF stimulation in patients with femoral neck fractures (Garden 1–3) treated with screw fixation in a randomized double-blind study (Faldini et al. 2010). Fracture healing was achieved in 94% of active patients compared with 69% of the placebo group. Patients were instructed to use the device for at least 8 h/day for 90 days. The patients' compliance was monitored, and the success rate was found to be higher in compliant patients, more than 6 h/day (Figure 3.6). Similar results were reported by Murray et al. in a follow-up study of 1,382 patients treated with PEMF in the management of nonunion fractures. Patients treated for 9 h/day or more had a significant reduction of healing time than patients treated with PEMF for an average of 3 h/day or less (Murray and Pethica 2016).

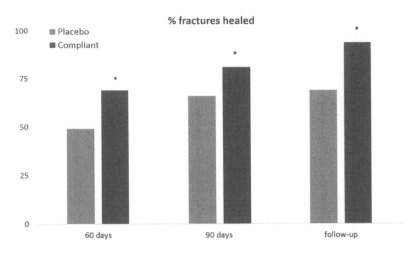

FIGURE 3.6   Femoral neck fracture healing is enhanced by PEMF exposure and depends on daily length of treatment (>6 h/day). *, p < 0.05.

Fontanesi et al. (1983) in a controlled study of 40 recent tibia fractures treated with plaster cast showed that PEMF-stimulated fractures healed in 85 days compared to 109 for the control group (p < 0.005). Hinsenkamp et al. (1984) estimated a shortening in the time to union of PEMF-stimulated recent tibia fractures. He also noted a shortening in the time to union of PEMF-stimulated tibial recent fractures treated with external fixation. Benazzo et al., using CCEF technique, observed earlier recovery in athletes with stress fractures (Benazzo et al. 1995).

Ultrasound stimulation has been used to enhance the healing of forearm and tibia fractures with good results. Heckman et al. (1994) performed a randomized, double-blind, placebo-controlled trial of 67 closed or grade-I open tibial fractures to evaluate the effect of LIPU on time to healing. LIPU treatment led to a significant reduction in the time to clinical healing (86 ± 5.8 days in the active-treatment group compared with 114 ± 10.4 days in the control group, p = 0.01), as well as to a 38% decrease in the time to overall healing (clinical and radiographic healing: 96 ± 4.9 days in the active-treatment group compared with 154 ± 13.7 days in the control group, p=0.0001). By contrast, Emami, in Sweden, used ultrasound in a double-blind study in patients with tibia fractures treated with endomedullary nailing but did not observe any positive effect of the ultrasound (Emami et al. 1999). The effects of this internal synthesis device on ultrasound stimulation have yet to be clarified.

Osteotomies

The study of electromagnetic stimulation on osteotomies represents an original approach in an attempt to quantify the effects of PEMF. Three double-blind studies have been performed: human femoral intertrochanteric osteotomies (Borsalino et al. 1988), tibial osteotomies (Mammi et al. 1993), and osteotomies in patients undergoing massive bone allograft (Capanna et al. 1994).

The use of PEMF stimulation favored rapid healing of the osteotomic line and, in the case of femur osteotomy, an early mineralization of the bone callus. As regards the effects on massive bone allografts, a significant shortening of the healing time (6.7 months in active group versus 9.4 months in the control group, p < 0.01) was observed for patients not undergoing chemotherapy after the surgery.

In the UK, researchers quantified with DEXA (dual-energy X-ray absorptiometry) a positive effect of PEMF stimulation on osteoporosis occurring distally in limbs undergoing lengthening (Eyres et al. 1996). In patients undergoing bilateral limb lengthening, it was shown that, in the limb subjected to PEMF stimulation, the external fixator could be removed 30 days earlier compared to unstimulated limbs (Luna et al. 1999).

Hip Prostheses

PEMF is an effective treatment for improving bone ingrowth in the presence of biomaterials and to prevent complications resulting from the failure of the implant, such as osteolysis. Rispoli et al., in painful noncemented press-fit primary or revision hip prostheses, reported a clinical success rate equal to good/excellent in 91.2% of patients who used the PEMF device more than 360 h, compared to only 12.5% of noncompliant patients (<360 h) (Rispoli et al. 1988). These data suggest a dose-related effect of PEMF treatment. A few years later, Kennedy et al. in a double-blind trial reported a 53% success rate in patients with femoral component loosening treated with PEMFs, compared to 11% of control patients (Kennedy et al. 1993). These data suggest that, for loosened cemented hip prostheses, the use of PEMF is an optional treatment to delay revision surgery. Dallari et al., in a prospective, randomized, double-blind study in subjects undergoing hip revision using the Wagner SL stem, demonstrate that treatment with PEMF eases pain, aids in clinical healing, and favors the restoration of bone mass following revision total hip replacement (Dallari et al. 2009).

Vertebral Fractures

Electrical stimulation has been used for over 30 years to enhance spinal fusions. Positive results have been widely reported on the use of PEMF to promote bone healing. Mooney et al., in the first multicenter randomized double-blind prospective study on 195 patients with anterior or posterior lumbar fusion, report 92% success rate in the PEMF active group, compared to 65% in the placebo group (Mooney 1990). The effectiveness of bone graft stimulation with PEMF is thus established. A 64% bone fusion rate in the active group compared to 43% in the placebo group (p < 0.003) was reported, few years later, by Linovitz et al., in 201 patients undergoing noninstrumented posterolateral spinal fusions (Linovitz et al. 2002). The authors conclude that the adjunctive use of the PEMF device significantly increased the 9-month success of radiographic spinal fusion and showed an acceleration of the healing process. Stimulation with CCEF has been proven much more comfortable than inductive stimulation, due to the ease of use of the applicators. The largest and most comprehensive study evaluating the use of CCEF as an adjunct to spinal fusion reported a success rates of 84% versus 65% in the control group (Goodwin et al. 1999). The authors suggest that CCEF is an effective adjunct to primary spine fusion, especially for patients with posterolateral fusion and those with internal fixation. Rossini et al. randomized 51 postmenopausal

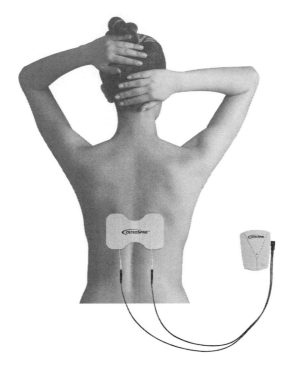

FIGURE 3.7   Representation of spinal stimulation CCEF device, Osteospine®.

women with multiple fractures and chronic pain in a prospective double-blind study with the use of CCEF by Osteospine® (IGEA SpA, Carpi, Italy) (Figure 3.7). The proportion of patients able to discontinue nonsteroidal anti-inflammatory drugs (NSAIDs) owing to elimination or reduction of pain was greater in the active group than in the control group. The authors suggested that CCEF stimulation controls pain in some patients and reduces the use of NSAIDs (Rossini et al. 2010). Piazzolla et al. demonstrated a significant improvement in pain relief and functional recovery and a significant reduction in the area of vertebral bone marrow edema in compression fractures, with clinical symptoms' resolution times reduced by half (Piazzolla et al. 2015).

META-ANALYSIS

Several meta-analyses have been conducted to evaluate the evidence supporting the use of biophysical stimulation in clinical practice (Schmidt-Rohlfing et al. 2011; Mollon et al. 2008) (Table 3.2). Overall a positive effect of the use of biophysical stimulation on reparative osteogenesis is recognized in the different analyses.

However, the limits of meta-analysis in this area should be recognized. For analysis, pooling together all treatment techniques is not appropriate. We have discussed above the importance of the physical parameters for the biological effects; this implies that the results of one technology cannot be extended to other ones. Furthermore, the

TABLE 3.2
Results of two different meta analysis to evaluate the efficacy of biophysical stimulation

Article	Control	Stimulated	Odds ratio (95% CI)
Mollon B et al.	59	57	1.76 (0.81, 3.80)
Schmidt-Rohlfing B et al.	396	379	3.50 (1.94, 6.30)

treatment conditions may be quite different (congenital pseudoarthrosis, nonunion, recent fracture, osteotomy) and difficult to be considered as a single entity.

WHEN TO USE BIOPHYSICAL STIMULATION

The clinical use of biophysical stimulation as proposed by Bassett initially entailed immobilization in plaster of the bone to be treated and subsequent application of the stimulator to be maintained until healing. Some patients underwent treatment even for 9–12 months and beyond. Now, all authors concur on the need to employ biophysical stimulation exclusively in combination with correct orthopedic treatment: immobilization, alignment, and fracture gap <50% of the diameter of the bone treated (Cadossi et al. 2005) (Figure 3.8).

In a recent randomized study of PEMF stimulation of scaphoid fractures, the authors could not demonstrate a benefit from the use of PEMF; nevertheless when they analyzed separately fractures properly immobilized, the use of PEMF stimulation resulted in a shorter healing time compared to controls (Hannemann et al. 2014).

It has been felt necessary to pose the problem of the differential diagnosis, i.e., to identify the causes underlying the failed union, in a prejudicial way. Following the observations of Frost, it can be recognized that 50% of pseudoarthrosis cases are due to mechanical failure, 20% are due to biological failure, namely inadequate activation and finalization of the reparative osteogenetic process, and 30% of cases combine mechanical and biological causes (Frost 1989). Failed union can be ascribed to a biological failure when, even in presence of adequate mechanical conditions, the fracture does not consolidate.

With these premises, differential diagnosis enables us to adopt the best solution therapy. Surgical solution in the case of mechanical failure, biophysical stimulation in the case of biological failure, and the adoption of both (surgery and stimulation) have been suggested in cases where the mechanical condition, as well as a biological deficiency, is capable of hindering healing (Figure 3.9).

CHOICE OF THE BIOPHYSICAL TECHNIQUE

The choice of technique should aim at the highest patient compliance. Ultrasound stimulation only lasts 20 min/day, compared to the 6–8 h requested by the other technologies (inductive and capacitive). However, an inverse relationship between daily treatment time and time to healing has been reported.

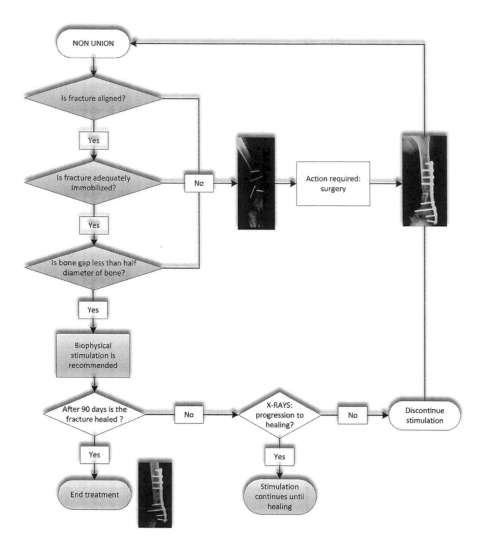

FIGURE 3.8 Indication for use of biophysical stimulation: decisional algorithm.

PEMF should always be used when the skin over the fractured bone is not accessible, in presence of a plaster, cast, or brace. Their use in proximal humerus and in clavicular bone can be difficult.

CCEF requires direct access to the skin as adhesive gels are used to deliver the electric field. Their use is particularly indicated in the upper limbs and clavicular bone. Also small bones, in the hand and foot, can easily be treated as the electrodes can be shaped to the bone size.

LIPU requires direct access to the skin. The active area of the ultrasound beam is small. Focusing the beam can be difficult in the case of the femur or in the presence of long fracture margins or comminuted fractures. Treatment of proximal humerus and clavicular bone can be difficult (Figure 3.10).

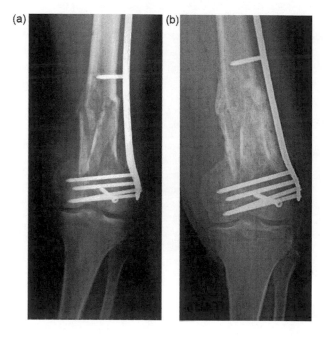

FIGURE 3.9 Biophysical stimulation of surgically treated fracture: (a) after surgery and (b) after 2 months of PEMF treatment.

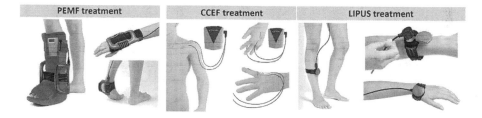

FIGURE 3.10 How PEMF, CCEF, and LIPUS devices are to be used.

CONCLUSIONS

The orthopedic surgeon needs to develop an increasing sensitivity toward the biological environment where the repair activity of a fracture takes place. While the mechanical instability may impede union, the healing of a fracture cannot but be the result of the local cellular activity only. Nowadays the orthopedic surgeon has available various physical, biological (bone marrow cells, platelets), and chemical (growth factors) methods capable of enhancing endogenous repair, reducing healing times, and enabling swifter functional and working recovery.

Biophysical stimulation is the result of solid scientific research. During the last century, treatment with a different physical energy has been successfully used to favor tissue healing, and its clinical use is based on sound scientific evidence.

The orthopedic community has undoubtedly played a central role in the development and understanding of the importance of physical stimuli to control biological activities. Orthopedic research has demonstrated that the effects are dependent on physical parameters and, mediated from pharmacology, has introduced the concept of physical dynamics.

REFERENCES

Aaron RK, Ciombor MD, Jolly G. Stimulation of experimental endochondral ossification by low-energy pulsing electromagnetic fields. *J Bone Miner Res.* 1989; 4(2): 227–33.

Aaron RK, Ciombor DM. Acceleration of experimental endochondral ossification by biophysical stimulation of the progenitor cell pool. *J Orthop Res.* 1996; 14(4): 582–9.

Aaron RK, Ciombor DM, Keeping H, Wang S, Capuano A, Polk C. Power frequency fields promote cell differentiation coincident with an increase in transforming growth factor-beta(1) expression. *Bioelectromagnetics.* 1999; 20(7): 453–8. Erratum in: *Bioelectromagnetics* 2000; 21(1): 73.

Aaron RK, Boyan BD, Ciombor DM, Schwartz Z, Simon BJ. Stimulation of growth factor synthesis by electric and electromagnetic fields. *Clin Orthop Relat Res.* 2004; 419: 30–7.

Assiotis A, Sachinis NP, Chalidis BE. Pulsed electromagnetic fields for the treatment of tibial delayed unions and nonunions. A prospective clinical study and review of the literature. *J Orthop Surg Res.* 2012; 7: 24.

Bassett CA, Schinkascani M. Long-term pulsed electromagnetic field results in congenital pseudoarthrosis. *Calcif Tissue Int.* 1991; 49(3): 3–12.

Bassett CAL. Therapeutic uses of electric and magnetic fields in orthopedics. In: *Biological Effects of Electric and Magnetic Fields*, Eds. DO Carpenter, S Ayrapetyan. Academic Press, San Diego, CA, Vol. II, 13–48, 1994.

Bassett CA, Pilla AA, Pawluk RJ. A non-operative salvage of surgically-resistant pseudarthroses and non-unions by pulsing electromagnetic fields. A preliminary report. *Clin Orthop Relat Res.* 1977; 124: 128–43.

Benazzo F, Mosconi M, Beccarisi G, Galli U. Use of capacitive coupled electric fields in stress fractures in athletes. *Clin Orthop Relat Res.* 1995; 310: 145–9.

Bianco B, Chiabrera A. From the Langevin-Lorentz to the Zeeman model of electromagnetic effects on ligand-receptor binding. *Bioelectrochem Bioenerg.* 1992; 28: 355–65.

Bisceglia B, Zirpoli H, Caputo M, Chiadini F, Scaglione A, Tecce MF. Induction of alkaline phosphatase activity by exposure of human cell lines to a low-frequency electric field from apparatuses used in clinical therapies. *Bioelectromagnetics.* 2011; 32(2): 113–9.

Black J. *Electrical Stimulation.* Praeger, New York, 1987.

Bodamyali T, Bhatt B, Hughes FJ, Winrow VR, Kanczler JM, Simon B, Abbott J, Blake DR, Stevens CR. Pulsed electromagnetic fields simultaneously induce osteogenesis and upregulate transcription of bone morphogenetic proteins 2 and 4 in rat osteoblasts in vitro. *Biochem Biophys Res Commun.* 1998; 250(2): 458–61.

Borsalino G, Bagnacani M, Bettati E, Fornaciari G, Rocchi R, Uluhogian S, Ceccherelli G, Cadossi R, Traina G. Electrical stimulation of human femoral intertrochanteric osteotomies: Double blind study. *Clin Orthop Relat Res.* 1988; 237: 256–63.

Brighton CT, Hozack WJ, Brager MD, Windsor RE, Pollack SR, Vreslovic EJ, Kotwick JE. Fracture healing in the rabbit fibula when subjected to various capacitively coupled electrical fields. *J Orthop Res.* 1985; 3(3): 331–40.

Brighton CT, Wang W, Seldes R, Zhang G, Pollack SR. Signal transduction in electrically stimulated bone cells. *J Bone Joint Surg Am.* 2001; 83-A(10): 1514–23.

Cadossi R, Traina CG, Massari L. Electric and magnetic stimulation of bone repair: Review of the European experience. In: *Symposium of Physical Regulation of Skeletal Repair*, Eds. RK Aaron, ME Bolander. American Academy of Orthopaedic Surgeons, Rosemont, IL, 37–51, 2005.

Canè V, Botti P, Farneti D, Soana S. Electromagnetic stimulation of bone repair: A histomorphometric study. *J Orthop Res.* 1991; 9(6): 908–17.

Canè V, Botti P, Soana S. Pulsed magnetic fields improve osteoblast activity during the repair of an experimental osseous defect. *J Orthop Res.* 1993; 11(5): 664–70.

Canè V, Zaffe D, Botti B, Cavani F, Soana S. Correlation between PEMF exposure time and new bone formation. *Paper Presented at the IV Congresso Nazionale Società Italiana Chirurgia Veterinaria*, Napoli, Maggio, 1997.

Capanna R, Donati D, Masetti C, Manfrini M, Panozzo A, Cadossi R, Campanacci M. Effect of electromagnetic fields on patients undergoing massive bone graft following bone tumor resection: A double-blind study. *Clin Orthop Relat Res.* 1994; 306: 213–21.

Carina V, Costa V, Raimondi L, Pagani S, Sartori M, Figallo E, Setti S, Alessandro R, Fini M, Giavaresi G. Effect of low-intensity pulsed ultrasound on osteogenic human mesenchymal stem cells commitment in a new bone scaffold. *J Appl Biomater Funct Mater.* 2017; 15(3): e215–22. doi: 10.5301/jabfm.5000342.

Cebrìan, JL, Gallego P, Francès A, Sànchez P, Manrique E, Marco F, Lòpez-Duràn L. Comparative study of the use of electromagnetic fields in patients with pseudoarthrosis of tibia treated by intramedullary nailing. *Int Orthop.* 2010; 34: 437–40.

Ceccarelli G, Bloise N, Mantelli M, Gastaldi G, Fassina L, De Angelis MG, Ferrari D, Imbriani M, Visai L. A comparative analysis of the in vitro effects of pulsed electromagnetic field treatment on osteogenic differentiation of two different mesenchymal cell lineages. *Biores Open Access.* 2013; 2(4): 283–94.

Ciombor DM, Aaron RK. The role of electrical stimulation in bone repair. *Foot Ankle Clin.* 2005; 10(4): 579–93.

Clark CC, Wang W, Brighton CT. Up-regulation of expression of selected genes in human bone cells with specific capacitively coupled electric fields. *J Orthop Res.* 2014; 32(7): 894–903.

Corradi C, Cozzolino A. The action of ultrasound on the evolution of an experimental fracture in rabbits. *Minerva Ortop.* 1952; 55: 44–5.

Dal Monte A, Fontanesi G, Giancecchi F, Poli G, Cadossi R. Treatment of congenital pseudarthrosis and acquired nonunion with pulsing electromagnetic fields (PEMFs). *Orthop Trans JBJS.* 1986; 10(3): 452.

Dallari D, Fini M, Giavaresi G Del Piccolo N, Stagni C, Amendola L, Rani N, Gnudi S, Giardino R. Effects of pulsed electromagnetic stimulation on patients undergoing hip revision prostheses: A randomized prospective double-blind study. *Bioelectromagnetics.* 2009; 30(6): 423–30.

Emami A, Petren-Mallmin M, Larsson S. No effect of low-intensity ultrasound on healing time of intramedullary fixed tibial fractures. *J Orthop Trauma.* 1999; 13(4): 252–7.

Esposito M, Lucariello A, Riccio I, Riccio V, Esposito V, Riccardi G. Differentiation of human osteoprogenitor cells increases after treatment with pulsed electromagnetic fields. *In Vivo.* 2012; 26(2): 299–304.

Eyres KS, Saleh M, Kanis JA. Effect of pulsed electromagnetic fields on bone formation and bone loss during limb lengthening. *Bone.* 1996; 18(6): 505–9.

Faldini C, Cadossi M, Luciani D, Betti E, Chiarello E, Giannini S. Electromagnetic bone growth stimulation in patients with femoral neck fractures treated with screws: Prospective randomized double-blind study. *Curr Orthop Pract.* 2010; 21(3): 282–7.

Fassina L, Visai L, Benazzo F, Benedetti L, Calligaro A, De Angelis MG, Farina A, Maliardi V, Magenes G. Effects of electromagnetic stimulation on calcified matrix production by SAOS-2 cells over a polyurethane porous scaffold. *Tissue Eng.* 2006; 12(7): 1985–99.

Fassina L, Saino E, De Angelis MG, Magenes G, Benazzo F, Visai L. Low-power ultrasounds as a tool to culture human osteoblasts inside cancellous hydroxyapatite. *Bioinorg Chem Appl.* 2010: 456240. doi: 10.1155/2010/456240.

Fini M, Giavaresi G, Giardino R, Cavani F, Cadossi R. Histomorphometric and mechanical analysis of the hydroxyapatite-bone interface after electromagnetic stimulation: An experimental study in rabbits. *J Bone Joint Surg Br.* 2006; 88(1): 123–8.

Fontanesi G, Giancecchi F, Rotini R, e Cadossi R. Terapia dei Ritardi di Consolidazione e Pseudoartrosi con Campi Elettromagnetici Pulsati a Bassa Frequenza. *GIOT.* 1983; IX: 319–33.

Fontanesi G, Traina GC, Giancecchi F, Tartaglia I, Rotini R, Virgili B, Cadossi R, Ceccherelli G, Marino AA. La lenta evoluzione del processo riparativo di una frattura può essere prevenuta? *GIOT.* 1986; XII(3): 389–404.

Fontijne WPJ, Konings PC. Botgroeistimulatie met PEMF bij gestoorde fractuurgenezing. *Ned Tijoschr Traum.* 1998; 5: 114–9.

Friedenberg ZB, Brighton CT. Bioelectric potentials in bone. *J Bone Joint Surg Am.* 1966; 48(5): 915–23.

Friedenberg ZB, Dyer R, Brighton CT. Electro-osteograms of long bones of immature rabbits. *J Dent Res.* 1971; 50(3): 635–9.

Friedenberg ZB, Harlow MC, Heppenstall R, Brighton CT. The cellular origin of bioelectric potentials in bone. *Calcif Tissue Res.* 1973; 13(1): 53–62.

Frost HM. The biology of fracture healing: An overview for clinicians. Part I and II. *Clin Orthop Relat Res.* 1989; 248: 283–309.

Goodwin CB, Brighton CT, Guyer RD Johnson JR, Light KI, Yuan HA. A double-blind study of capacitively coupled electrical stimulation as an adjunct to lumbar spinal fusions. *Spine.* 1999; 24(13): 1349–56.

Govey PM, Jacobs JM, Tilton SC, Loiselle AE, Zhang Y, Freeman WM, Waters KM, Karin NJ, Donahue HJ. Integrative transcriptomic and proteomic analysis of osteocytic cells exposed to fluid flow reveals novel mechano-sensitive signaling pathways. *J Biomech.* 2014; 47(8): 1838–45.

Green J, Kleeman CR. Role of bone in regulation of systemic acid-base balance. *Kidney Int.* 1991; 39: 9–26.

Guerkov HH, Lohmann CH, Liu Y, Dean DD, Simon BJ, Heckman JD, Schwartz Z, Boyan BD. Pulsed electromagnetic fields increase growth factor release by nonunion cells. *Clin Orthop.* 2001; 384: 265–79.

Guzelsu N. Piezoelectric and electrokinetic effects in bone tissue. *Electro Magnet Biol.* 1993; 12(1): 51–82.

Hannemann PF, Mommers EH, Schots JP, Brink PR, Poeze M. The effects of low-intensity pulsed ultrasound and pulsed electromagnetic fields bone growth stimulation in acute fractures: A systematic review and meta-analysis of randomized controlled trials. *Arch Orthop Trauma Surg.* 2014; 134(8): 1093–106.

Hartig M, Joos U, Wiesmann HP. Capacitively coupled electric fields accelerate proliferation of osteoblast-like primary cells and increase bone extracellular matrix formation in vitro. *Eur Biophys J.* 2000; 29(7): 499–506.

Hauser J, Hauser M, Muhr G, Esenwein S. Ultrasound-induced modifications of cytoskeletal components in osteoblast-like SAOS-2 cells. *J Orthop Res.* 2009; 27(3): 286–94.

Heckman JD, Ryaby JP, McCabe J, Frey JJ, Kilcoyne RF. Acceleration of tibial fracture-healing by non-invasive, low-intensity pulsed ultrasound. *J Bone Joint Surg Am.* 1994; 76(1): 26–34.

Hernandez Vaquero D, Suarez Vazquez A, Midura Blanco L. La EEM en traumatologia. In: *La estimulacion electromagnetica en la patologia osea*, Eds. D Hernandez Vaquero, L Lopez-Duran Stern. San Martin IG, Madrid, Vol. 8, 125–34, 1999.

Hinsenkamp M, Bourgois R, Bassett CA, Chiabrera A, Burny F, Ryaby J. Electromagnetic stimulation of fracture repair. Influence on healing of fresh fracture. *Acta Orthop Belg.* 1978; 44(5): 671–98.

Hinsenkamp M, Burny F, Donkerwolcke M, Coussaert E. Electromagnetic stimulation of fresh fractures treated with Hoffmann external fixation. *Orthopedics.* 1984; 7(3): 411–6.

Hinsenkamp M, Ryaby J, Burny F. Treatment of nonunion by pulsing electromagnetic field: European multicenter study of 308 cases. *Reconstr Surg Traumatol.* 1985; 19: 147–56.

Huang AJ, Gemperli MP, Bergthold L, Singer SS, Garber A. Health plans' coverage determinations for technology-based interventions: The case of electrical bone growth stimulation. *Am J Manag Care.* 2004; 10(12): 957–62.

Impagliazzo A, Mattei A, Spurio Pompili GF, Setti S, Cadossi R. Treatment of nonunited fractures with capacitively coupled electric field. *J Orthop Traumatol.* 2006; 7: 16–22.

Jansen JH, van der Jagt OP, Punt BJ, Verhaar JA, van Leeuwen JP, Weinans H, Jahr H. Stimulation of osteogenic differentiation in human osteoprogenitor cells by pulsed electromagnetic fields: An in vitro study. *BMC Musculoskelet Disord.* 2010; 23(11): 188.

Jingushi S, Azuma V, Ito M, Harada Y, Takagi H, Ohta T, Komoriya K. Effect of noninvasive pulsed low-intensity ultrasound on rat femoral fracture. *Paper Presented at the Third World Congress of Biomechanics*, Sapporo, Japan, 1998.

Kennedy WF, Roberts CG, Zuege RC, et al. Use of pulsed electromagnetic fields in treatment of loosened cemented hip prostheses. A double-blind trial. *Clin Orthop.* 1993; 286: 198–205.

Kort J, Schink MM, Mitchell SN, Bassett CA. Congenital pseudarthrosis of the tibia: Treatment with pulsing electromagnetic fields. *Clin Orthop Relat Res.* 1982; 165: 124–37.

Kraus W, Lechner F. Die Heilung von Pseudarthrosen und Spontanfrakturen durch strukturbildende elektrodynamische Potentiale. *Munchen Mudizinische Woch.* 1972; 114: 1814–7.

Lavine LS, Lustrin I, Shamos MH. Treatment of congenital pseudoathrosis of the tibia with direct current. *Clin Orthop Relat Res.* 1977; 124: 69–74.

Lim K, Hexiu J, Kim J, Seonwoo H, Cho WJ, Choung PH, Chung JH. Effects of electromagnetic fields on osteogenesis of human alveolar bone-derived mesenchymal stem cells. *Biomed Res Int.* 2013; 2013: 296019. doi: 10.1155/2013/296019.

Lin HY, Lin YJ. In vitro effects of low frequency electromagnetic fields on osteoblast proliferation and maturation in an inflammatory environment. *Bioelectromagnetics.* 2011; 32(7): 552–60.

Linovitz RJ, Pathria M, Bernhardt M et al. Combined magnetic fields accelerate and increase spine fusion: A double-blind, randomized, placebo controlled study. *Spine (Phila Pa 1976).* 2002; 27(13): 1383–9.

Lohmann CH, Schwartz Z, Liu Y, Guerkov H, Dean DD, Simon B, Boyan BD. Pulsed electromagnetic field stimulation of MG63 osteoblast-like cells affects differentiation and local factor production. *J Orthop Res.* 2000; 18(4): 637–46.

Luna GF, Arevalo RL, Labajos UV. La EEMen las elongaciones y transportes oseos. In: *La estimulacion electromagnetica en la patologia osea*, Eds. D Hernandez Vaquero, L Lopez-Duran Stern. San Martin IG, Madrid, 236–46, 1999.

Mammi GI, Rocchi R, Cadossi R, Traina GC. The electrical stimulation of tibial osteotomies. Double-blind study. *Clin Orthop Relat Res.* 1993; 288: 246–53.

Marcer M, Musatti G, Bassett CA. Results of pulsed electromagnetic fields (PEMFs) in ununited fractures after external skeletal fixation. *Clin Orthop Relat Res.* 1984; 190: 260–5.

Mayr E, Wagner S, Ecker M, Rüter A. Ultrasound therapy for nonunions. Three case reports. *Unfallchirurg.* 1999; 102(3): 191–6.

Midura RJ, Ibiwoye MO, Powell KA, Sakai Y, Doehring T, Grabiner MD, Patterson TE, Zborowski M, Wolfman A. Pulsed electromagnetic field treatments enhance the healing of fibular osteotomies. *J Orthop Res.* 2005; 23(5): 1035–46.

Mollon B, da Silva V, Busse JW, Einhorn TA, Bhandari M. Electrical stimulation for long-bone fracture-healing: A meta-analysis of randomized controlled trials. *J Bone Joint Surg Am.* 2008; 90(11): 2322–30.

Mooney V. A randomized double-blind prospective study of the efficacy of pulsed electromagnetic fields for interbody lumbar fusions. *Spine.* 1990; 15(7): 708–12.

Murray HB, Pethica BA. A follow-up study of the in-practice results of pulsed electromagnetic field therapy in the management of nonunion fractures. *Orthop Res Rev.* 2016; 8: 67–72.

Otter MW, Vincent R, Palmieri VR, DD Wu, Seiz KG, Mac Ginitie LA, Cochran GVB. A comparative analysis of streaming potentials in vivo and in vitro. *J Orthop Res.* 1992; 10: 710–19.

Piazzolla A, Solarino G, Bizzoca D Garofalo N, Dicuonzo F, Setti S, Moretti B. Capacitive coupling electric fields in the treatment of vertebral compression fractures. *J Biol Regul Homeost Agents.* 2015; 29(3): 637–46.

Poli G, Verni E, Dal Monte A. A double approach to the treatment of congenital pseudo-arthrosis: Endomedullary nail fixation and stimulation with low frequency pulsing electromagnetic fields (PEMFs). *Bioelectrochem Bioenerget.* 1985; 14: 151.

Pollack SR. Bioelectrical properties of bone. Endogenous electrical signals. *Orthop Clin North Am.* 1984; 15: 3–14.

Rijal KP, Kashimoto O, Sakurai M. Effect of capacitively coupled electric fields on an experimental model of delayed union of fracture. *J Orthop Res.* 1994; 12: 262–7.

Rispoli FP, Corolla FM, Mussner R. The use of low frequency pulsing electromagnetic fields in patients with painful hip prosthesis. *J Bioelectricity.* 1988; 7: 181.

Romanò CL, Romanò D, Logoluso N. Low-intensity pulsed ultrasound for the treatment of bone delayed union or nonunion: A review. *Ultrasound Med Biol.* 2009; 35(4): 529–36.

Rossini M, Viapiana O, Gatti D, de Terlizzi F, Adami S. Capacitively coupled electric field for pain relief in patients with vertebral fractures and chronic pain. *Clin Orthop Relat Res.* 2010; 468(3): 735–40.

Rubin C, Bolander M, Ryaby JP, Hadjiargyrou M. The use of low-intensity ultrasound to accelerate the healing of fractures. *J Bone Joint Surg Am.* 2001; 83: 259–70.

Ryaby JT, Mathew J, Duarte-Alves P. Low intensity pulsed ultrasound affects adenylate cyclase activity and TGF-β synthesis in osteoblastic cells. *Trans Orthop Res Soc.* 1992; 7: 590.

Schmidt-Rohlfing B, Silny J, Gavenis K, Heussen N. Electromagnetic fields, electric current and bone healing: What is the evidence? *Z Orthop Unfall.* 2011; 149(3): 265–70.

Scott G, King JB. A prospective double blind trial of electrical capacitive coupling in the treatment of non-union of long bones. *J Bone Joint Surg Am.* 1994; 76(6): 820–6.

Sedel L, Christel P, Duriez J, Duriez R, Evard J, Ficat C, Cauchoix J, Witvoet J. Acceleration of repair of non-unions by electromagnetic fields (author's transl). *Rev Chir Orthop Reparatrice Appar Mot.* 1981; 67(1): 11–23.

Sharrard WJ. Treatment of congenital and infantile pseudoarthrosis of the tibia with pulsing electromagnetic fields. *Biol Eff Non-ionizing Electromagn Radiat.* 1985; IX(3): 42.

Sharrard WJW. A double-blind trial of pulsed electromagnetic field for delayed union of tibial fractures. *J Bone Joint Surg Br.* 1990; 72(3): 347–55.

Sollazzo V, Massari L, Caruso A, De Mattei M, Pezzetti F. Effect of low-frequency pulsed electromagnetic fields on human osteoblast-like cells in vitro. *Electro Magnetobiol.* 1996; 15(1): 75–83.

Sollazzo V, Scapoli L, Palmieri A, Fanali S, Girardi A, Farinella F, Pezzetti F, Carinci F. Early effects of pulsed electromagnetic fields on human osteoblasts and mesenchymal stem cells. *Eur J Inflammation.* 2011; 1(S): 95–100.

Sutcliffe ML, Goldberg AAJ. The treatment of congenital pseudoarthrosis of the tibia with pulsing electromagnetic fields. A survey of 52 cases. *Clin Orthop Relat Res.* 1982; 166: 45–57.

Traina GC, Fontanesi G, Costa P, Mammi GI, Pisano F, Giancecchi F, Adravanti P. Effect of electromagnetic stimulation on patients suffering from nonunion. A retrospective study with a control group. *J Bioelectricity*. 1991; 10: 101–17.

Tschon M, Veronesi F, Contartese D, Sartori M, Martini L, Vincenzi F, Ravani A, Varani K, Fini M. Effects of pulsed electromagnetic fields and platelet rich plasma in preventing osteoclastogenesis in an in vitro model of osteolysis. *J Cell Physiol*. 2018; 233(3): 2645–56.

Varani K, Vincenzi F, Ravani A, Pasquini S, Merighi S, Gessi S, Setti S, Cadossi M, Borea PA, Cadossi R. Adenosine receptors as a biological pathway for the anti-inflammatory and beneficial effects of low frequency low energy pulsed electromagnetic fields. *Mediators Inflammation*. 2017; 2017: 2740963. doi: 10.1155/2017/2740963.

Veronesi F, Fini M, Sartori M, Parrilli A, Martini L, Tschon M. Pulsed electromagnetic fields and platelet rich plasma alone and combined for the treatment of wear-mediated periprosthetic osteolysis: An in vivo study. *Acta Biomater*. 2018; 77: 106–15.

Vincenzi F, Targa M, Corciulo C, Gessi S, Merighi S, Setti S, Cadossi R, Goldring MB, Borea PA, Varani K. Pulsed electromagnetic fields increased the anti-inflammatory effect of A2A and A3 adenosine receptors in human T/C-28a2 chondrocytes and hFOB 1.19 osteoblasts. *PLoS One*. 2013; 8(5): e65561. doi: 10.1371/journal.pone.0065561.

Wang SJ, Lewallen DG, Bolander ME, Chao EY, Ilstrup DM, Greenleaf JF. Low intensity ultrasound treatment increases strength in a rat femoral fracture model. *J Orthop Res*. 1994; 12: 40–7.

Wang Z, Clark CC, Brighton CT. Up-regulation of bone morphogenetic proteins in cultured murine bone cells with use of specific electric fields. *J Bone Joint Surg Am*. 2006; 88(5): 1053–65.

Xue H, Zheng J, Cui Z, Bai X, Li G, Zhang C, He S, Li W, Lajud SA, Duan Y, Zhou H. Low-intensity pulsed ultrasound accelerates tooth movement via activation of the BMP-2 signaling pathway. *PLoS One*. 2013; 8(7): e68926. doi: 10.1371/journal.pone.0068926.

Yuan J, Xin F, Jiang W. Underlying signaling pathways and therapeutic applications of pulsed electromagnetic fields in bone repair. *Cell Physiol Biochem*. 2018; 46(4): 1581–94.

Zhou J, Wang JQ, Ge BF, Ma XN, Ma HP, Xian CJ, Chen KM. Different electromagnetic field waveforms have different effects on proliferation, differentiation and mineralization of osteoblasts in vitro. *Bioelectromagnetics*. 2014; 35(1): 30–8.

Zhuang H, Wang W, Seldes RM, Tahernia AD, Fan H, Brighton CT. Electrical stimulation induces the level of TGF-β1 mRNA in osteoblastic cells by a mechanism involving calcium/calmodulin pathway. *Biochem Biophys Res Comm*. 1997; 237: 225–9.

4 Biophysical Stimulation of Articular Cartilage for Chondroprotection and Chondroregeneration

Simona Salati, Stefania Setti, and Ruggero Cadossi

CONTENTS

BIOPHYSICAL STIMULATION OF CARTILAGE REPAIR

Articular cartilage is a hypocellular, avascular tissue composed of a dense collagen and proteoglycan (PG) matrix. The complex structure of articular cartilage enables this tissue to perform its biomechanical role providing low friction and highly wear-resistant surface to both shear and compressive stress (Ulrich-Vinther et al. 2003). On the other side, the hypocellular and avascular nature of cartilage appears to hamper repair; thus lesions without further treatment often progress to tissue degeneration and osteoarthritis (OA) (Holland et al. 2004). OA is one of the most common degenerative diseases of the joints, characterized by progressive and permanent degradation of the articular cartilage, synovial hypertrophy, and change in the underlying bone. OA is a degenerative disorder that is prevalent in the aged population. In the United States, between 6% and 12% of the population complain of a painful knee condition (Ramsey and Russell 2009); more than 300,000 total knee replacements are performed each year, and this number is expected to increase to 3.48 million in 2030 (Kurtz et al. 2007). At present, the most advanced frontier for the treatment of inflammatory and degenerative diseases of the cartilage is represented by treatment modalities that aim to accelerate cartilage tissue repair and to reconstruct the functional properties of the damaged tissues (Redman et al. 2005). Currently several strategies have been proposed for cartilage repair, including chondroabrasion, microfractures that recruit medullary stromal cells from subchondral bone, or techniques which fill the cartilage damage with autologous *ex vivo* cultured chondrocytes, enriched bone marrow mesenchymal stem cells (MSCs), or osteochondral grafts (Ahmed and Hincke 2010). Surgical procedures that can replicate the biological composition and biomechanical characteristics of native cartilage haven't been identified yet. Moreover, the joint environment is severely modified after even a minimal invasive surgical procedure such as arthroscopy; in particular, the proinflammatory cytokine concentration increases in the synovial fluid.

In the articular environment, the activity of inflammatory cells and proinflammatory cytokines can lead to degradation of the extracellular matrix and loss of PGs. In response to tissue damage, local release of catabolic cytokines and enzymes, such as interleukin-1b (IL-1β), tumor necrosis factor-a (TNF-α), interleukin-6 (IL-6), matrix metalloproteinases (MMPs), and inflammation cause a depletion of glycosaminoglycans (GAGs) and suppress type II collagen synthesis (Berg et al. 2001).

During inflammation, the physiological maintenance of the cartilage, which is guaranteed by the balance between anabolic and catabolic activity, is lost. Once cartilage loss has begun, it progresses through a mixture of mechanical and biological events. Owing to the low reparative capacity of the cartilage, it is of utmost importance to develop chondroprotective treatments aimed at limiting the damage of cartilage following injury, increasing the reparative response and finally preventing degeneration.

Biophysical stimulation (BS), through the action of nonionizing physical stimuli, i.e., pulsed electromagnetic fields (PEMFs), is currently applied to promote and reactivate the formation of bone tissue to control joint inflammation and to stimulate anabolic activities of cartilage cells. The scientific bases of the BS technique for bone healing lie on the pioneer works performed by Fukada and Yasuda (1957) and by Bassett and Becker (1962), who identified the relationship between mechanical loading and electrical activity in the bone. Later, the physical signal has been shown to trigger a series of intracellular events through the cell membrane resulting in a specific biological response. In 2002, Varani et al. described for the first time the mechanism of action of low-frequency low-energy PEMFs in inflammatory cells (Varani et al. 2002), identifying the A_{2A} adenosine receptor (AR) as the main target of PEMF stimulation.

The rationale for using PEMF stimulation for chondroprotection and chondroregeneration is based on two important findings: (i) PEMF stimulation increases the anabolic activity of chondrocytes and cartilage explants and (ii) PEMF stimulation antagonizes the catabolic effects of inflammation through its agonistic activity on the A_{2A} AR. In this view, BS represents a new powerful tool available in the physicians' arsenal to treat degenerative joint disease in human beings.

MECHANISMS OF ACTION

Signaling Pathways Activated by PEMF Stimulation

Studies on signal transduction identified increase in intracellular Ca^{2+} levels and activation of membrane receptors as the main signaling pathways triggered by PEMFs stimulation (Brighton et al. 2001; Varani et al. 2002; Uzieliene et al. 2018).

Brighton et al. demonstrated in a murine osteoblastic cell line that inductive coupling and combined electromagnetic fields (EMFs) activate intracellular release of Ca^{2+} leading to increase in cytosolic Ca^{2+} and activated cytoskeletal calmodulin (Uzieliene et al. 2018). Activated calmodulin is known to promote nucleotide synthesis and cellular proliferation (Tomlinson et al. 1984). However, the precise mechanism by which the EMFs are transduced to the intracellular Ca^{2+} storage hasn't been clarified yet.

Recently, Varani et al. demonstrated the interaction between PEMFs and ARs (Varani et al. 2008). Adenosine acts through the interaction with four cell surface ARs, A_1, A_{2A}, A_{2B}, and A_3 which are coupled to G proteins. A_1 and A_3 ARs inhibit adenylate cyclase activity and decrease cyclical AMP (cAMP) production, while A_{2A} and A_{2B} ARs are coupled to G stimulatory protein (G_s) thus activating adenylate cyclase and inducing increase in cAMP intracellular levels (Borea et al. 2015). Modulation of ARs has an important role in the regulation of inflammatory processes suggesting their involvement in different pathologies resulting from inflammation (Gessi et al. 2011). The ARs are expressed on chondrocytes, synoviocytes, and osteoblasts with affinity in the nanomolar range and variable density depending on the cell type (Figure 4.1). Varani et al. demonstrated that exposure to PEMFs increases the expression of A_{2A} and A_{3A} ARs in chondrocytes and synoviocytes. Notably, A_1 and A_{2B} receptors were not influenced by the same exposure conditions (Varani et al. 2008).

Worth of notice, in an animal model of septic arthrosis, it was found that a specific A_{2A} AR agonist significantly reduced cartilage damage, synovial inflammation, and infiltration of leukocytes (Cohen et al. 2004).

Anabolic Effect of PEMFs on Chondrocytes and Cartilage

Extensive efforts have been made to characterize the *in vitro* effects of PEMFs on articular cells (Table 4.1). Most of these studies showed that PEMFs affect chondrocytes in different experimental settings (cell cultures, cartilage explants, and 3D scaffolds) by increasing cell proliferation and synthesis of cartilage extracellular matrix (ECM) components, such as PGs and collagen type II. Exposure to PEMFs increased H^3-thymidine incorporation in human articular chondrocytes cultured *in vitro* (Pezzetti et al. 1999). PEMFs also stimulate anabolic activities *ex vivo* in full-thickness cartilage explants: PG synthesis is increased when cartilage explants are exposed to PEMFs (De Mattei et al. 2003). Moreover, PEMFs are able to counteract the IL-1β-triggered cartilage ECM degradation in healthy and OA-joint-derived cells (Ongaro et al. 2011, 2015). Worth of notice, De Mattei showed that the increase in PGs synthesis induced by PEMF stimulation in human cartilage explants is of the same magnitude as that induced by the main cartilage anabolic growth factor, insulin-like growth factor-1 (IGF-1) (De Mattei et al. 2004).

(a)

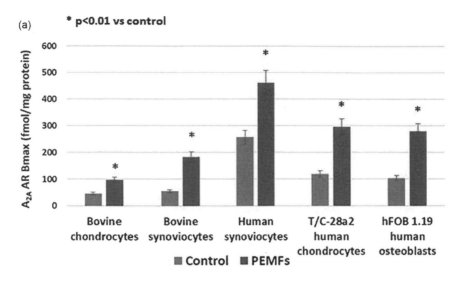

(b)

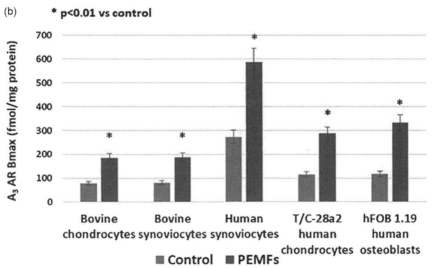

FIGURE 4.1 A_{2A} AR (a) and A_{3A} receptor (b) density in bovine chondrocytes and synoviocytes, human synoviocytes, T/C-28a2 human chondrocytes, and hFOB 1.19 human osteoblasts in the absence and in the presence of PEMFs.

Anti-Inflammatory Effect of PEMFs on Chondrocytes and Cartilage

Several pieces of evidence from different cellular models, including human synoviocytes, demonstrated that A_{2A} or A_{3A} receptor stimulation mediates the inhibition of the NF-kB pathway (Lee et al. 2006; Martin et al. 2006). In chondrocytes and synoviocytes, NF-kB is a key regulator of the expression of cyclooxygenase type 2 (COX-2) and MMPs, together with a large number of genes involved in response to inflammation (Shakibaei et al. 2007). PEMF stimulation induces a

TABLE 4.1
PEMF Effects on Articular Cells

Cell Type	Effect of PEMFs	References
Bovine chondrocytes and synovial fibroblasts	Increase of A_{2A} and A_3 receptors Increase of cellular proliferation Inhibition of PGE2 release	De Mattei et al. (2009)
Bovine articular cartilage explants	Increase of PG synthesis Chondroprotective effect	Fini et al. (2005)
Human synovial fibroblasts	Inhibition of PGE2 IL-6, IL-8, and TNF-α release Stimulation of IL-10 release	Ongaro et al. (2012)
Human articular cartilage explants	Increase of PG synthesis Counteracting of the catabolic activity of IL-1b Increase of cartilage explant anabolic activities	De Mattei et al. (2003)
Human T/C-28a2 chondrocytes and hFOB 1.19 osteoblasts	Increase of A_{2A} and A_3 receptors Inhibition of PGE2 IL-6, IL-8, and VEGF release Increase of cellular proliferation Increase of osteoprotegerin (OPG) production Inhibition of NF-κB activation	Vincenzi et al. (2013)

strong adenosine-agonist effect, mediated by the increase in the number of receptors themselves, which determines a significant increase in cAMP and a reduction in the release of superoxide anion (O^{2-}), thus supporting a strong anti-inflammatory effect (Vincenzi et al. 2013).

It has been shown that A_{2A} and A_3 AR activation in the presence of PEMFs reduces the release of prostaglandin E2 (PGE2) and the expression of COX-2 in bovine synoviocytes, thus reducing inflammation and cartilage degradation associated with joint disease (Figure 4.2) (De Mattei et al. 2009). These effects could be attributed to the capability of PEMFs to potentiate the activation of ARs, which, in turn, inhibit the NF-kB signaling pathway, resulting in decreased synthesis of inflammatory molecules. Specifically, PEMFs stimulation of A_{2A} and A_3 ARs in human synoviocytes has been shown to inhibit p38 MAPK and NF-kB pathways, thus leading to decreased synthesis of TNF-α and IL-1β and other mediators involved in joint inflammation and bone diseases (Vincenzi et al. 2013).

In human synovial fibroblasts taken from OA patients, Ongaro et al. demonstrated that PEMF stimulation reduces the synthesis of inflammatory mediators such as PGE2 and proinflammatory cytokines (IL-6 and IL-8), while stimulating the release of the anti-inflammatory cytokine interleukin-10 (IL-10) (Ongaro et al. 2012). Finally, Ongaro et al. demonstrated that PEMFs are able to counteract the IL-1β-induced inhibition of chondrogenesis in mesenchymal stem cells suggesting BS as a therapeutic strategy for improving the clinical outcome of cartilage engineering repair procedures (Ongaro et al. 2015).

Altogether, these data support the use of PEMFs in promoting the anabolic activity of chondrocytes and as anti-inflammatory agents acting via the upregulation of ARs (Table 4.1).

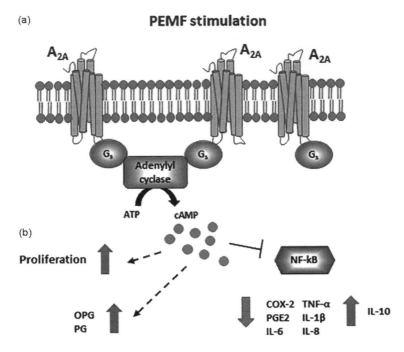

FIGURE 4.2 Effects of PEMF on A_{2A} ARs: (a) A_{2A} ARs on the cell membrane of PEMF-unstimulated cells and (b) up-regulation of A_{2A} ARs after PEMF exposure.

In Vivo Chondroprotective Effect of PEMFs

These *in vitro* effects of PEMFs translate *in vivo* into chondroprotection of cartilage. The capability of PEMFs to modify OA progression was first described by Ciombor et al. (2003). The authors reported that PEMFs prevent cartilage degeneration measured by Mankin score. By immunohistochemistry, the authors also demonstrated a decreased level of IL-1β and increased levels of transforming growth factor-1 beta (TGF-1β). TGF-1β is of extreme importance in cartilage healing: TGF-1β stimulates collagen type II synthesis and counteracts the action of proinflammatory cytokines (Martel-Pelletier and Pelletier 2010). More recently, Fini et al. treated Dunkin Hartley guinea pigs with PEMFs and demonstrated their ability in halting the progression of osteoarthrosis, limiting cartilage surface clefts and fibrillation, preserving cartilage thickness, and preventing sclerosis of the subchondral bone (Fini et al. 2005, 2008).

After autologous osteochondral grafts, PEMFs favor graft integration and prevent its reabsorption. The synovial liquid in the stimulated animals contained significantly lower levels of proinflammatory cytokines IL-1β and TNF-α and a higher concentration of TGF-β1 compared to the untreated animals (Benazzo et al. 2008). PEMFs have been proven effective in rabbits with osteochondral lesions: BS significantly improved the quality of the regenerated cartilage and bone in joint defects in the presence of collagen scaffold and bone marrow concentrate (Veronesi et al. 2015). Finally, Boopalan studied PEMFs on full-thickness articular osteochondral defects

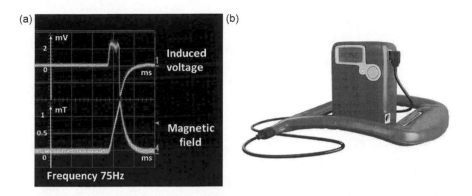

FIGURE 4.3 (a) Signal waveform of I-ONE therapy and (b) clinical device for I-ONE therapy.

in rabbit knees. The defect was filled with calcium phosphate scaffold, which provides the structural support for cells from the marrow to migrate and regenerate. The authors reported complete restoration of osteochondral defects in the animals treated with PEMFs (Boopalan et al. 2011).

In order to translate these results into clinical practice, dose-response effects of PEMF stimulation (amplitude, frequency, and exposure time) on cells and cartilage explants have been performed (De Mattei et al. 2007). This work of De Mattei et al. identified the optimal exposure conditions in terms of exposure time (h/day), field peak value (mT), and frequency (Hz) to be subsequently used in clinical studies: 1.5 mT, 75 Hz, 4 h/day (I-ONE therapy, IGEA, Italy) (Figure 4.3).

CLINICAL STUDIES

Preclinical research shows that treatment with PEMFs is antidegenerative, helping to control local inflammatory phenomena and supporting cartilage repair processes in clinical settings.

PEMFs IN CONSERVATIVE TREATMENTS

OA is a progressive chronic disease associated with damage to the cartilage and surrounding tissues of the joint. Additionally, OA has a chronic cycle of aberrant attempts to repair joints leading to inflammation and tissue degradation. In early stages of OA, every effort should be made to prevent further damage to the articular cartilage from the catabolic effects of the proinflammatory cytokines. Based on the proven *in vitro* and *in vivo* anti-inflammatory and anabolic effects of PEMFs, studies on BS on OA patients have been performed. PEMFs stimulation has been shown to reduce stiffness in OA knee already at 2 weeks after treatment.

The positive effects of PEMFs on knee have been reported especially in early OA and younger patients (age <65 years) (Thamsborg et al. 2005). In agreement with these findings, a study conducted on patients in the early stages of osteoarthritic degeneration of the knee, treated conservatively with I-ONE therapy rather than surgically, showed significantly better results in terms of functional recovery and pain resolution at 12 months and at 2 years follow-up (Gobbi et al. 2014). In particular, the authors reported a progressive decline in outcome scores at 2 years, despite having a significant improvement at the end of the first year. Thus, the authors recommend to repeat the treatment annually to maintain sustained improvement.

Similar improvements in joint function, pain resolution, and time to return to sport activity were obtained with I-ONE therapy in patients with patellofemoral pain (Servodio Iammarrone et al. 2016).

The effects of PEMFs have also been studied in the context of edema, Complex Regional Pain Syndrome (CRPS), and osteonecrosis. These pathologies originate from an ischemic damage to the bone. In edema and primarily in CRPS, there is an inflammatory process in a localized osteoporosis zone with pain. The local release of proinflammatory cytokines seems to be the pathway that triggers and maintains the disease. PEMFs have been shown to be effective in controlling various painful and inflammatory disorders; therefore their application in CRPS is interesting and promising (Pagani et al. 2017). PEMF stimulation has been successfully applied in combination with surgery and rehabilitation for simple and complex wrist fractures associated with CRPS. The use of I-ONE therapy was able to shorten the time of functional recovery of the wrist, demonstrating full resolution of pain and a complete functional recovery (Borelli 2017). In the same way, bone marrow edema (BME) of the talus has been successfully treated with PEMF stimulation. The authors reported reduction in BME area associated with a significant decrease in pain within 3 months from the beginning of treatment (Martinelli et al. 2015). This study confirms reports by other authors that suggest an improved healing process after I-ONE therapy for the early phase of osteonecrosis. Working on patients in the initial stages of spontaneous osteonecrosis of the knee, Marcheggiani Muccioli et al. reported that I-ONE therapy significantly reduces knee pain and necrosis area in the first 6 months, saving 86% of knees from prosthetic surgery (Marcheggiani Muccioli et al. 2013).

Altogether, these results demonstrate a strong chondroprotective effect for PEMFs in conservative clinical settings.

PEMFs in Chondroregenerative Treatments

Joint environment is modified following a trauma or a local surgical procedure; in particular increased IL-1β concentration has been documented after joint surgery, and its levels have been found to correlate with the severity of cartilage damage (Lotz et al. 2010).

The effects of PEMF stimulation have also been assessed in chondroregenerative clinical settings, including patients suffering from cartilage injuries treated with microfractures, with autologous chondrocyte implant, and with scaffold seeded with bone-marrow-derived cells (BMDCs) (Table 4.2). Regardless of the surgical procedure applied, it has been reported in literature that I-ONE therapy favors pain relief (Figure 4.4) and enhances functional recovery (Figure 4.5), leading to a better quality of life in long term follow-up (Figure 4.6).

TABLE 4.2

Summary of the Main Clinical Studies Using PEMF Stimulation on Joint Diseases

Disease/Treatment	Design of the Study	Results	References
		Conservative Treatments	
Knee OA	Randomized, controlled, double-blind add-on study	PEMF significantly improved stiffness after 2 weeks in patients <65 years	Thamsborg et al. (2005)
Early OA	Case series	Improvement in symptoms, knee function, and activity. At 2 years follow-up, 80% of patients were satisfied and willing to repeat the treatment	Gobbi et al. (2014)
Patellofemoral pain	Prospective comparative study	Pain relief, better functional recovery, and lower NSAID use	Iammarrone et al. (2016)
Wrist fractures associated with algodystrophy	Case series	Pain relief, better functional recovery	Borelli (2017)
BME of the talus	Case series	Reduction in BME area, pain relief	Martinelli et al. (2015)
Spontaneous osteonecrosis of the knee	Case series	Pain relief, better functional recovery, and reduction in the necrosis area 86% of knees preserved from prosthetic surgery at 2 years follow-up	Marcheggiani Muccioli et al. (2013)
		Chondroregenerative Treatments	
Cartilage knee lesions, chondroabrasion/ perforation	Randomized, controlled, double-blind study	Pain relief, better functional recovery, and lower NSAID use Complete functional recovery higher in the active group at 3 years follow-up	Zorzi et al. (2007)
Grade III–IV cartilage knee lesions/partial medial meniscectomy and microfractures	Randomized, controlled, double-blind study	Clinical and functional outcomes were better in the PEMF-treated group at 5 years follow-up	Osti et al. (2015)
Osteochondral lesions in talar/BMDC transplantation	Prospective comparative study	Pain relief, better functional recovery	Cadossi et al. (2014)
Chondral knee lesions/ MACI	Prospective comparative study	Pain relief, better functional recovery. Better clinical outcome up to 5 years of follow-up ($p < 0.05$)	Collarile et al. (2018)

(Continued)

TABLE 4.2 (*Continued*)
Summary of the Main Clinical Studies Using PEMF Stimulation on Joint Diseases

Disease/Treatment	Design of the Study	Results	References
	Postsurgical Treatments		
Anterior cruciate ligament lesion/ reconstruction and meniscectomy	Randomized, controlled, double-blind study	Pain relief, better functional recovery, and lower NSAID use. Complete functional recovery, no knee pain, and return to sport activity higher in the active group at 2 years follow-up	Benazzo et al. (2008)
Grade 4 OA/total knee arthroplasty	Prospective comparative study	Pain, joint swelling, and knee score were significantly better Lower NSAID use	Moretti et al. (2012)
Grade 4 OA/total knee arthroplasty	Prospective comparative study	Pain, knee swelling, and functional score were significantly better. Severe pain and occasional walking limitations were reported in a lower number at 3 years follow-up	Adravanti et al. (2014)

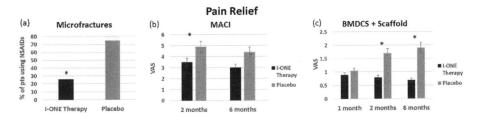

FIGURE 4.4 Pain relief induced by PEMF treatment after chondroregenerative treatments: microfractures, data from Zorzi et al. (2007) (a), MACI, data from Collarile et al. (2018) (b), and BMDCS + Scaffold, data from Cadossi et al. (2014) (c). *p < 0.05, **p < 0.01.

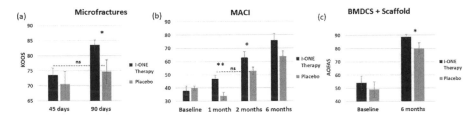

FIGURE 4.5 Functional recovery induced by PEMF treatment after chondroregenerative treatments: microfractures, data from Zorzi et al. (2007) (a); MACI, data from Collarile et al. (2018) (b); and BMDCs + scaffold, data from Cadossi et al. (2014) (c). *p < 0.05, **p < 0.01. ns – not significant.

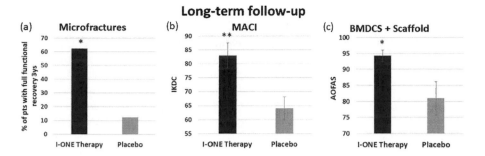

FIGURE 4.6 Effects of PEMF treatment in long-term follow-up after chondroregenerative treatments: microfractures, data from Zorzi et al. (2007) (a); MACI, data from Collarile et al. (2018) (b); and BMDCs + scaffold, data from Cadossi et al. (2014) (c). *p < 0.05, **p < 0.01.

Zorzi et al., in a prospective, randomized double-blind study, evaluated the effect of PEMF stimulation after arthroscopic treatment of cartilage lesions in the knee (Zorzi et al. 2007). The authors showed that the percentage of patients assuming nonsteroidal anti-inflammatory drugs (NSAIDs) was significantly lower in those patients treated with the active device compared to the placebo group. Functional scores were significantly better in the I-ONE therapy group in the short term (90 days after surgery) and at 3 years follow-up. Worth of notice, the Knee injury and Osteoarthritis Score (KOOS) at 45 days in the I-ONE therapy group was equal to the score of untreated patients at 90 days, suggesting that PEMF stimulation allows to halve the time of functional recovery. In agreement with this study, Osti et al. combined microfractures and PEMF stimulation for the treatment of knee OA and reported better clinical and functional outcomes in PEMF-treated patients at 5 years follow-up (Osti et al. 2015).

Similar results have been obtained on talar osteochondral lesions treated with collagen scaffold seeded with BMDCs: Cadossi et al. reported significantly higher American Orthopaedic Foot and Ankle Society (AOFAS) score and lower pain in the I-ONE therapy group both at 6 and 12 months follow-up (Cadossi et al. 2014).

Moreover, PEMF stimulation has been proven effective following matrix-assisted autologous chondrocyte implantation (MACI) in the treatment of chondral lesions of the knee (Collarile et al. 2018). The authors reported that the I-ONE therapy group achieves significantly better pain relief (2 and 6 months follow-up) and clinical outcome at the time of 60 months follow-up.

These studies suggest that PEMF stimulation improves the effectiveness of tissue engineering approaches for cartilage repair by favoring the anabolic activity of the implanted cells and periarticular tissues and protecting the construct from the catabolic effects of inflammation, regardless of the surgical procedure applied.

PEMFs in Postsurgical Treatments

Anterior cruciate ligament (ACL) injuries cause an increase in proinflammatory cytokines which alter the cartilage metabolism favoring catabolic phenomena leading to the development of OA (Bigoni et al. 2013).

The anti-inflammatory and chondroprotective effects of PEMFs on the joint environment have been extensively described, laying the ground for the application of PEMF stimulation as postsurgical treatment. The ultimate goal of BS after surgery is to quickly control the inflammatory response due to the surgical procedure, thus allowing the maintenance in the long term of the mechanical and biological properties of the cartilage and preventing the development of OA.

Benazzo, in a prospective, randomized, double-blind study showed that I-ONE therapy reduces the recovery time of patients undergoing ACL reconstruction and allows an early return to sport activity (Benazzo et al. 2008). In particular, the percentage of patients using NSAIDs following ACL reconstruction was significantly lower in the patients treated with I-ONE therapy compared to the placebo group. Two years after surgery, a complete functional recovery was achieved by 86% of the patients in the active group compared with 75% of the patients in the placebo group.

In two recent Italian studies on patients undergoing total knee arthroplasty (TKA), the authors reported that pain, knee swelling, and functional score were significantly better in PEMF-treated compared with the control group (Moretti et al. 2012; Adravanti et al. 2014). In addition to these findings, the study by Adravanti included also an evaluation of patients 36 months after the procedure. At this time point, only 7% of the patients in the PEMF group reported a level of persistent pain requiring the use of anti-inflammatory drugs, compared to 33% of control patients. Furthermore, no patient in the PEMF group expressed the need for walking aids compared to ~20% of the control group.

These pieces of clinical evidence highlight how early resolution of local inflammatory response after surgery enables patients to attain faster functional recovery and achieve better long-term results, most probably by favoring faster physiological healing.

BS FOR CARTILAGE REGENERATION: FUTURE PERSPECTIVES

It is commonly accepted that articular cartilage is a tissue with little or no regenerative potential and thus undergoes degradation over time. In 1853, James Paget reported that there are "no instances in which a lost portion of cartilage has been restored, or a wounded portion repaired with new and well-formed cartilage". For centuries, the dogma has been that cartilage repair is an unattainable goal. The notion that it is impossible to reverse the degeneration of articular cartilage has been questioned recently by the growing body of basic-research findings concerning the pathophysiology of articular cartilage. Recently, the spontaneous repair of full-thickness cartilage defects was noted in humans following joint offloading, either by realignment osteotomy (Koshino et al. 2003) or by total joint distraction techniques (Intema et al. 2011; Wiegant et al. 2013). These reparative events did not depend on cell expansion protocols but instead relied on endogenous reparative capabilities of native joint-resident or periarticular cells. Spontaneous cartilage regeneration suggests potentially overlapping roles for stem cells from different niches and also for mature chondrocytes. MSCs are adult stromal cells that are highly proliferative, clonogenic, and capable of multilineage differentiation into mesenchymal tissues including bone, cartilage, and adipose tissue. Joint-resident MSCs in humans were first described in an adult synovial membrane in 2001 by De Bari et al. (2001b). MSCs have also been detected in the synovial fluid compartment. Interestingly, the number of recoverable MSCs is much greater in synovial fluid samples from patients with rheumatoid arthritis or OA, as well as following ligament injury, than in samples from healthy joints (Jones et al. 2008). Synovial fluid MSCs and synovium-derived MSCs may represent the cartilage reservoir of chondrogenic precursors responsible for tissue maintenance. Several pieces of evidence from mouse models showed that adult mammalian synovium MSCs contained chondroprogenitor cells that participate in postinjury cartilage repair *in vivo* (Roelofs et al. 2017; Decker et al. 2017).

BS has already been shown to exert a positive effect on human articular chondrocyte proliferation (De Mattei et al. 2007; Chang et al. 2010) and on cartilage ECM synthesis (Ongaro et al. 2011, 2012). Moreover, PEMF stimulation has been reported to increase proliferation and osteogenic differentiation of adult MSCs (Esposito et al. 2012; Ongaro et al. 2014).

Based on these premises, the application of PEMF stimulation in cartilage regeneration mediated by resident MSCs can be foreseen. In this view, several efforts need to be taken in order to understand whether and how BS is able to enhance synovial MSCs proliferation, chondrogenic differentiation, and anabolic activities.

An increasing body of evidence suggests that aberrant or defective activity of MSCs might contribute to the development of OA. Murphy et al. demonstrated that MSCs isolated from patients with end-stage OA are functionally deficient for *in vitro* proliferation and differentiation (Murphy et al. 2002). Results from other studies have pointed to both age and disease as factors that influence the phenotype of MSCs. For example, De Bari et al. showed that human periosteal MSCs from young donors exhibit spontaneous chondrogenic activity in culture, whereas this activity is absent in cells from older donors (De Bari et al. 2001a).

Senescent chondrocytes have been described in cartilage tissue isolated from patients with posttraumatic and age-related OA (Martin et al. 2004). Although Senescent Cells (SnCs) are nondividing, their persistence can negatively impact surrounding cells and tissues, primarily through their senescence-associated secretory phenotype (SASP) (Campisi 2013). The SASP involves secretion of several extracellular proteases, proinflammatory cytokines, chemokines, and growth factors, which in turn cause inflammation and tissue degeneration. Recent studies provide some insight into the role of SnCs in OA development. Using a transgenic mouse model that allows selective imaging and removal of SnCs, Jeon OH et al. found that SnCs increased in the articular cartilage and synovium after anterior cruciate ligament transection (ACLT). The selective removal of SnCs reduced the development of posttraumatic OA, decreased pain, and promoted cartilage development (Jeon et al. 2017).

The instrumental role of SnCs and the SASP in OA opens the way to new strategies for treatment. It has been recently reported that the activation of A_{2A} AR inhibits cellular senescence in hepatic stellate cells (Ahsan and Mehal 2014).

PEMFs have been shown to induce the upregulation of A_{2A} and A_3 ARs on the cell membrane of several cell types (Varani et al. 2010). We can then hypothesize that PEMFs might stimulate ARs on articular cells, thus inhibiting cellular senescence. This in turn will decrease the levels of SASP proteins in the joint microenvironment, thus reducing inflammation and pain. The reduction in SASP proteins together with the well-known anti-inflammatory effect of PEMF stimulation will be able to significantly reduce the proinflammatory and catabolic environment associated with OA.

We can then anticipate three levels of action for BS in cartilage regeneration: (i) chondrogenic effect, acting on joint-resident MSCs to increase cellular proliferation, chondrogenic differentiation, and anabolic activities; (ii) anti-inflammatory effect mediated by the activation of A_{2A} and A_3 ARs; and (iii) antisenescence effect mediated by the activation of A_{2A} AR.

CONCLUSIONS

In consideration of the sensitivity of cartilage tissue to physical stimuli, the orthopedic community has now focused its interest on the use of BS to prevent cartilage degeneration, to enhance cartilage repair, and to favor patients' functional recovery.

A systematic review of the literature on the influence of PEMFs on cartilage concluded that PEMFs have a beneficial effect on chondrocyte proliferation, ECM synthesis, and chondrogenic differentiation. Moreover PEMFs decrease proinflammatory cytokines via A_{2A} and A_3 ARs leading to beneficial effects on pain and functional recovery of OA knees (Iwasa and Reddi 2018). These well-documented mechanisms of action of PEMFs allow us to hypothesize the application of PEMF stimulation not only for chondroprotective purposes but also for cartilage regenerative treatments.

Compared to the pharmacological treatments, BS has the advantage of being a local therapy; thus it can reach maximum concentration and therapeutic efficacy at the treatment site, without general negative side effects. BS has been proven suitable for prolonged treatments in the presence of local chronic degenerative diseases, whereas at present, there is no evidence for its application to treat systemic disorders.

In the last 30 years, the field of BS applied in clinical settings made impressive progress. The bases of this success lie on sound basic research performed following the principles and method of pharmacology. The remarkable amount of data obtained from *in vitro* and *in vivo* studies allowed the successful translation of such knowledge to the clinic.

Further development of the clinical use of physical agents involves facing numerous and complex issues. The first step in this process is recognizing and defining a therapeutic development area as clinical biophysics and thus creating a common ground of reference for researchers in different fields. This step is fundamental to point the way for future research in several other clinical areas.

REFERENCES

Adravanti P, Nicoletti S, Setti S, Ampollini A, de Girolamo L. Effect of pulsed electromagnetic field therapy in patients undergoing total knee arthroplasty: A randomised controlled trial. *Int Orthop* 2014;38(2):397–403.

Ahmed TA, Hincke MT. Strategies for articular cartilage lesion repair and functional restoration. *Tissue Eng Part B* 2010;16:305.

Ahsan MK, Mehal WZ. Activation of adenosine receptor A2A increases HSC proliferation and inhibits death and senescence by down-regulation of p53 and Rb. *Front Pharmacol* 2014;5:69.

Bassett CAL, Becker RO. Generation of electric potentials in bone in response to mechanical stress. *Science* 1962;137:1063–1064.

Benazzo F, Zanon G, Pederzini L, Modonesi F, Cardile C, Falez F, Ciolli L, La Cava F, Giannini S, Buda R, Setti S, Caruso G, Massari L. Effects of biophysical stimulation in patients undergoing arthroscopic reconstruction of anterior cruciate ligament: Prospective, randomized and double blind study. *Knee Surg Sports Traumatol Arthrosc* 2008;16:595.

Berg EE, Pollard ME, Kang Q. Interarticular bone tunnel healing. *Arthroscopy* 2001;17:189.

Bigoni M, Sacerdote P, Turati M, Franchi S, Gandolla M, Gaddi D, Moretti S, Munegato D, Augusti CA, Bresciani E, Omeljaniuk RJ, Locatelli V, Torsello A. Acute and late changes in intraarticular cytokine levels following anterior cruciate ligament injury. *Orthop Res* 2013;31(2):315–321.

Boopalan PRJVC, Arumugam S, Livingston A, Mohanty, M, Chittaranjan S. Pulsed electromagnetic field therapy results in healing of full thickness articular cartilage defect. *Int Orthop* 2011;35:143.

Borea PA, Varani K, Vincenzi F, Baraldi PG, Tabrizi MA, Merighi S, Gessi S. The A3 adenosine receptor: History and perspectives. *Pharmacol Rev* 2015;67(1):74–102.

Borelli PP. Aspetti di riabilitazione nelle fratture di polso: Ruolo della stimolazione biofisica e del tutore modulare. *Chirurgia della Mano* 2017;54(3):72–90.

Brighton CT, Wang W, Seldes R, Zhang G, Pollack SR. Signal transduction in electrically stimulated bone cells. *J Bone Joint Surg Am* 2001;83:1514–1523.

Cadossi M, Buda RE, Ramponi L, Sambri A, Natali S, Giannini S. Bone marrow derived cells and biophysical stimulation for talar osteochondral lesions: A randomized controlled study. *Foot Ankle Int* 2014;35(10):981–987.

Campisi J. Aging, cellular senescence, and cancer. *Annu Rev Physiol* 2013;75:685–705.

Chang CH, Loo ST, Liu HL, Fang HW, Lin HY. Can low frequency electromagnetic field help cartilage tissue engineering? *J Biomed Mater Res A* 2010;92(3):843–851.

Ciombor DM, Aaron RK, Wang S, Simon B. Modification of osteoarthritis by pulsed electromagnetic field: A morphological study. *Osteoarthritis Cartilage* 2003;11(6):455–462.

Cohen SB, Gill SS, Baer GS, Leo BM, Scheld WM, Diduch DR. Reducing joint destruction due to septic arthrosis using an adenosine2A receptor agonist. *J Orthop Res* 2004;22(2):427–435.

Collarile M, Sambri A, Lullini G, Cadossi M, Zorzi C. Biophysical stimulation improves clinical results of matrix-assisted autologous chondrocyte implantation in the treatment of chondral lesions of the knee. *Knee Surg Sports Traumatol Arthrosc* 2018;26(4):1223–1229.

De Bari C, Dell'Accio F, Luyten FP. Human periosteum-derived cells maintain phenotypic stability and chondrogenic potential throughout expansion regardless of donor age. *Arthritis Rheum* 2001a;44:85–95.

De Bari C, Dell'Accio F, Tylzanowski P, Luyten FP. Multipotent mesenchymal stem cells from adult human synovial membrane. *Arthritis Rheum* 2001b;44:1928–1942.

De Mattei M, Fini M, Setti S, Ongaro A, Gemmati D, Stabellini G, Pellati A, Caruso A. Proteoglycan synthesis in bovine articular cartilage explants exposed to different low-frequency low-energy pulsed electromagnetic fields. *Osteoarthritis Cartilage* 2007;15:163.

De Mattei M, Pasello M, Pellati A, Stabellini G, Massari L, Gemmati D, Caruso A. Effects of electromagnetic fields on proteoglycan metabolism of bovine articular cartilage explants. *Connect Tissue Res* 2003;44(3–4):154–159.

De Mattei M, Pellati A, Pasello M, Ongaro A, Setti S, Massari L, Gemmati D, Caruso A. Effects of physical stimulation with electromagnetic field and insulin growth factor-I treatment on proteoglycan synthesis of bovine articular cartilage. *Osteoarthritis Cartilage* 2004;12(10):793–800.

De Mattei M, Varani K, Masieri FF, Pellati A, Ongaro A, Fini M, Cadossi R, Vincenzi F, Borea PA, Caruso A. Adenosine analogs and electromagnetic fields inhibit prostaglandin E(2) release in bovine synovial fibroblasts. *Osteoarthritis Cartilage* 2009;17(2):252–262.

Decker RS, Um HB, Dyment NA, Cottingham N, Usami Y, Enomoto-Iwamoto M, Kronenberg MS, Maye P, Rowe DW, Koyama E, Pacifici M. Cell origin, volume and arrangement are drivers of articular cartilage formation, morphogenesis and response to injury in mouse limbs. *Dev Biol* 2017;1:56–68.

Esposito M, Lucariello A, Riccio I, Riccio V, Esposito V, Riccardi G. Differentiation of human osteoprogenitor cells increases after treatment with pulsed electromagnetic fields. *In Vivo* 2012;26(2):299–304.

Fini M, Giavaresi G, Carpi A, Nicolini A, Setti S, Giardino R. Effects of pulsed electromagnetic fields on articular hyaline cartilage: Review of experimental and clinical studies. *Biomed Pharmacother* 2005;59(7):388–394.

Fini M, Torricelli P, Giavaresi G, Aldini NN, Cavani F, Setti S, Nicolini A, Carpi A, Giardino R. Effect of pulsed electromagnetic field stimulation on knee cartilage, subchondral and epyphiseal trabecular bone of aged Dunkin Hartley guinea pigs. *Biomed Pharmacother* 2008;62(10):709–715.

Fukada E, Yasuda I. On the piezoelectric effect of bone. *J Phys Soc Japan* 1957;12:121–128.

Gessi S, Merighi S, Fazzi D, Stefanelli A, Varani K, Borea PA. Adenosine receptor targeting in health and disease. *Expert Opin Invest Drugs* 2011;20(12):1591–1609.

Gobbi A, Lad D, Petrera M, and Karnatzikos G, Symptomatic Early Osteoarthritis of the Knee Treated With Pulsed Electromagnetic Fields: Two-Year Follow-up. *Cartilage* 2014;5(2):78–85.

Holland TA, Tessmar JKV, Tabata Y, Mikos AG. Transforming growth factor-b1 release from oligo(poly(ethylene glycol) fumarate) hydrogels in conditions that model the cartilage wound healing environment. *J Controlled Release* 2004;94:101–114.

Intema F, Van Roermund PM, Marijnissen AC, Cotofana S, Eckstein F, Castelein RM, Bijlsma JW, Mastbergen SC, Lafeber FP. Tissue structure modification in knee osteoarthritis by use of joint distraction: An open 1-year pilot study. *Ann Rheum Dis* 2011;70:1441–1446.

Iwasa K, Reddi AH. Pulsed electromagnetic fields and tissue engineering of the joints. *Tissue Eng Part B Rev* 2018;24(2):144–154.

Jeon OH, Kim C, Laberge RM, Demaria M, Rathod S, Vasserot AP, Chung JW, Kim DH, Poon Y, David N, Baker DJ, van Deursen JM, Campisi J, Elisseeff JH. Local clearance of senescent cells attenuates the development of post-traumatic osteoarthritis and creates a pro-regenerative environment. *Nat Med* 2017;23(6):775–781.

Jones EA, Jones EA, Crawford A, English A, Henshaw K, Mundy J, Corscadden D, Chapman T, Emery P, Hatton P, McGonagle D. Synovial fluid mesenchymal stem cells in health and early osteoarthritis: Detection and functional evaluation at the single-cell level. *Arthritis Rheum* 2008;58:1731–1740.

Koshino T, Wada S, Ara Y, Saito, T. Regeneration f degenerated articular cartilage after high tibial valgus osteotomy for medial compartmental osteoarthritis of the knee. *Knee* 2003;10:229–236.

Kurtz S, Ong K, Lau E, Mowat F, Halpern M. Projections of primary and revision hip and knee arthroplasty in the United States from 2005 to 2030. *J Bone Joint Surg Am* 2007;89:780.

Lee JY, Jhun BS, Oh YT, Lee JH, Choe W, Baik HH, Ha J, Yoon KS, Kim SS, Kang I. Activation of adenosine A3 receptor suppresses lipopolysaccharide- induced TNF-alpha production through inhibition of PI 3-kinase/Akt and NF-kappaB activation in murine BV2 microglial cells. *Neurosci Lett* 2006;396(1):2006.

Lotz MK, Kraus VB. New developments in osteoarthritis. Posttraumatic osteoarthritis: Pathogenesis and pharmacological treatment options. *Arthritis Res Ther* 2010;12:211.

Marcheggiani Muccioli GM, Grassi A, Setti S, Filardo G, Zambelli L, Bonanzinga T, Rimondi E, Busacca M, Zaffagnini S. Conservative treatment of spontaneous osteonecrosis of the knee in the early stage: Pulsed electromagnetic fields therapy. *Eur J Radiol* 2013;82(3):530–537.

Martel-Pelletier J, Pelletier JP. Is osteoarthritis a disease involving only cartilage or other articular tissues? *Eklem Hastalik Cerrahisi* 2010;21:2.

Martin JA, Brown T, Heiner A, Buckwalter JA. Post-traumatic osteoarthritis: The role of accelerated chondrocyte senescence. *Biorheology* 2004;41:479–491.

Martin L, Pingle SC, Hallam DM, Rybak LP, Ramkumar V. Activation of the adenosine A3 receptor in RAW 264.7 cells inhibits lipopolysaccharide-stimulated tumor necrosis factor-alpha release by reducing calcium dependent activation of nuclear factor kappaB and extracellular signal-regulated kinase 1/2. *J Pharmacol Exp Ther* 2006;316:71.

Martinelli N, Bianchi A, Sartorelli E, Dondi A, Bonifacini C, Malerba F. Treatment of bone marrow edema of the talus with pulsed electromagnetic fields. *J Am Podiatr Med Assoc* 2015;105(1):27–32.

Moretti B, Notarnicola A, Moretti L, Setti S, De Terlizzi F, Pesce V, Patella V. I-ONE therapy in patients undergoing total knee arthroplasty: A prospective, randomized and controlled study. *BMC Musculoskelet Disord* 2012;13(1):88.

Murphy JM, Dixon K, Beck S, Fabian D, Feldman A, Barry F. Reduced chondrogenic and adipogenic activity of mesenchymal stem cells from patients with advanced osteoarthritis. *Arthritis Rheum* 2002;46:704–713.

Ongaro A, Pellati A, Bagheri L, Fortini C, Setti S, De Mattei M. Pulsed electromagnetic fields stimulate osteogenic differentiation in human bone marrow and adipose tissue derived mesenchymal stem cells. *Bioelectromagnetics* 2014;35(6):426–436.

Ongaro A, Pellati A, Masieri FF, Caruso A, Setti S, Cadossi R, Biscione R, Massari L, Fini M, De Mattei M. Chondroprotective effects of pulsed electromagnetic fields on human cartilage explants. *Bioelectromagnetics* 2011;32:543.

Ongaro A, Pellati A, Setti S, Masieri FF, Aquila G, Fini M, Caruso A, De Mattei M. Electromagnetic fields counteract IL-1β activity during chondrogenesis of bovine mesenchymal stem cells. *J Tissue Eng Regen Med* 2015;9(12):E229–E238.

Ongaro A, Varani K, Masieri FF, Pellati A, Massari L, Cadossi R, Vincenzi F, Borea PA, Fini M, Caruso A, De Mattei M. Electromagnetic fields (EMFs) and adenosine receptors modulate prostaglandin E(2) and cytokine release in human osteoarthritic synovial fibroblasts. *J Cell Physiol* 2012;227(6):2461–2469.

Osti L, Del Buono A, Maffulli N. Application of pulsed electromagnetic fields after microfractures to the knee: A mid-term study. *Int Orthop* 2015;39:1289.

Pagani S, Veronesi F, Aldini NN, Fini M. Complex regional pain syndrome type I, a debilitating and poorly understood syndrome. *Possible Role Pulsed Electromagn Fields Narrative Rev Pain Phys* 2017;20(6):807–822.

Pezzetti F, De Mattei M, Caruso A, Cadossi R, Zucchini P, Carinci F, Traina GC, Sollazzo V. Effects of pulsed electromagnetic fields on human chondrocytes: An in vitro study. *Calcif Tissue Int* 1999;65(5):396–401.

Ramsey DK, Russell ME. Unloader braces for medial compartment knee osteoarthritis: Implications on mediating progression. *Sports Health* 2009;1:416.

Redman SN, Oldfield SF, Archer CW. Current strategies for articular cartilage repair. *Eur Cell Mater* 2005;9:23.

Roelofs AJ, Zupan J, Riemen AHK, Kania K, Ansboro S, White N, Clark SM, De Bari C. Joint morphogenetic cells in the adult mammalian synovium. *Nat Commun* 2017;8:15040.

Servodio Iammarrone C, Cadossi M, Sambri A, Grosso E, Corrado B, Servodio Iammarrone F. Is there a role of pulsed electromagnetic fields in management of patellofemoral pain syndrome? Randomized controlled study at one year follow-up. *Bioelectromagnetics* 2016;37(2):81–88.

Shakibaei M, John T, Schulze-Tanzil G, Lehmann I, Mobasheri, A. Suppression of NF-kappaB activation by curcumin leads to inhibition of expression of cyclo-oxygenase-2 and matrix metalloproteinase-9 in human articular chondrocytes: Implications for the treatment of osteoarthritis. *Biochem Pharmacol* 2007;73(9):1434–1445.

Thamsborg G, Florescu A, Oturai P, Fallentin E, Tritsaris K, Dissing S. Treatment of knee osteoarthritis with pulsed electromagnetic fields: A randomized, double-blind, placebo-controlled study. *Osteoarthritis Cartilage* 2005;13(7):575–581.

Tomlinson S, MacNeil S, Walker SW, Ollis CA, Merrit JE, Brown BL. Calmodulin and cell function. *Clin Sci (Colch)* 1984;66:497–507.

Ulrich-Vinther M, Maloney MD, Schwarz EM, Rosier R, O'Keefe RJ. Articular cartilage biology. *J Am Acad Orthop Surg* 2003;11(6):421–30.

Uzieliene I, Bernotas P, Mobasheri A, Bernotiene E. The role of physical stimuli on calcium channels in chondrogenic differentiation of mesenchymal stem cells. *Int J Mol Sci* 2018;19(10):2998.

Varani K, Gessi S, Merighi S, Iannotta V, Cattabriga E, Spisani S, Cadossi R, Borea PA. Effect of low frequency electromagnetic fields on A2A adenosine receptors in human neutrophils. *Brit J Pharmacol* 2002;136:57–66.

Varani K, De Mattei M, Vincenzi F. et al. Characterization of adenosine receptors in bovine chondrocytes and fibroblast like synoviocytes exposed to low frequency low energy pulsed electromagnetic fields. *Osteoarthritis Cartilage* 2008;16(3):292–304.

Varani K, Padovan M, Govoni M, Vincenzi F, Trotta F, Borea PA. The role of adenosine receptors in rheumatoid arthritis. *Autoimmun Rev* 2010;10(2):61–64.

Veronesi F, Fini M, Giavaresi G, Ongaro A, De Mattei M, Pellati A, Setti S, Tschon M. Experimentally induced cartilage degeneration treated by pulsed electromagnetic field stimulation; an in vitro study on bovine cartilage. *BMC Musculoskelet Disord* 2015;16(1):308.

Vincenzi F, Targa M, Corciulo C, Gessi S, Merighi S, Setti S, Cadossi R, Borea PA, Varani K. Pulsed electromagnetic fields increased the anti-inflammatory effect of A2A and A3 adenosine receptors in human T/C-28a2 chondrocytes and hFOB 1.19 osteoblasts. *PLoS One* 2013;8(5):e65561.

Wiegant K, van Roermund PM, Intema F, Cotofana S, Eckstein F, Mastbergen SC, Lafeber FP. Sustained clinical and structural benefit after joint distraction in the treatment of severe knee osteoarthritis. *Osteoarthritis Cartilage* 2013;21:1660–1667.

Zorzi C, Dall'Oca C, Cadossi R, Setti S. Effects of pulsed electromagnetic fields on patients' recovery after arthroscopic surgery: Prospective, randomized and double blind study. *Knee Surg Sports Traumatol Arthrosc* 2007;15:830.

5 Electromagnetic Field Effects on Soft Tissues – Muscles and Tendons

Erik I. Waldorff, Nianli Zhang,
James T. Ryaby, and Andrew F. Kuntz

CONTENTS

INTRODUCTION

Pulsed electromagnetic field (PEMF) therapy has been shown to contain a wide range of spectral components allowing for potential coupling to a variety of possible biochemical signaling pathways (Pilla 2015). The possibility of treatment using electromagnetic fields (EMFs) for various disorders has therefore drawn interest in part due to the ability to noninvasively induce an electric current in the target tissue. While electromagnetic studies have included disorders such as Major Depressive Disorder (using transcranial magnetic stimulation (TMS)) (Carpenter, Janicak et al. 2012), fibromyalgia (Thomas, Graham et al. 2007), and osteoarthritis of the knee (Nicolakis, Kollmitzer et al. 2002), the only Class III EMF devices currently approved by the U.S. Food & Drug Administration (FDA) have been within the category of bone growth simulation/ostegenesis stimulation.

As mentioned, common for all the approved commercial EMF devices for osteogenesis stimulation in the U.S. is their classification by the FDA as a Class III device requiring the establishment of safety and effectiveness of the device through valid scientific evidence before approval from the FDA can be obtained. This is done through the initial FDA approval of an investigational device exemption (IDE) allowing for the device to be used in a clinical study collecting safety and effectiveness data. This data is required to support a premarket approval (PMA) application which upon approval enables the device to enter the market. This process ensures that safety issues such as hardware failure, inadvertent exposure of incorrect target tissues, incorrect exposure (amplitude, duration, etc.), and unanticipated adverse events are considered and evaluated.

As it has been shown in other chapters of this book, PEMF effects vary significantly depending on alterations to the PEMF parameters such as pulse period, burst period, amplitude, and number of pulses/burst. Combining this with PEMF field parameter limitations due to engineering considerations such as battery life and device portability emphasizes the need to gain enough *in-vitro/in-vivo* knowledge of specific PEMF signals and their target tissue interaction to enable a high success rate in clinical trials.

Hence, while PEMF application for soft tissues have been examined for diseases such as intervertebral disc degeneration (Tang, Coughlin et al. 2015a,b, Miller, Coughlin et al. 2016, Tang, Coughlin et al. 2016, Tang, Coughlin et al. 2017a,b), this chapter will focus on the application of PEMF on muscles and tendons as more clinical studies have been conducted for these soft tissue types. This will enable a better presentation of the full translational perspective from preclinical investigations to clinical studies.

PRECLINICAL INVESTIGATIONS (*IN VITRO*): TENOGENIC AND MYOGENIC *IN-VITRO* EXPERIMENTS

MYOCYTE *IN-VITRO* EXPERIMENTS

Feng, He et al. (2011) examined the effect of duration (10, 15, and 20 days) of PEMF exposure (50 Hz, 1 mT, 30 min/day) on rat bone marrow mesenchymal stem cell (rBMSC) differentiation into cardiomyocyte-like cells. Positive control groups were treated with 5-azacytidine (10 μmol/L for 1 day), a DNA methylation inhibitor which stimulates cardiac differentiation of stem cells (Kaur, Yang et al. 2014), while standard controls were cultured with standard media. Outcome measures included mRNA expression of cardiac troponin T (TNNT2) and alpha-actinin (ACTN2) as determined by RT-PCR. Results indicated that, as early as 10 days, PEMF induced a 15.78-fold and 4.92-fold increase in TNNT2 mRNA and ACTN2 mRNA expression, respectively, relative to control groups. This increase was similar to that of the positive control groups at 10 days.

Norizadeh-Abbariki, Mashinchian et al. (2014) used dextran-coated superparamagnetic iron oxide nanoparticles (SPIONs, ~20 nm size, 20 μg/mL) to label embryonic stem cells to determine if the interaction of PEMF with the particles could enhance skeletal muscle formation/differentiation which is controlled by transcription factors such as MyoD, Myf5, Myf6, myogenin (MyoG), and sarcomeric myosin heavy chain (myHC). Using a Helmholtz coil system, PEMF treatment (0.1 mT, 12 Hz) was done 6 h/day for 5 days during differentiation period. Cells were cultured in myogenic differentiation medium (1% dimethyl sulfoxide (DMSO) and 5% horse serum). Reverse transcription polymerase chain reaction (RT-PCR) showed that Myh2 expression was enhanced by PEMF (with or without SPIONs) relative to the differentiation media treated groups. However, for MyoG expression, PEMF acted synergistically with differentiation media and SPIONs (36- and 85-fold increase relative to control, respectively). Performing the same experiment in serum-free media, it was furthermore shown that the combined effect of PEMF, differentiation media, and SPIONs could enhance myogenic differentiation of stem cells as compared to each stimuli by itself.

Using rodent neonatal ventricular myocytes in serum-free medium, Wei, Sun et al. (2015) assessed cell viability and intracellular free calcium concentration when exposed to caffeine (10 nM for 60 s) or EMF (0, 15, 50, 75, and 100 Hz at 2.0 mT for 180 s) using Fura-2-acetoxymethyl ester (Fura-2/AM) and spectrofluorometry. The former served as a control for the Na^+/Ca^{2+} exchanger (NCX) function and activity of the sarco/endoplasmic reticulum Ca^{2+}-ATPase (SERCA). While caffeine application expectedly caused a rapid increase in intracellular Ca^{2+}, EMF (all frequencies) was able to increase the intracellular Ca^{2+} transient baseline and reduce the amplitude of Ca^{2+} transients (i.e., a reduction in half-decay times for both caffeine-induced intracellular Ca^{2+} transients and spontaneous intracellular Ca^{2+} transients) and Ca^{2+} levels in the sarco/endoplasmic reticulum. Although the authors indicate that the mechanisms behind the effect of PEMF on NCX and SERCA activity are unknown, the data suggests that EMF can change intracellular Ca^{2+} signaling in cardiomyocytes.

In a subsequent study, Wei, Tong et al. (2016) also examined the effect of EMF on cardiomyocytes following hypoxia-induced injury. Neonatal ventricular myocytes were cultured under hypoxic (1% O_2, 5% CO_2, 37°C) or normoxic (21% O_2, 5% CO_2, 37°C) conditions for 12 h. The hypoxic conditions were also done for both control conditions (no EMF) and 30 min of EMF exposure (15 Hz, 2 mT) prior to hypoxic conditions and with or without KNK437 (100 μM), a heat shock protein 70 (HSP70) inhibitor, which was added 1 h before EMF exposure. Outcome measures included morphological analysis (cell surface area) as determined by microscopy; protein content; cytotoxicity assessment; cell proliferation using DNA bromodeoxyuridine (BrdU) incorporation assay; intracellular calcium assessment using Fura-2 fluoroscopy; and gene expression of BCL-2, Bax, Caspase-3, β-MHC, HSP70, and 18S using RT-PCR. It was shown that EMF exposure reduced hypoxia-induced injury in cardiomyocytes through a reduction in cell death, increase in cell viability, and BrdU incorporation and reduction of mRNA expression of genes related to apoptosis. In addition, it was shown that EMF inhibited the hypoxia-induced cardiomyocytes hypertrophy. Finally EMF was shown to promote cardioprotection through activation of HSP70 in hypoxic cardiomyocytes.

Xu, Zhang et al. (2016) exposed C2C12 myoblasts grown in standard media to PEMF for 4 days (4 h/day, sinusoidal waveform, 100 Hz, 1 mT) and found an increase in proliferation without a change in apoptosis rate relative to non-PEMF controls. The MAPK/ERK pathway was furthermore shown to be activated by PEMF. Specifically it was also shown that ERK activation was necessary for the induced cell proliferation caused by PEMF. Although proliferation was shown to increase with PEMF, myogenesis also involves the differentiation of myoblast which, as the authors discuss, has not been examined yet.

At the same time as Xu, Zhang et al. (2016), Liu, Lee et al. (2016) examined the effect of PEMF (15 Hz burst frequency, 3.85 kHz pulse frequency, 1.5 mT) on both myocyte proliferation and differentiation using C2C12 murine myoblasts. Liu et al. showed that daily 3-h PEMF exposure for 2 weeks enhanced gene expression of the MyoD growth factor in myocytes under inflammatory conditions (10 ng/mL of IL-1) but not under normal conditions. In addition, it was found that myotube formation was increased under both normal and inflammatory conditions (10 ng/mL of IL-1). The implications from these results may be the potential use of PEMF as a nonoperative treatment to promote muscle regeneration after injury.

TENOCYTE *IN-VITRO* EXPERIMENTS

Other researchers have investigated the effect of PEMF on tenocytes under various conditions. Specifically, Denaro, Ruzzini et al. (2011) exposed human tenocyte cultures (from human supraspinatus and quadriceps tendons) to continuous PEMF (50 Hz, 0.4 mT) for 3, 5, and 7 days. In addition they examined the effect of PEMF on *in-vitro* microcuts on confluent tenocyte cultures 12, 24, and 35 h following injury. PEMF did not significantly change tenocyte cell growth or collagen accumulation relative to controls at any time points. Furthermore flow-cytometry analysis indicated that tenocyte cell cycle was not affected by PEMF either. However, wound size following the *in-vitro* microcuts was significantly smaller for the PEMF group at 12 and 24 h with wound closure for both control and PEMF groups at 36 h. These results indicate that while PEMF did not change the individual tenocytes in terms of maturity or proliferation, tenocyte migration rates were enhanced by PEMF.

Interestingly de Girolamo, Stanco et al. (2013) found contradictory results when they exposed primary tendon cells from semitendinosus and gracilis tendons from young patients to PEMF (75 Hz, 1.5 mT) for 4, 8, or 12 h. While no changes were seen in cell viability or DNA content due to PEMF, qPCR showed a significant increase in tendon-specific gene transcription for scleraxis and type-1 collagen which was also positively correlated with PEMF exposure time. Furthermore, anti-inflammatory factors (IL-6 (membrane bound), IL-10, and TGF-β) were increased with PEMF (8 and 12 h) while proinflammatory factors (IL-1β and TNF-α) were not affected by PEMF. Finally, VEGF-A expression was also increased with PEMF (8 and 12 h).

Subsequently de Girolamo, Vigano et al. (2015) repeated their experiment while varying the PEMF field intensity (1.5 and 3.0 mT, 75 Hz), duration (8 and 12 h), and number of durations (single or three (only for 1.5 mT)). Results indicated that, similarly to their first study (de Girolamo, Stanco et al. 2013), none of the PEMF treatments affected tendon cell apoptosis. However, cell viability was increased with the 3mT PEMF for 12 h of exposure. A single 1.5mT PEMF exposure, whether 8 or 12 h, resulted in the largest increase in scleraxis, type-1 collagen, and VEGF-A. Unlike the previous study, slight increase was found in IL-1β and TNF-α; however, anti-inflammatory factors (IL-6 (membrane bound), IL-10, and TGF-β) were increased as before with PEMF (8 and 12 h). Noticeably the single 1.5 mT exposure led to an even greater IL-10 expression than other PEMF treatments. This indicates that PEMF treatment of tendon cells *in vitro* is indeed parameter dependent requiring a need to investigate specific PEMF waveforms prior to using them in an *in-vivo* setting whether preclinically or clinically.

Marmotti, Peretti et al. (2018) used a similar PEMF system (1.5 mT, 75 Hz) as de Girolamo, Stanco et al. (2013), de Girolamo, Vigano et al. (2015) to treat mesenchymal stem cells from human umbilical cord *in vitro* with/without PEMF (2, 4, or 8 h/day) for 7, 14, or 21 days while culturing them in differentiation medium (FGF-2). Similarly to other studies, cell apoptosis was not affected by PEMF treatment. Treatment with both FGF-2 and PEMF showed the greatest production of scleraxis and collagen type-1 for all time points and was positively correlated with daily exposure time. The proliferative marker PCNA was also greatest when FGF-2 was combined with PEMF for day 7 and 14. However, a treatment of 2 hours/day resulted in greater expressions of PCNA than 4 or 8 h/day.

Randelli, Menon et al. (2016) on the other hand exposed human tendon stem cells to PEMF for 1 h utilizing a commercially available PEMF system (PST® by Global Munich Germany). Unlike other PEMF waveforms, this system delivered a quasi-rectangular waveform with continuously varying frequency (10–30 Hz) and field strength (0.5–1.5 mT). Following PEMF exposure, PST and control cells were grown for another 10, 24, and 48 h. In addition, cell migration was also assessed 5, 20, 24, and 30 h following a wound-healing assay similar to Denaro, Ruzzini et al. (2011). Randelli et al. found no change in cell proliferation, cell morphology, wound closure rate, apoptosis, or expression of tendon markers (Tenascin C and collagen type-1) at any time points. However, a higher expression of stem cell markers (Oct4 and KLF4) was maintained following PEMF exposure indicating the preservation of a more undifferentiated cellular state.

Liu, Lee et al. (2016) also examined the effect of PEMF (15 Hz burst frequency, 3.85 kHz pulse frequency, 1.5 mT) on tenocyte proliferation and differentiation *in vitro* using human rotator cuff tenocytes. Interestingly it was found that daily 3-h PEMF exposure for 2 weeks enhanced gene expression of growth factors in human rotator cuff tenocytes (COL1, TGFβ-1, PDGFβ, BMP12, and TIMP4) under inflammatory conditions (10 ng/mL of IL-1) but not under normal conditions.

The implications from these results may be the potential use of PEMF as a nonoperative treatment to improve clinical outcomes following rotator-cuff repair.

PRECLINICAL INVESTIGATION (*IN VIVO*): MUSCLE AND TENDON ANIMAL MODELS

MUSCLE ANIMAL MODELS

In a rodent model, Smith, Wong-Gibbons et al. (2004) showed that PEMF (15 Hz burst frequency, 3.85 kHz pulse frequency, 1.5 mT) applied to the arteriolar microvessels in the cremaster muscle could lead to an immediate increase (9%) in arteriolar diameter with both 2 and 60 min of exposure. This effect was not accompanied by any changes in heart rate or systemic arterial pressure. In addition, the vasodilation effects were shown to be accumulative between the two exposure times, which were done in series. The authors suggested that, since the rapid effect was within 2 min, the effect did not require a specific expression of a gene for a specific protein to achieve the said effect.

In a crush-injury rodent model of the extensor digitorum longus muscle, Cheon, Park et al. (2012) showed that PEMF (80–600 Hz burst frequency, 27.12 MHz pulse frequency, 0.2 mT, 20 min exposure) applied once a day for 5 days following injury led to a significant increase in HSP70. Delayed PEMF application (once a day starting 3 days after injury) showed a significant increase in HSP70, too, relative to controls albeit lower than the immediate PEMF application group. Qualitative histological evidence furthermore showed an improvement of atrophy and irregular muscle fiber arrangement caused by the crush injury. These results parallel the previously mentioned *in-vitro* work by Wei, Tong et al. (2016) who showed cardioprotection promoted by EMF through activation of HSP70 in hypoxic cardiomyocytes.

Tendon Animal Models

Robotti, Zimbler et al. (1999) examined the effect of PEMF on flexor tendon healing in chickens. Specifically, following mid-section flexor tendon transection and repair, animals were divided into groups in which the operative extremity was either immobilized or not, with animals further divided into groups that either received or did not receive 8 h of daily PEMF treatment (15 Hz burst frequency, 3.73 kHz pulse frequency, 0.02 mT) for 3 weeks. The study found a small but significant decrease in tensile strength due to PEMF but no changes in joint flexion. However, the interactive effect due to casting could not be determined. The authors noted that the utilized PEMF signal had previously been shown to successfully increase collagen formation and proteoglycan synthesis in endochondral bone formation indicating the dependency on signal specificity for various indications.

Lee, Maffulli et al. (1997) examined the effect of pulsed magnetic fields (PMFs) and PEMFs in a rodent Achilles tendon inflammation model. Specifically, inflammation was caused through direct mass impact to the Achilles tendon with PMF (17 or 50 Hz, 4.95 mT) or PEMF (15 or 46 Hz burst frequency, 27.12 MHz pulse frequency, 3.6 or 12.9 W mean power) being applied subsequently for 15 min, 5 times a week, for up to 4 weeks. Sacrifice occurred 2 h and 1, 3, 7, 14, and 28 day(s) after injury. Outcome measures included tendon weight, water content, and histological appearance. Results showed a significant increase in water content and weight for PEMF at early time points but a significantly sustained lower water content for PMF at early and later time points. In addition while collagen alignment was improved qualitatively for all treatment groups across the 28 days, PMF (17 Hz) showed the best return to normal physiological appearance with an accompanying decrease in inflammation.

Tucker, Cirone et al. (2016) showed that daily 3-h PEMF exposure (15 Hz burst frequency, 3.85 kHz pulse frequency, 1.5 mT) improved tendon-to-bone healing in an acute rotator cuff repair model in rats. Specifically the tendon modulus increased significantly at early time periods (100% and 60% at 4 and 8 weeks, respectively) with increased maximum stress (4 weeks) and subsequent improved bone quality at 16 weeks (increased bone volume fraction, trabecular thickness, and bone mineral density). The results indicated a potential new usage for PEMF as an adjunct treatment to surgical rotator cuff repair to prevent post-op retears.

Using the same acute rotator cuff repair model in rats, further investigations by Huegel, Choi et al. (2017) revealed that using PEMFs (15 Hz burst frequency, 1.5 mT) with varying fundamental pulse frequencies (3.85–40 kHz) or exposure durations (1, 3, or 6 h/day) led to improvements in tendon properties at the earlier time points and at lower fundamental frequencies.

The same laboratory also investigated the structural/functional effects of the same PEMFs (15 Hz burst frequency, 3.85–40 kHz pulse frequency, 1.5 mT) on tendon-to-tendon healing in a full- and partial-tear Achilles tendon repair model with and without immobilization (Boorman-Padgett, Huegel et al. 2018, Huegel, Boorman-Padgett et al. 2018, Boorman-Padgett, Huegel et al. 2019). Specifically male rats underwent acute, complete transection and repair of the Achilles tendon (FULL, n = 144) or full thickness, partial width injury (PART, n = 160) followed by immobilization for 1 week. In addition, a third group also underwent full thickness, partial width injury (PART-NI, n = 144) but without subsequent immobilization. All animals received PEMF exposure for either 1 or 3 h/day until sacrifice at 1, 3, and 6 weeks. Outcome measures included passive joint mechanics, gait analysis, biomechanical assessments, histological analysis of the repair site, and mCT (humerus) assessment (FULL only). A decrease in stiffness and limb-loading rate was observed for the PEMF groups for the FULL groups, whereas an increase in stiffness with no change in range of motion (ROM) was seen for the PART groups. An increase in limb loading rate and speed was found at early time points for the PART-NI groups, but no other changes were found for any other outcome measures.

CLINICAL APPLICATIONS: MUSCLE AND TENDON CLINICAL STUDIES

MUSCLE CLINICAL APPLICATIONS

Jeon, Kang et al. (2015) applied PEMF (1 Hz burst frequency, 2.08 kHz pulse frequency, 200 mT) for 10 min to the most painful area of the bicep brachii of healthy young subjects in South Korea at 0, 24, 48, and 72 h following exercise-induced delayed onset muscle soreness (DOMS). Outcome measures at each time point included muscle soreness using a visual analog scale (VAS) scale, peak torque using an isokinetic dynamometer, and median frequency (MDF) and electromechanical delay using EMG measurements. Overall, it was found that PEMF improved perceived muscle soreness (i.e., reduced VAS scores), MDF (increased), EMD (increased), while isometric peak torque was unaffected. This indicates that PEMF could reduce DOMS symptoms which is also supported by the fact that HSP70 has been shown to increase significantly during intensive training sessions and may play a role in muscle recovery and remodeling/adaptation following high-intensity exercise (Liu, Mayr et al. 1999, Thompson, Clarkson et al. 2002, Paulsen, Vissing et al. 2007). Since, as it was discussed earlier, PEMF increases HSP70 (Cheon, Park et al. 2012, Wei, Sun et al. 2015), a reduction in DOMS could be expected.

On the other hand, Szemerszky, Szabolcs et al. (2018) found no effect of PEMF (2.05 Hz burst frequency, 25.3 µT) in a randomized, double-blind, placebo-controlled study in Hungary investigating acute ischemic muscle pain in the forearm.

Specifically, PEMF application following pain induction in the forearm using a tourniquet technique did not lead to a change in pain threshold, pain tolerance, subject heart rate, or perceived decrease in pain. However, a significant placebo effect was found for pain tolerance and perceived change in pain which reduced any potential effects from PEMF.

TENDON CLINICAL APPLICATIONS

PEMF has been clinically investigated for the adjunctive treatment of rotator cuff repair showing early reduction in pain and increased ROM. Specifically Osti, Buono et al. (2015) performed a randomized, controlled study in Italy in 66 patients who underwent arthroscopic repair of small to medium rotator cuff tears. Thirty-two patients were treated with PEMF (I-ONE, IGEA, Carpi, Italy; 1.5 mT at 75 Hz) at the shoulder with the remaining being treated with a placebo device. Patients were treated with PEMF for 6–8 h/day for the first 6 weeks post-op. All patients underwent the same post-operative rehabilitation protocols and had follow-up visits at 3 months and at minimum 2 years. At 3 months post-op, patients' VAS, ROM, and University of California at Los Angeles (UCLA) and Constant-Murley shoulder scores were significantly better for the PEMF group relative to the placebo group. However, at 2 years, no difference was seen between the groups. The only adverse events noted were related to capsulitis which was mild to moderate for three PEMF and seven placebo patients. Severe capsulitis occurred in one PEMF and two placebo patients. All of these patients responded well to physical therapy and were doing well as other patients at the last follow-up (>2 years). The adverse events related to capsulitis are not likely related to PEMF as these side effects were observed in both active and placebo groups with the possibility of occurrence even higher in placebo group.

In Turkey, Aktas, Akgun et al. (2007) investigated PEMF for adjunctive use for conservative treatment of shoulder impingement syndrome which showed signs of early improvements in shoulder strength and pain but ultimately was inconclusive. Specifically a double-blind, randomized, and controlled study examined 46 patients with unilateral shoulder pain due to impingement. All patients received a 3-week conservative treatment regimen of Codman's pendulum exercise and cold applications. In addition, half the patients received PEMF treatment (Magnetoterapia model MG/3P, Elettromed, Roma, Italy; 3 mT at 50 Hz) of the shoulder 5 times/week (25 min/session) for the entire 3 weeks with the other half receiving a placebo PEMF treatment. No adverse events were noted for any patients. Shoulder pain and function was recorded using VAS and Constant score, respectively. In addition, daily living activities were evaluated using a shoulder disability questionnaire (SDQ). All assessments were done before and after treatment. Although significant improvements occurred over the 3 weeks for both the placebo and PEMF group, no differences were found between the placebo and PEMF group for any of the outcome measures.

In parallel to Aktas, Akgun et al.'s (2007) study, Galace de Freitas, Marcondes et al. (2014) performed another double-blind, randomized, and placebo-controlled study of 56 patients diagnosed with shoulder impingement syndrome in Brazil. These patients were initially treated with PEMF (n = 26; Magnetherp 330, Meditea, Buenos Aires, Argentina; 20 mT at 50 Hz) or placebo (n = 30) at the shoulder for

30 min, 3 times/week, for 3 weeks. Subsequently all patients underwent similar shoulder exercise protocols for 6 weeks and were then followed up for an additional 3 months. No adverse events were noted for any patients. At baseline, 3 weeks (post active), at 9 weeks (post exercises), and 3 months post treatment, VAS, handheld dynamometry strength measurements, and the UCLA and Constant-Murley shoulder scores were collected. While the PEMF group had less pain and higher shoulder function at all follow-ups relative to baseline, the placebo group only saw these improvements at 9 weeks and 3 months post follow-up. PEMF increased shoulder strength relative to baseline (at 9 weeks in lateral rotation and 9 weeks/3 months for medial rotation). However while the placebo group did not see an increase in strength, the difference between placebo and PEMF was not significant either.

Lastly, in England, Binder, Parr et al. (1984) examined the use of PEMFs for the treatment of rotator cuff tendinitis which showed early improvements in pain and function relative to placebo treatment. Specifically, Binder et al. carried out a double-blind, controlled study examining the effect of PEMF on 29 patients whose symptoms did not respond to conventional treatment of rotator cuff tendonitis. One group (n = 15) received PEMF treatment (5–9 h/day, 2.7 mT at 73 Hz, custom-made investigational coil) for 8 weeks while the second group (n = 14) were treated with a placebo PEMF device for 4 weeks and subsequently for 4 weeks with an active PEMF device. Following this, both groups received no treatment for 8 weeks. VAS, pain on resisted movement, painful arc on active abduction, and total ROM were collected at baseline, 2, 4, 6, 8, 12, and 16 weeks. During the initial 4 weeks, the PEMF group saw a greater pain reduction and increase in the ROM than the placebo group. However for the next 4 weeks and the subsequent 8 weeks (when both groups received PEMF and then placebo, respectively), no differences were seen between the two groups. No adverse events were noted. The study does mention that, in their pilot study, one patient with known cervical spondylosis treated with cervical laminectomy had recurrence of neck pain and neurological deficit in both upper limbs following successful treatment of shoulder pain, but the authors concluded that the PEMF treatment was considered safe.

CONCLUSIONS

While *in-vitro* research has shown that PEMF had a positive effect on tenocyte and myocyte proliferation and differentiation, contradictory results exists. Animal models, on the other hand, have shown some translation of the *in-vitro* results of the myocyte protective activation of HSP70 with results indicating improved muscle recovery following injury. Preclinical *in-vivo* works on tendons, however, also show varying results, with tendon-to-tendon healing being compromised by PEMF for both flexor and Achilles tendon, while tendon-to-bone healing following injury and repair was improved significantly.

The preclinical *in-vitro* and *in-vivo* work has shown some translation clinically as it supports the evidence for a reduction in delayed onset muscle soreness although no effect was found for acute ischemic muscle pain in forearm. In addition, other clinical studies showed that adjunctive PEMF treatment for rotator cuff repair, tendinitis, and impingement syndrome led to early improvements in pain and function with no associated severe side effects.

The combined studies show that PEMF can be effective for soft tissue repair in muscles and tendons but is dependent on the location of application, the PEMF waveform parameters, and daily PEMF exposure duration.

This heavy dependency of outcomes on the many PEMF parameters is evident in the small amount of translation from the several *in-vitro* studies to the few preclinical *in-vivo* animal models examining PEMF effects on muscles and tendons. Furthermore, even fewer studies have been done clinically worldwide and none clinically in the U.S. due to additional factors such as PEMF devices being regulated by the FDA as a Class III device requiring adherence to the expensive and lengthy IDE/PMA clinical trial pathway.

In general, it is therefore advisable to gain enough preclinical (*in-vitro/in-vivo*) knowledge of the specific PEMF signal and its target tissue interaction to enable a high success rate in any future clinical trials investigating the effect of PEMF on soft tissues such as muscles or tendons.

REFERENCES

Aktas, I., K. Akgun and B. Cakmak (2007). "Therapeutic effect of pulsed electromagnetic field in conservative treatment of subacromial impingement syndrome." *Clin Rheumatol* **26**(8): 1234–1239.

Binder, A., G. Parr, B. Hazleman and S. Fitton-Jackson (1984). "Pulsed electromagnetic field therapy of persistent rotator cuff tendinitis. A double-blind controlled assessment." *Lancet* **1**(8379): 695–698.

Boorman-Padgett, J. F., J. Huegel, C. A. Nuss, M. C. C. Minnig, A. F. Kuntz, E. I. Waldorff, N. Zhang, J. T. Ryaby and L. J. Soslowsky (2018). *Effect of Pulsed Electromagnetic Field Therapy on Healing in a Rat Achilles Tendon Partial Tear Model.* Orthopaedic Research Society (ORS). New Orleans, LA.

Boorman-Padgett, J. F., J. Huegel, C. A. Nuss, M. C. C. Minnig, A. F. Kuntz, E. I. Waldorff, N. Zhang, J. T. Ryaby and L. J. Soslowsky (2019). *Effects of Pulsed Electromagnetic Field Therapy on Healing in a Rat Achilles Tendon Partial Width Injury Model Without Immobilization.* Orthopaedic Research Society (ORS). Austin, TX.

Carpenter, L. L., P. G. Janicak, S. T. Aaronson, T. Boyadjis, D. G. Brock, I. A. Cook, D. L. Dunner, K. Lanocha, H. B. Solvason and M. A. Demitrack (2012). "Transcranial magnetic stimulation (TMS) for major depression: a multisite, naturalistic, observational study of acute treatment outcomes in clinical practice." *Depress Anxiety* **29**(7): 587–596.

Cheon, S., I. Park and M. Kim (2012). "Pulsed electromagnetic field elicits muscle recovery via increase of HSP 70 expression after crush injury of rat skeletal muscle." *J Phys Ther Sci* **24**: 589–592.

de Girolamo, L., D. Stanco, E. Galliera, M. Vigano, A. Colombini, S. Setti, E. Vianello, M. M. Corsi Romanelli and V. Sansone (2013). "Low frequency pulsed electromagnetic field affects proliferation, tissue-specific gene expression, and cytokines release of human tendon cells." *Cell Biochem Biophys* **66**(3): 697–708.

de Girolamo, L., M. Vigano, E. Galliera, D. Stanco, S. Setti, M. G. Marazzi, G. Thiebat, M. M. Corsi Romanelli and V. Sansone (2015). "In vitro functional response of human tendon cells to different dosages of low-frequency pulsed electromagnetic field." *Knee Surg Sports Traumatol Arthrosc* **23**(11): 3443–3453.

Denaro, V., L. Ruzzini, S. A. Barnaba, U. G. Longo, S. Campi, N. Maffulli and A. Sgambato (2011). "Effect of pulsed electromagnetic fields on human tenocyte cultures from supraspinatus and quadriceps tendons." *Am J Phys Med Rehabil* **90**(2): 119–127.

Feng, X., X. He, K. Li, W. Wu, X. Liu and L. Li (2011). "The effects of pulsed electromagnetic fields on the induction of rat bone marrow mesenchymal stem cells to differentiate into cardiomyocytes-like cells in vitro." *Sheng Wu Yi Xue Gong Cheng Xue Za Zhi (Journal of Biomedical Engineering)* **28**: 676–682.

Galace de Freitas, D., F. B. Marcondes, R. L. Monteiro, S. G. Rosa, P. Maria de Moraes Barros Fucs and T. Y. Fukuda (2014). "Pulsed electromagnetic field and exercises in patients with shoulder impingement syndrome: a randomized, double-blind, placebo-controlled clinical trial." *Arch Phys Med Rehabil* **95**(2): 345–352.

Huegel, J., J. F. Boorman-Padgett, C. A. Nuss, M. C. C. Minnig, A. F. Kuntz, E. I. Waldorff, N. Zhang, J. T. Ryaby and L. J. Soslowsky (2018). *Effects of Pulsed Electromagnetic Field Therapy on Healing of Complete Achilles Tendon Tears in a Rat Model.* Orthopaedic Research Society (ORS). New Orleans, LA.

Huegel, J., D. Choi, C. A. Nuss, M. C. Minnig, J. J. Tucker, C. D. Hillin, A. F. Kuntz, E. I. Waldorff, N. Zhang, J. T. Ryaby and L. J. Soslowsky (2017). *Effects of Pulsed Electromagnetic Field Therapy at Different Frequencies and Durations on Rotator Cuff Tendon-to-Bone Healing in a Rat Model.* Orthopaedic Research Society (ORS). San Diego, CA.

Jeon, H. S., S. Y. Kang, J. H. Park and H. S. Lee (2015). "Effects of pulsed electromagnetic field therapy on delayed-onset muscle soreness in biceps brachii." *Phys Ther Sport* **16**(1): 34–39.

Kaur, K., J. Yang, C. A. Eisenberg and L. M. Eisenberg (2014). "5-azacytidine promotes the transdifferentiation of cardiac cells to skeletal myocytes." *Cell Reprogram* **16**(5): 324–330.

Lee, E. W., N. Maffulli, C. K. Li and K. M. Chan (1997). "Pulsed magnetic and electromagnetic fields in experimental achilles tendonitis in the rat: a prospective randomized study." *Arch Phys Med Rehabil* **78**(4): 399–404.

Liu, M., C. Lee, D. Laron, N. Zhang, E. I. Waldorff, J. T. Ryaby, B. Feeley and X. Liu (2016). "Role of pulsed electromagnetic fields (PEMF) on tenocytes and myoblasts-Potential application for treating rotator cuff tears." *J Orthop Res* **35**(5): 956–964.

Liu, Y., S. Mayr, A. Opitz-Gress, C. Zeller, W. Lormes, S. Baur, M. Lehmann and J. M. Steinacker (1999). "Human skeletal muscle HSP70 response to training in highly trained rowers." *J Appl Physiol (1985)* **86**(1): 101–104.

Marmotti, A., G. M. Peretti, S. Mattia, L. Mangiavini, L. de Girolamo, M. Vigano, S. Setti, D. E. Bonasia, D. Blonna, E. Bellato, G. Ferrero and F. Castoldi (2018). "Pulsed electromagnetic fields improve tenogenic commitment of umbilical cord-derived mesenchymal stem cells: a potential strategy for tendon repair-an in vitro study." *Stem Cells Int* **2018**: 9048237.

Miller, S. L., D. G. Coughlin, E. I. Waldorff, J. T. Ryaby and J. C. Lotz (2016). "Pulsed electromagnetic field (PEMF) treatment reduces expression of genes associated with disc degeneration in human intervertebral disc cells." *Spine J* **16**(6): 770–776.

Nicolakis, P., J. Kollmitzer, R. Crevenna, C. Bittner, C. B. Erdogmus and J. Nicolakis (2002). "Pulsed magnetic field therapy for osteoarthritis of the knee--a double-blind sham-controlled trial." *Wien Klin Wochenschr* **114**(15–16): 678–684.

Norizadeh-Abbariki, T., O. Mashinchian, M. A. Shokrgozar, N. Haghighipour, T. Sen and M. Mahmoudi (2014). "Superparamagnetic nanoparticles direct differentiation of embryonic stem cells into skeletal muscle cells." *J Biomater Tissue Eng* **4**: 1–7.

Osti, L., A. D. Buono and N. Maffulli (2015). "Pulsed electromagnetic fields after rotator cuff repair: a randomized, controlled study." *Orthopedics* **38**(3): e223–e228.

Paulsen, G., K. Vissing, J. M. Kalhovde, I. Ugelstad, M. L. Bayer, F. Kadi, P. Schjerling, J. Hallen and T. Raastad (2007). "Maximal eccentric exercise induces a rapid accumulation of small heat shock proteins on myofibrils and a delayed HSP70 response in humans." *Am J Physiol Regul Integr Comp Physiol* **293**(2): R844–R853.

Pilla, A. A. (2015). Pulsed electromagnetic fields: From signaling to healing. In M. Markov, *Electromagnetic fields in Biology and Medicine*. CRC Press. Boca Raton, FL: pp. 29–47.

Randelli, P., A. Menon, V. Ragone, P. Creo, U. Alfieri Montrasio, C. Perucca Orfei, G. Banfi, P. Cabitza, G. Tettamanti and L. Anastasia (2016). "Effects of the pulsed electromagnetic field PST(R) on human tendon stem cells: a controlled laboratory study." *BMC Complement Altern Med* **16**: 293.

Robotti, E., A. G. Zimbler, D. Kenna and J. A. Grossman (1999). "The effect of pulsed electromagnetic fields on flexor tendon healing in chickens." *J Hand Surg Br* **24**(1): 56–58.

Smith, T. L., D. Wong-Gibbons and J. Maultsby (2004). "Microcirculatory effects of pulsed electromagnetic fields." *J Orthop Res* **22**(1): 80–84.

Szemerszky, R., Z. Szabolcs, T. Bogdany, G. Janossy, G. Thuroczy and F. Koteles (2018). "No effect of a pulsed magnetic field on induced ischemic muscle pain. A double-blind, randomized, placebo-controlled trial." *Physiol Behav* **184**: 55–59.

Tang, X., D. G. Coughlin, A. Ouyang, E. Liebenberg, S. Dudli, E. I. Waldorff, N. Zhang, J. T. Ryaby and J. C. Lotz (2016). *Anti-Inflammatory Effects of Pulsed Electromagnetic Fields (PEMF) Treatment in Rat Intervertebral Disc Degeneration Model*. International Combined Orthopaedic Research Societies (ICORS). Xian.

Tang, X., D. G. Coughlin, A. Ouyang, E. Liebenberg, S. Dudli, E. I. Waldorff, N. Zhang, J. T. Ryaby and J. C. Lotz (2017a). *Pulsed Electromagnetic Fields (PEMF) Reduce Acute Inflammation in the Injured Rat Intervertebral Disc*. Orthopaedic Research Society. San Diego, CA.

Tang, X., D. G. Coughlin, E. I. Waldorff, J. T. Ryaby, T. Alliston and J. C. Lotz (2015a). *Dynamic Imaging Demonstrates the Effect of Pulsed Electromagnetic Fields (PEMF) Treatment on IL-6 Transcription in Bovine Nucleus Pulposus Cells*. The International Society for the Study of the Lumbar Spine (ISSLS). San Francisco, CA.

Tang, X., D. G. Coughlin, E. I. Waldorff, J. T. Ryaby, T. Alliston and J. C. Lotz (2015b). *Dynamic Tracking of IL-6 Transcription in Nucleus Pulposus Cells In Vitro*. Orthopaedic Research Society (ORS). Las Vegas, NV.

Tang, X., D. G. Coughlin, E. I. Waldorff, N. Zhang, J. T. Ryaby, T. Alliston and J. C. Lotz (2017b). *NF-kβ Involved in PEMF Treatment Effects on Pro-Inflammatory Cytokine IL-6 Expression in Nucleus Pulposus*. Orthopaedic Research Society (ORS). San Diego, CA.

Thomas, A. W., K. Graham, F. S. Prato, J. McKay, P. M. Forster, D. E. Moulin and S. Chari (2007). "A randomized, double-blind, placebo-controlled clinical trial using a low-frequency magnetic field in the treatment of musculoskeletal chronic pain." *Pain Res Manag* **12**(4): 249–258.

Thompson, H. S., P. M. Clarkson and S. P. Scordilis (2002). "The repeated bout effect and heat shock proteins: intramuscular HSP27 and HSP70 expression following two bouts of eccentric exercise in humans." *Acta Physiol Scand* **174**(1): 47–56.

Tucker, J. J., J. M. Cirone, T. R. Morris, C. A. Nuss, J. Huegel, E. I. Waldorff, N. Zhang, J. T. Ryaby and L. J. Soslowsky (2016). "Pulsed electromagnetic field therapy improves tendon-to-bone healing in a rat rotator cuff repair model." *J Orthop Res* 35(4): 902–909.

Wei, J., J. Sun, H. Xu, L. Shi, L. Sun and J. Zhang (2015). "Effects of extremely low frequency electromagnetic fields on intracellular calcium transients in cardiomyocytes." *Electromagn Biol Med* **34**(1): 77–84.

Wei, J., J. Tong, L. Yu and J. Zhang (2016). "EMF protects cardiomyocytes against hypoxia-induced injury via heat shock protein 70 activation." *Chem Biol Interact* **248**: 8–17.

Xu, H., J. Zhang, Y. Lei, Z. Han, D. Rong, Q. Yu, M. Zhao and J. Tian (2016). "Low frequency pulsed electromagnetic field promotes C2C12 myoblasts proliferation via activation of MAPK/ERK pathway." *Biochem Biophys Res Commun* **479**(1): 97–102.

6 Clinical Use of Pulsed Electromagnetic Fields (PEMFs)

William Pawluk

CONTENTS

THE NEED

Society in general and medicine in particular continue to struggle and strive to develop and provide better, innovative, low-cost, low-risk and cost-effective healthcare strategies across a spectrum of health conditions and diseases and for disease prevention and health maintenance. The practice of medicine is still hugely reliant on pharmaceuticals (House of Commons, 2005) and procedures to address the health needs of our societies. Relying mostly on these approaches, which have significant limitations and risks, leaves large gaps in obtaining the best health levels and resolving health problems. Innovative technologies such as gene therapies (Kaemmerer, 2018) and stem cell therapies (Poltavtseva et al., 2018) offer major new pathways to approaching these health issues, but they are very early in development, and their true potential and value are unknown.

There is a greater societal emphasis and an explosion of resources available for lifestyle strategies including behavioral management, nutrition, and exercise and fitness. Pulsed electromagnetic field (PEMF) therapy is a largely neglected technology, already in therapy use for over 70 years (Farinas and Martinez Farinas, 1951;

Markov, 2015). PEMFs, used alone or in a complementary fashion, have the potential for significantly changing and improving therapeutic options available to clinicians, their patients/clients, and the consumers directly. Their unique actions are supported by a large and ever-growing database of magnetic field effects on biology and a growing scientific literature of their clinical benefits across a spectrum of health conditions and diseases (Shupak, 2003).

THE CHALLENGES

While there are a number of devices with government approvals for a limited set of indications, some of which have been in use for decades (Bassett et al., 1974; Markov, 2015), the cost of entry for approvals is prohibitive. Medical education about the opportunities for the use of PEMF technologies is grossly lacking. Funding for the PEMF therapy research necessary to obtain regulatory approvals is almost nonexistent. There are major challenges for protecting the investment in developing PEMF technologies. Existing research literature is scattered across the spectrum of available published sources. Where it is concentrated in some limited publications, such as the *Bioelectromagnetics* journal, the emphasis is heavily biomedical and not clinical.

There is a huge gap between the information available from the bench and the information available and needed for clinical application. While there is some emerging large-scale commercial interest in developing and introducing applied electromagnetic tools – some call this electroceuticals – this is still in the nascent stages. As a result, consumer interest in and adoption of PEMF technology is outpacing applied research data for cost-effectiveness. Consumers, and even clinicians, often acquire PEMF technologies for specific health conditions without the benefit of adequate education about the scientific principles, appropriate device options, and guidance for application. Most sources of information for consumers and clinicians are from the Internet and the PEMF vendors.

THE CLINICIAN

In medicine, PEMF technology is most commonly used in the areas of orthopedics, wound healing, pain, brain tumors, and depression as approvals have been granted and investment has been made for its application in these areas. The electromagnetic devices that are regularly used include lasers and devices based on radio frequency, diathermy, electrical stimulation, high-intensity transcranial magnetic stimulation, and photobiomodulation. Clinical electromagnetic applications in disciplines other than medicine include chiropractic, naturopathy, physical therapy, acupuncture, biofeedback, psychology, and cosmetology, among others. Clinical disciplines within medicine, besides orthopedics and psychiatry, that are increasingly using electromagnetic technologies include physical medicine, pain medicine, neurosurgery, and the growing field called complementary, integrative, holistic, or functional medicine. The number of clinical disciplines in medicine getting exposed to the potential uses of PEMFs is increasing with the increase in commercial availability and promotion of these devices.

CLINICAL USE VERSUS CONSUMER USE

The growth of complementary, holistic, and alternative practices; explosion of information available on the Internet to the consumer; and the medical community not adopting and supporting these alternative practices have led consumers to increasingly seek care outside the conventional medical care system (Barnes et al., 2008). Restrictive changes in health insurance policy coverage and increase in costs of health insurance premiums are also a contributing factor. Consumer access to health information in the last three decades has exploded, allowing consumers to often be more informed than their doctors about their health conditions, their therapeutic options, and the risks and benefits of these options. As a result of these factors, consumers are increasingly self-directing their care. These societal changes contribute to a widening gap between the adoption of electromagnetic technologies in conventional medicine and self-adoption by the public.

THE REGULATORY ENVIRONMENT FOR PEMFS

PEMF and other electromagnetic technology approved by regulatory bodies, such as the Food and Drug Administration (FDA) of the United States of America, can typically only be accessed with a prescription from medical doctors or other clinicians licensed to prescribe use of these approved devices. With regard to selling and applying for off-label uses, happening routinely in the case of approved pharmaceuticals, these approved devices cannot be sold or applied for such off-label uses because of tight regulatory controls. As a result, the vast majority of PEMF devices, particularly lower-cost systems, are accessed directly by consumers and clinicians, without formal approval by regulators. Increasingly, clinicians are acquiring nonapproved devices for use in their practices or recommending or selling them directly to their patients/clients. An Internet search using the term PEMFs will reveal a large number of devices and vendors. The current FDA position regarding most PEMF devices, beyond those that are FDA-approved, limits their use for wellness purposes (FDA, 2016).

FDA approval is required for devices marketed for specific medical indications. This is a very costly and onerous process, and only entities with adequate financial, development, marketing, or other resources can seek approval. In addition, approvals generally restrict the use of these technologies for specific "label" indications and applications, even though PEMF technology has much wider utility across a spectrum of applications (see "Clinical Opportunities for PEMF Therapies section" below). FDA approval is not required for devices that are used or marketed primarily for wellness (FDA, 2016). This has resulted in a dramatic proliferation of relatively lower-cost, easily accessible, commercially available "wellness" PEMF systems. This position of FDA most likely results from the perspective that low-intensity, low-frequency PEMF systems are generally regarded as safe (GRAS).

CLINICAL DECISION-MAKING ABOUT PEMF USE

Clinicians need to be aware of the many different PEMF device options available, with their respective advantages and disadvantages. The choice of a device should be based on the conditions to be treated and guided as much as possible by available research data, especially clinical studies. The technical specifications and limitations of any given PEMF system need to be understood to achieve desired and expected results.

For any given PEMF device and clinical application, the following technical parameters should be known (Markov, 2017):

a. type of field
b. component (electrical or magnetic)
c. intensity or induction
d. gradient (dB/dt)
e. vector (dB/dx)
f. frequency
g. pulse shape
h. localization
i. duration of the exposure
j. depth of penetration.

Practical considerations in the clinical setting will have to include the cost of treatments to the patient/client and the probable benefit for any particular device. Most clinicians will use available technologies across a range of conditions appropriate to their areas of specialization. In other words, a given technology, including PEMFs, is used more based on its functional value than on a specific indication or health/medical condition.

Clinicians deciding to use or recommend PEMFs in their practices need to consider the following:

- for the patient/client:
  a. condition being treated, including the pathology and physiologic aspects
  b. severity of the condition
  c. functional status of the patient/client
  d. acute versus chronic condition/s
  e. depth in the body of the lesion/s to be targeted
  f. likely cooperation of the patient/client
  g. other therapies being used or intended to be used
  h. outcome measurement tools
  i. whether to be used for in office and/or home treatment
- for the PEMF to be used or recommended:
  j. whole body versus local application
  k. intensity needed
  l. frequency/frequencies needed
  m. ease of application
  n. treatment time
  o. periodicity of treatments
  p. cost of equipment
  q. clinician or self-administration.

Fortunately, most PEMF treatments can be applied without significant patient or client preparation. Since PEMFs penetrate clothing, casts or bandages without attenuation, they can be applied without having to expose or have direct contact with the skin or target tissue. This allows direct targeted application to organs within the body usually without concern for harm or negative effects to intervening tissue.

CLINICAL PEMF DOSIMETRY

A critical consideration for clinicians in selecting and applying PEMF therapy is the total "dose" or dosimetry of the magnetic field. The clinician needs to determine the goal magnetic field intensity, length of treatment, and the number of sessions needed to achieve desired results in treating the target tissue. There are no clear consensus guidelines for the myriad of conditions and circumstances that clinicians are likely to encounter.

Their use is further complicated because manufacturers may or may not provide maximum intensities for their devices and applicators. They rarely provide dB/dT values or induced current estimates or measurements. Many devices have multiple selectable frequencies or programs with various frequencies without listing the intensity values at each of the frequencies. Clinicians do not understand that the stated maximum intensity for a given PEMF system will vary with the frequency used. In addition, there are often multiple applicators available, and maximum intensity is not usually given for every applicator. So, the clinician is often left guessing what the intensity or the dosimetry at the target tissue/s would be.

PEMF dosimetry at the adenosine receptor and inflammation. It may be helpful to provide a dosimetry example using a specific receptor target, based on available science. The receptor target to be illustrated is the adenosine receptor (AR). The AR was chosen because of its role in reducing inflammation (Varani et al., 2017) and because of the ubiquity of the receptor and multitude of demonstrated actions. Because of the wide range of physiologic actions and roles in numerous pathologic conditions, stimulating the adenosine nucleoside and its receptor target would have a broad range of potential applications in a broad-based clinical setting.

Adenosine is a building block for RNA/DNA and a part of the energy molecule ATP (Chen et al., 2013). In addition, adenosine is a signaling molecule. Adenosine acts through four types of ARs – A_1, A_{2A}, A_{2B}, and A_3. These receptors are widely distributed throughout the body and have been found to be part of both physiological and pathological biological functions. They affect, at the least, cardiac rhythm and circulation, breakdown of fat, kidney blood flow, immune function, regulation of sleep, development of new blood vessels, inflammatory diseases/inflammation, blood flow, and neurodegenerative disorders.

The role of ARs and adenosine in modifying inflammation is well accepted (Varani et al., 2017). Neutrophils have a major role in inflammation. There are A_{2A} receptors in the membranes of neutrophils. PEMFs applied *in vitro* at the surface of neutrophils have been found to significantly increase the binding of adenosine to the A_{2A} receptor in the human neutrophils exposed to PEMFs. This effect was time, intensity, and temperature dependent. PEMF dose-response studies found that a PEMF effect was detectable after 30 min of exposure, and the receptors became saturated with a 1.5 mT magnetic field (Massari et al., 2007). The effect plateaued with intensities >1.5 mT (Figure 6.1).

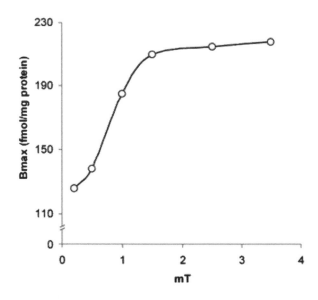

FIGURE 6.1   Saturation binding of A_{2A} AR agonist ((3H)-ZM 241385) as a function of magnetic field peak amplitude (millitesla) in human neutrophil membranes Bmax = receptor-binding capacity. (Adapted from Massari et al., 2007.)

The PEMF used had an intensity range from 0.1 to 4.5 mT, frequency range from 10 to 120 Hz. The signal had a pulse duration of 1.3 ms, frequency of 75 Hz, 10% duty cycle. The most used peak intensity of the magnetic field was 1.5 mT. The corresponding peak amplitude of the induced electric voltage was 2.0 ± 0.5 mV.

Armed with this information, the clinician might reasonably assume that a 1.5 mT "dose" for at least 30 minutes at the inflammatory target tissue would be the optimized amplitude of a magnetic field to help with reducing inflammation, at least as far as neutrophil involvement is concerned. The clinician is faced routinely with the need to reduce inflammation at various depths in the body, depending on the target organ and tissue, and it is not unreasonable to consider the role of the AR on the neutrophil membrane as a worthy and useful approach.

To achieve the 1.5 mT goal PEMF intensity acting on neutrophils at various distances from the applicator, that is, at various depths into the body, the clinician must be aware of the inverse square law governing the loss of magnetic field intensity with distance from the applicator.

Values of Table 6.1 were calculated at 1.5 mT goal intensity at various depths in the body. Using Newton's inverse square law, the following formula applies: $I_1 \times d_1^2 = I_2 \times d_2^2$, where I = intensity, d = distance (depth), I_1 = initial intensity, d_1 = depth at the surface of the PEMF applicator, I_2 = intensity at the target tissue, and d_2 = depth at the target tissue. Therefore $I_2 = \left(I_1 \times d_1^2\right)/d_2^2$ (Pawluk, 2017). The starting depth (d_1) used was 0 cm and equals unity. Therefore, for I_2, d_2 at 0 cm is $0 + 0$ cm; at 2 cm d_2 is $0 + 2$ cm, etc.

TABLE 6.1

Initial Intensity (millitesla) Needed to Obtain 1.5 mT at the Target Tissue (Depth D_2) in Centimeters

D_2 (cm)	0	1	2	3	4	5	6	7	8	9	10
I_1 (mT)	2	6	14	24	38	54	74	96	122	150	182
D_2 (cm)	11	12	13	14	15	16	17	18	19	20	21
I_1 (mT)	216	254	294	338	384	434	486	542	600	662	726
D_2 (cm)	22	23	24	25	26	27	28	29	30	31	32
I_1 (mT)	794	864	938	1,014	1,094	1,176	1,262	1,350	1,442	1,536	1,634

To use the table, determine the goal depth; then after that, determine the intensity needed to achieve 1.5 mT in the target tissue. From the table, it can be seen that to deliver the goal 1.5 mT at the target tissue 2 cm from the applicator, a 14 mT intensity magnetic field would be required. At 20 cm, 662 mT would be required to deliver 1.5 mT at the target tissue.

For example, the kidneys may be a target for PEMF treatment. Inflammation in the kidney is common, and the kidneys have been found to have ARs. Assume that neutrophils are present within the kidney circulation and tissue when there is inflammation. The depth of the center of the kidneys in the body is typically between 5 and 7 cm from the front of the abdomen (Xue et al., 2017). The thickness of the kidneys from front to back of the kidney is typically about 5 cm or about 2.5 cm from the center of the kidney to the back of the kidney (Moorthy and Venugopal, 2011). If a PEMF applicator is placed over the anterior abdomen and the expected depth at the center of the kidney is 9.5 cm (or rounding up, 10 cm), with the goal intensity being 1.5 mT, the maximum PEMF intensity would need to be 182 mT. This means that a PEMF system that can deliver at least this magnetic field intensity to adequately target the kidneys would need to be selected.

From Figure 6.1, it can be seen that intensities lower than 1.5 mT can still be beneficial in stimulating ARs. One millitesla will deliver about 88% of the optimal benefit. One-half millitesla will deliver about 75% of the optimal benefit. In our experience, an intensity less than the optimal intensity will still produce results clinically, although it usually takes a longer treatment time to get optimal results.

Targeting the anti-inflammatory effects of AR stimulation is only one possible consideration for the selection of magnetic field intensity in judging dosimetry needs. Since there are so many different physiologic effects and actions of PEMFs (Pawluk and Layne, 2017), dosimetry calculations for each of these effects are not available to the clinician. It is challenging to extrapolate from *in-vitro* or *in-vivo* research dosimetric experience in lower-order mammals to humans. In addition, it is unlikely that any specific physiologic action, for example, enhanced circulation, accelerated healing, pain reduction, etc., can be uniquely and specifically selected as a target effect when considering actual applications in the clinical environment. Experience suggests that multiple actions are at play any time a PEMF is used. At the moment at least, neutrophil AR research is a helpful starting point.

CELL AND TISSUE INJURY AS A PEMF TARGET

All diseases start with micromolecular or structural alterations in individual cells. Injury to sufficient numbers of cells and to the matrix between cells ultimately leads to tissue and organ dysfunction. The increasing cumulative burden of these unrecovered cells and cell functions, leads, at the very least, to aging, if not ultimately, cumulatively, to organ dysfunction, diseases, and death.

While clinicians may be aware of some of the physiologic and symptomatic benefits of PEMFs, resulting from their effects on tissues and organs, they may not be aware that many of these benefits are derived from the impact of PEMFs directly at the level of injured (dysfunctional) cells. The number of dysfunctional and irreversibly damaged cells determines the vitality and functionality of tissues and organs. The goal of clinical practice is to impact the process of cellular and tissue dysfunction early enough to either reverse or halt the progression of damage and dysfunction.

Cell injury causes. Cell injury results when the limits of adaptive responses of cells are exceeded, whatever the cause, including but not limited to:

- oxygen deprivation
- physical agents
- chemical agents and drugs
- infectious agents
- immunologic reactions
- genetic derangements
- nutritional imbalances.

Physical agents causing cell injury include mechanical trauma, extremes of temperature (burns and deep cold), sudden changes in atmospheric pressure, radiation, and electric shock.

Mechanical trauma, which we most commonly associate with injury, including lacerations, sprains, dislocations, muscle tears, fractures, etc., is a fraction of the causes of cell injury (Figure 6.2).

Cell injury effects. Some examples of changes that occur as a result of various injurious stimuli are as follows.

At the reversible stage of injury, the hallmark changes are

- reduced oxidative phosphorylation with depletion of ATP
- cellular swelling caused by changes in ion concentrations and water influx
- mitochondrial and cell skeleton alterations
- inflammation
- DNA damage.

For instance, in response to increased hemodynamic loads, the heart muscle becomes enlarged, a form of adaptation, and can even undergo injury. If the blood supply to the myocardium is compromised or inadequate, the muscle first suffers reversible injury, manifested by certain cell changes. If this is not reversed, some percent of the cells suffer irreversible injury and die, and, if enough die, organ or tissue function can be reduced.

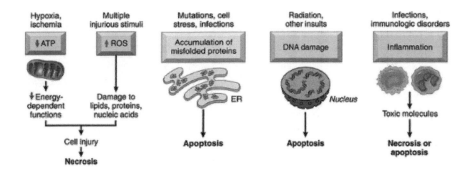

FIGURE 6.2   The principal biochemical mechanisms and sites of damage in cell injury. Note that causes and mechanisms of cell death by necrosis and apoptosis are shown as being independent of each other, but there may be overlap; for instance, both may contribute to cell death caused by ischemia, oxidative stress, or radiation. ATP, adenosine triphosphate; ROS, reactive oxygen species; ER, endoplasmic reticulum; DNA, deoxyribonucleic acid (Kumar, 2007).

The end results of unresolved genetic, biochemical, or structural changes in cells and tissues are functional abnormalities, which lead to clinical manifestations (symptoms and signs) which then may become diagnosed as disease.

Stages of cell and tissue injury. At the earliest stages of cell and tissue injury, the damaging reactions of the cells may be reversible. It is at this stage that PEMFs may have their most dramatic and rapid responses, with the least amount of therapeutic effort. If the progression of damage is not halted, cells enter into an irreversible stage, leading to cell death, by either necrosis or apoptosis (Figure 6.3).

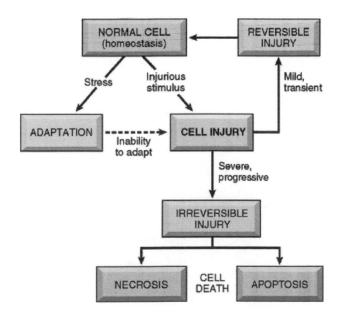

FIGURE 6.3   Stages of the cellular response to stress and injurious stimuli. (From Kumar, 2007.)

Within limits, cells can repair these potentially reversible cell function derangements through their own homeostatic and repair processes and, if the injury stimulus is removed, cell function can return to normal. With continuing damage, the injury becomes irreversible, and cells and associated tissues cannot recover and may obviously manifest dysfunction or damage. The dysfunction or damage may then reach the stage of being detectable by measuring biochemical changes, imaging studies, and/or pathologic techniques depending on the stage.

Most stresses and noxious influences exert their effects first at the molecular or biochemical level. There is a time lag between the stress and the physical changes of cell injury or death. Cell function is lost long before cell death occurs (Figure 6.4). Note that varying degrees and amounts of cell death typically precede ultrastructural, light microscopic, and grossly visible morphologic changes. As the number of cells damaged in a tissue increases, detectable physical changes become gradually more evident. For example, cardiomyocytes may cease contracting after 1–2 min of ischemia but may not die for another 20–30 min. Myocyte death may be evident within 2–3 h with electron microscopy but would take 6–12 h to be evident with light microscopy (Kumar, 2007).

In clinical practice, the degree of accumulation of cell dysfunction and death determines the degree of loss of function of an organ. Even in failing organs, certain degrees of repair and reversal of cell injury and stabilization of damage are possible, even before complete organ failure, if adequately facilitated, either naturally or by the external application of various treatments.

Cell injury and aging. The consequences of cell and tissue injury depend on the type, state, and adaptability of the injured cells and tissues, including nutritional and hormonal status, and vulnerability to, e.g., hypoxia, degree of toxic exposure, etc. The more depleted the cells and tissues and the less adaptable and regenerative, for example, in the aged, the more challenging will be their ability to repair themselves and recover.

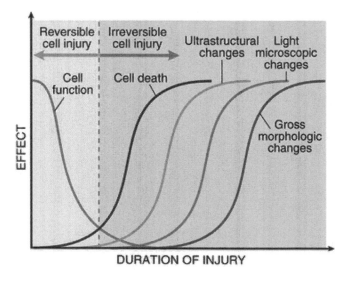

FIGURE 6.4 Reversible versus irreversible cell injury, the relationship among cellular function, cell death, and the morphologic changes of cell injury (Kumar, 2007).

One of the most important points of this section is that intervention will more likely produce benefits early in the process of cellular or tissue stress, before it reaches the "point of no return" or irreversibility. Since tissue stress occurs throughout life, preventive maintenance is a rational approach. Optimizing cell and tissue function regularly, if not daily, before quantifiable cell and tissue damage occurs, may significantly decrease the inevitable effects of cumulative lifetime cell and tissue stress that results in aging.

Cellular and tissue aging also likely represents the progressive accumulation over the years of chronic sublethal cell injury that may or may not lead to cell death. This process can lead to a diminished capacity of cells to respond to even mild injury, thus making cells and tissues more vulnerable to the cumulative effects of borderline stress. While there are many possible examples, one may be the impact of multiple upper respiratory infections, leading to chronic bronchitis. The chronic bronchitis (chronic bronchial inflammation) then predisposes the bronchi and lungs to hypoxemia and recurrent infections, which may then, with repetitive insults, progress to chronic obstructive pulmonary disease.

PEMFs and cell and tissue injury. Clinical treatment and prevention approaches should address the multiple causes and mechanisms of cell and tissue injury. The classic biologic effects of PEMFs touch almost all aspects of cell injury, to varying degrees, especially early in the injury process.

PEMFs can be used to improve or optimize many cellular and body functions and rapidly and safely reduce the effects of cell and tissue injury. In addition to acting to reverse or slow the progression of loss of cell function, where injury has progressed to the stage of irreversible damage, PEMFs have been shown to help to varying degrees with tissue repair (Dindar et al., 1993). PEMFs have extensive physiologic and biologic effects. About 25 of these effects of PEMFs are reviewed in Pawluk and Layne, 2017. Based on even this subset of PEMF effects, the potential uses and value of PEMFs extends significantly beyond known studies of PEMFs for individual disease conditions.

Many of the biologic effects of PEMFs are seen in both similar and dissimilar health or disease conditions, including but not limited to anti-inflammatory action, improved circulation, tissue healing and pain reduction, among many others. This means that improvements in symptoms and functions can be significant, across many different disease conditions, even though the underlying cause of the disease process itself may not be impacted dramatically. This is a clinically acceptable therapeutic target, in the absence of the ability to cure, and is typical for commonly applied pharmaceutical therapies.

Extremely low-frequency (ELF) PEMFs, at the appropriate field intensities, penetrate through the body with almost no attenuation, affecting every cell and tissue in their path. And, fortunately, it appears that healthy cells are minimally impacted by common therapeutic ELF PEMFs (Blank and Soo, 1992). There is no other technology, which I'm aware of, that can, as a single modality, have both the range and depth of action that appropriate clinically directed ELF PEMFs can have, with little effect on healthy cells and an extraordinary safety profile.

PEMFs as early intervention. From the perspective of intervention at the earliest stages of cell and tissue injury, PEMFs could ideally be recommended as an almost daily prevention tool to address milder cell and tissue dysfunctions prior to irreversible cell injury. These early cell dysfunction events usually happen before clinical services are needed or accessed.

Per the cell injury model, much cell death happens so early that the healthcare system is not likely to be involved at that stage. The exception would be obvious events such as acute wounds, burns, trauma, poisoning, etc. In chronic disease, cell injury and cell death often progress slowly until the condition reaches a point requiring clinical attention. Clinical evaluation processes typically use personal history reporting, laboratory testing, imaging studies, or visual or physical examination to determine clinical status. So, the role of the healthcare system, including the clinician, is often limited to delaying the progression of damage (tertiary prevention), improving symptoms, and recovering as much function as possible.

Conventional medicine and cellular repair. Once clinical services are accessed, conventional medical approaches are used that rarely facilitate cellular repair or create other risks. Examples are as follows:

- Antibiotics reduce progression of infection but do nothing to repair damage from the infectious process.
- The use of anti-inflammatories limits or reduces inflammation but does not improve healing and repair. In fact, NSAIDs also cause their own problems, including gastric inflammation and bleeding, leading to about 30,000 gastrointestinal bleeding related deaths per year (Thomas and Prato, 2002).
- The current epidemic of opioid use is an attempt to manage pain, but it does nothing to limit or ease the pain. Opioid overdoses accounted for more than 42,000 deaths in 2016, more than any previous year on record. An estimated 40% of opioid overdose deaths involved a prescription opioid (HHS, 2018).

If PEMFs were more widely accessible and used more routinely in the clinical setting or recommended to patients/clients for their own more regular home use for prevention, early intervention, active complementary or sole treatment and disease maintenance, greater reduction of morbidity and recovery of function could be expected.

PEMF COMPLEMENTS OTHER THERAPIES AND TECHNOLOGIES

"To cure sometimes, to relieve often, and to comfort always" is quoted as a 15th century folk saying and also attributed to Dr. Edward Trudeau of the 1800s, founder of a tuberculosis sanatorium (Cayley, 2006; Shaw, 2009). It is still as true for clinical practice today as it was in the past. To achieve the best ends of helping people with their diseases and illnesses, clinicians often need to use multiple modalities or approaches in their practices and their patients/clients need to use other modalities as well. These additional modalities include medications, including opioids, supplements, herbs, acupuncture, chiropractic adjustments or manipulations, low-level laser, electrical stimulation, invasive procedures, and/or physical therapy, and PEMFs more recently, among others.

How do PEMFs interact with other modalities, and how can they be used in a complementary fashion?

Because PEMF's penetrate through all biologic tissues, equally well, unlike other technologies, such as laser, electrical stimulation, and ultrasound, they may be used to penetrate clothing, shoes, casts, wraps, bandages, and splints, decreasing preparation time and resources. This ability of PEMFs to penetrate all the way through an entire volume of tissue, including the whole body, is what makes them, not only safe but also practical and simple to use, alone or in conjunction with any other therapies.

In patients with pain, PEMF therapy is frequently performed concomitantly with medication management. PEMFs are commonly seen clinically to not only decrease pain, but also reduce dependence on pain medications. In one study using high-frequency PEMFs for the treatment of cervical dorsal root ganglion pain, pain relief was found to be satisfactory (Van Zundert et al., 2007). The need for pain medication was significantly reduced in the active group even after 6 months. One other study on knee pain (Pawluk et al., 2002) found that even after follow-up at 1 year, 85% claimed to have pain reduction beyond the time of stimulation. Medication consumption decreased from 39% at 8 weeks to 88% after 8 weeks. Rohde et al. (2010) found a 2.2-fold reduction in narcotic use in PEMF-treated acute post-traumatic patients.

There are few studies evaluating the effects of combining PEMFs with other modalities, other than pain medication, used concurrently. A number of studies combined usual care with PEMFs, comparing them to usual care alone. Wistar rats with induced acute or chronic pneumonia received conventional antibiotic treatment alone or in combination with a PEMF (2.5–35 mT, 10–180 min/exposure, 10–20 daily exposures). The most pronounced anti-inflammatory effect was observed in rats exposed to 2.5-mT PEMF for 180 min/day for 10 days. This same group then conducted clinical studies in 165 patients with chronic bronchitis or inflammatory-suppurative lung diseases. Adding PEMFs resulted in marked inhibition of lung inflammation (Iashchenko et al., 1988).

Adding ultraviolet B (UVB) radiation to a 100 Hz, 1 mT, PEMF produces greater inhibition of T-cell proliferation than UVB alone (Nindl et al., 2004). This observation may indicate that adding PEMFs to UVB treatment in psoriasis would extend the value of UVB treatment and reduce the risk of skin cancer caused by the UVB.

Osteomyelitis and wounds complicated by *Pseudomonas* or *Staphylococcus* infection treated with bone infusions of a mixture of antibiotics and other substances, combined with 100–150 Hz, 11–17 mT, 25–30 exposures of 25–30 min each per course, followed by 50 Hz, 25–30 mT; 10–15 exposures of 25–30 min each, reduced rehabilitation hospital stays by 4–5 weeks (Alyshev et al., 1988).

Posttraumatic, late-stage reflex sympathetic dystrophy (RSD), which is now called regional complex pain syndrome (CRPS), a form of neuropathy, is very painful and largely unsatisfactorily treatable by standard medical approaches. In one report, ten 30 min PEMF sessions at 50 Hz followed by a further ten sessions at 100 Hz plus physiotherapy and medication reduced edema and pain in 10 days (Saveriano and Ricci, 1989).

Hemiplegic stroke patients were recruited and randomly assigned to the experimental group (scalp acupuncture + low-frequency repetitive transcranial magnetic stimulation (rTMS) + routine rehabilitation treatment) or the control group (scalp acupuncture + routine rehabilitation treatment). Compared with pretreatment, the upper limb motor function score and quality of life score increased significantly in the two groups, but motor function improvement was much greater in the experimental group (Zhao et al., 2018).

One group evaluated pain and swelling of distal radius fractures after an immobilization period of 6 weeks (Cheing et al., 2005). Eighty-three patients were randomly allocated to receive 30 min of either ice plus PEMF (group A), ice plus sham PEMF (group B), PEMF alone (group C), or sham PEMF for five consecutive days (group D). All had a standard home exercise program. The end result was that the addition of PEMF to ice therapy produces better overall treatment outcomes than ice alone, or PEMF alone, in pain reduction and ulnar ROM.

Treatment of duodenal ulcers with a 50-Hz, 20- to 25-mT (200- to 250-G) magnetic field applied for 1 min to acupuncture points, commonly used for general adaptation and more specifically for gastrointestinal function, was compared to standard antiulcer medication and the combination of medication and acupuncture point stimulation (Kravtsova et al., 1994). Time to pain relief, reduction of dyspepsia symptoms, and ulcer healing were compared. Pain and dyspepsia were best controlled in the sole therapy group in 2.75 days and 3.08 days, respectively. Ulcer healing took 18.25 days, 8 days less than in the medication-only group (26.6 days). Combining PEMF and drug therapy resulted in pain control in 8.61 days; dyspepsia was relieved in 6.05 days, and also healing occurred in 19 days after the start of the treatment. So, compared to other studies, where medication plus active therapy was better, in this study, adding medication appears to delay improvement.

Hypnosis is a complementary therapeutic approach. A study was done to assess whether PEMFs might facilitate hypnosis (Healey et al., 1996). Weak (1 µT), burst-firing magnetic fields were applied for 20 min over the left or right temporoparietal lobe or both hemispheres and compared with sham treatment. Only the group that received the stimulation over the right hemisphere exhibited a marked increase in suggestibility following the treatment. PEMFs may directly affect the neurocognitive processes that are associated with hypnotizability.

THE PATIENT/CLIENT/CONSUMER AND PEMF THERAPY

While there is considerable information available on the Internet and in currently published books and other reference materials about the value and use of PEMFs, many patients/clients are still not aware of this technology as a safe and effective therapeutic option. Clinicians have a responsibility to educate and inform their patients about the availability of this technology; what it does; how it works; what's available; and likely benefits, risks, and costs.

THE CLINICAL OPPORTUNITIES FOR PEMF THERAPIES

Since clinicians can be confronted by a vast range of medical conditions, it would be helpful to know which conditions PEMFs may be most helpful for and which ones have been studied to, at least, a limited extent.

PEMFs are most often mentioned or considered for pain and musculoskeletal problems (Pawluk, 2015). However, significant benefits have been found in studies on PEMF therapy, either as a primary or complementary therapy, for a variety of conditions, including, among others,

- addiction
- anxiety
- asthma
- cancer
- chronic fatigue
- concussion
- depression
- enuresis
- keloids
- multiple sclerosis
- osteoporosis
- overactive bladder
- Parkinson's disease
- prostate hyperplasia
- sleep disorder
- stroke
- tremor
- vascular conditions
- wound healing.

References to these are provided in Pawluk and Layne, 2017.

COST BENEFIT OF PEMF THERAPIES

One of the things clinicians have to consider in regards to any therapy is whether it is cost effective. There is minimal formal research information available regarding the cost-effectiveness of PEMFs. Clinicians who routinely use or recommend PEMFs find them to be effective and low risk. Patients and clients are often able to reduce or stop medication. Recovery times after surgery and injuries are significantly faster with fewer risks of complications. In one woman with cancer pain, in my clinical experience, narcotics could be reduced because of improved pain control. People with anxiety have been found to maintain or reduce medication without dose escalation. Improvement in quality of life is routinely reported when PEMFs are used.

Evidence to support cost-effectiveness has been reported in one series of studies in France where PEMFs were used for chronic neck pain (Forestrier et al., 2007a,b). They used an improvement in pain scores of more than 20% after 6 months as a primary outcome measure. Patients were randomly subjected either to spa therapy or PEMFs. Seventy-nine percent improved in the PEMF group versus 55% in the spa therapy group (p = 0.02). In the cost effectiveness phase of their research, they compared the healthcare costs of the PEMF and spa therapy study participants for the 6 months before participation in the research to the costs for the 6 months after participation and to the same costs of individuals with standard care. Average healthcare costs increased for all groups. The cost increases for the PEMF group were the least, being 18% and 11% of the cost increases for the spa care and standard care groups, respectively. So, one might conclude that, in the long run, PEMF therapy is a better and more effective investment for the management of chronic neck pain than standard care. A major issue with a cost-benefit evaluation is whether a device is used in a clinical practice setting, where there is a limited opportunity for exposure, or a home setting where home therapy can be applied over longer periods of time, especially for chronic health issues.

More recently, rTMS for treatment-resistant depression has been found to be more cost effective than antidepressant medication (Nguyen and Gordon, 2015). The probability of rTMS being cost effective was 73%. A major cost-effectiveness analysis (Woods et al., 2017) was performed in the U.K. comparing 10 nonpharmacological interventions for osteoarthritis of the knee, including PEMFs, with usual care. Quality adjusted life years (QALY) saved was the cost-effectiveness measure used (Weinstein et al. 2009). A score of 1.0 is perfect health. Based on the five trials assessed using PEMFs, PEMFs produced positive QALYs averaging 0.56. There was a 12% improvement in QALYs compared to usual care. By way of comparison, acupuncture was found to have a QALY of only 0.11, compared to 0.56 for PEMFs. In this analysis, PEMFs had the second-highest incremental cost versus usual care, £577. Most of the PEMF studies were carried out in clinical settings, and the durations of the studies were 6 weeks or less. In summary, PEMFs improve the quality of life but with variable cost effectiveness, certainly relative to other interventions.

RESOURCES

Informational resources for clinicians are limited to research published in journals. PEMF research is found across a vast variety of publications and is mostly biomedical and not clinical, making it difficult for clinicians to gain access and understand. Research summaries, especially for clinical application, are also very limited. A useful curation of PEMF research is available at www.emf-portal.org. A clinically useful, albeit limited, review of PEMF information is found in Pawluk and Layne, 2017.

REFERENCES

Alyshev VA, Viaznikov AL, Gertsen IG, et al. Magnetotherapy in the complex treatment of patients with suppurative wounds and osteomyelitis. *Vestn Khir.* 1988; 140(4): 141–3.

Barnes PM, Bloom B, Nahin R. Complementary and alternative medicine use among adults and children: United States, 2007. Nat Health Stat Rep. 2008; 12: 1–23.

Bassett CA, Pawluk RJ, Pilla AA. Augmentation of bone repair by inductively coupled electromagnetic fields. *Science.* 1974; 184(4136): 575–7.

Blank M, Soo L. Threshold for inhibition of Na, K-ATPase by ELF alternating currents. *Bioelectromagnetics.* 1992; 13(4): 329–33.

Cayley WE Jr. Comfort always. *Fam Pract Manag.* 2006; 13(9): 74.

Cheing GL, Wan JW, Kai Lo S. Ice and pulsed electromagnetic field to reduce pain and swelling after distal radius fractures. *J Rehabil Med.* 2005; 37(6): 372–7.

Chen JF, Eltzschig HK, Fredholm BB. Adenosine receptors as drug targets: What are the challenges? *Nat Rev Drug Discovery.* 2013; 12(4): 265–86.

Dindar H, Renda N, Barlas M, et al. The effect of electromagnetic field stimulation on corticosteroids-inhibited intestinal wound healing. *Tokai J Exp Clin Med.* 1993; 18(1–2): 49–55.

Farinas PL, Martinez Farinas PO. Effect of electromagnetic field on the action of electron in the treatment of cancer. *Vida Nueva.* 1951; 67(6): 141–9.

FDA. General wellness: Policy for low risk devices: Guidance for industry and Food and Drug Administration Staff. July 29, 2016. www.fda.gov/downloads/medicaldevices/deviceregulationandguidance/guidancedocuments/ucm429674.pdf.

Forestier R, Françon A, Saint-Arromand F, et al. Are SPA therapy and pulsed electromagnetic field therapy effective for chronic neck pain? Randomised clinical trial I. First part: Clinical evaluation. *Ann Readapt Med Phys.* 2007a; 50(3): 140–7.

Forestier R, Françon A, Saint Arroman F, et al. Are SPA therapy and pulsed electromagnetic field therapy effective for chronic neck pain? Randomised clinical trial. II. Second part: Medicoeconomic approach. *Ann Readapt Med Phys.* 2007b; 50(3): 148–53.

Healey F, Persinger MA, Koren SA. Enhanced hypnotic suggestibility following application of burst-firing magnetic fields over the right temporoparietal lobes: A replication. *Int J Neurosci.* 1996; 87(3–4): 201–7.

HHS. What is the U.S. opioid epidemic? 2018.

House of Commons London. The Stationery Office Ltd. The influence of pharmaceutical industry 4th report of Session 2004–2005. 2005. www.hhs.gov/opioids/about-the-epidemic/index.html.

Iashchenko LV, Chistiakov IV, Gakh LM, et al. Low-frequency magnetic fields in the combined therapy of inflammatory lung diseases. *Probl Tuberk.* 1988; 3: 53–6.

Kaemmerer WF. How will the field of gene therapy survive its success? *Bioeng Transl Med.* 2018; 3(2): 166–77.

Kravtsova T, Rybolovlev EV, Kochurov AP. The use of magnetic puncture in patients with duodenal peptic ulcer. *Vopr Kurortol Fizioter Lech Fiz Kult* 1994; 1: 22–4.

Kumar V. Chapter 1: Cell injury, cell death and adaptations. In: *Robbins and Cotran Pathologic Basis of Disease.* Professional Edition, 8th edn. Elsevier, Philadelphia, 2007.

Markov M. Benefit and hazard of electromagnetic fields. In: Markov M (ed) *Electromagnetic Fields in Biology and Medicine.* CRC Press, Boca Raton, FL, 15–29, 2015.

Markov M. *Dosimetry in Bioelectromagnetics.* CRC Press, Boca Raton, FL, 1–23, 2017.

Massari L, Benazzo F, De Mattei M, et al. Effects of electrical physical stimuli on articular cartilage. *J Bone Joint Surg Am.* 2007; 89 (Suppl 3): 152–61.

Moorthy HK, Venugopal P. Measurement of renal dimensions in vivo: A critical appraisal. *Indian J Urol.* 2011; 27(2): 169–75.

Nguyen KH, Gordon LG. Cost-effectiveness of repetitive transcranial magnetic stimulation versus antidepressant therapy for treatment-resistant depression. *Value Health.* 2015; 18(5): 597–604.

Nindl G, Johnson MT, Balcavage WX. Low-frequency electromagnetic field effects on lymphocytes: Potential for treatment of inflammatory diseases. In: Ayrapetyan SN, Markov MS (eds) *Bioelectromagnetic Medicine.* Marcel Dekker, New York, 369–89, 2004.

Palmer TM, Trevethick MA. Suppression of inflammatory and immune responses by the A2A adenosine receptor: An introduction. *Br J Pharmacol.* 2008; 153: S27–34.

Pawluk W, Layne CJ. *Power Tools for Health: How Magnetic Fields (PEMFs) Help You.* Friesen Press, Victoria, 2017.

Pawluk W, Turk Z, Fischer G, et al. Treatment of osteoarthritis with a new broadband PEMF signal. *In 24th Annual Meeting of Bioelectromagnetics Society,* Quebec City, Canada, June 2002.

Pawluk W. Chapter 17: Clinical dosimetry of extremely low-frequency pulsed electromagnetic fields. In: Markov M (ed) *Dosimetry in Bioelectromagnetics.* CRC Press, Boca Raton, FL, 369–82, 2017.

Pawluk W. Chapter 17: Magnetic fields for pain control. In: Markov M (ed) *Electromagnetic Fields in Biology and Medicine*. CRC Press, Boca Raton, FL, 273–96, 2015.

Poltavtseva RA, Poltavtsev AV, Lutsenko GV, et al. Myths, reality and future of mesenchymal stem cell therapy. *Cell Tissue Res*. 2018; 375(3): 563–74.

Rohde C, Chiang A, Adipoju O, et al. Effects of pulsed electromagnetic fields on interleukin-1 beta and postoperative pain: A double-blind, placebo-controlled, pilot study in breast reduction patients. *Plast Reconstr Surg*. 2010; 125(6): 1620–1629.

Saveriano G, Ricci S. Experiences in treating secondary post-traumatic algodystrophy with low-frequency PEMFs in conjunction with functional rehabilitation. In International Symposium in Honor of Luigi Galvani, Bologna, Italy. *J Bioelectr*. 1989; 8(2): 320.

Shaw Q. On aphorisms. *Br J Gen Pract*. 2009; 59(569): 954–5.

Shupak NM. Therapeutic uses of pulsed magnetic-field exposure: a review. *The Radio Science Bulletin*. 2003; 307: 1–32.

Thomas AW, Prato FS. Magnetic field based pain therapeutics and diagnostics. Bioelectromagnetics Society, 24th Annual Meeting, Quebec City, PQ, Canada, June, 2002.

Van Zundert J, Patijn J, Kessels A, et al. Pulsed radiofrequency adjacent to the cervical dorsal root ganglion in chronic cervical radicular pain: A double blind sham controlled randomized clinical trial. *Pain* 2007; 127(1–2): 173–82.

Varani K, Vincenzi F, Ravani A, et al. Adenosine receptors as a biological pathway for the anti-inflammatory and beneficial effects of low frequency low energy pulsed electromagnetic fields. *Mediators Inflammation*. 2017; 2017: 2740963.

Weinstein M, Torrance G, McGuire A. QALYs: the basics. *Value Health*. 2009; 12(Suppl 1): S5–9.

Woods B, Manca A, Weatherly H, et al. Cost-effectiveness of adjunct non-pharmacological interventions for osteoarthritis of the knee. *PLoS One*. 2017; 12(3): e0172749.

Xue J, Deng H, Jia X, et al. Establishing a new formula for estimating renal depth in a Chinese adult population. *Medicine (Baltimore)*. 2017; 96(5): e5940.

Zhao N, Zhang J, Qiu M, et al. Scalp acupuncture plus low-frequency rTMS promotes repair of brain white matter tracts in stroke patients: A DTI study. *J Integr Neurosci*. 2018; 17(1): 61–9.

7 Electromagnetic Fields in Relation to Cardiac and Vascular Function

Harvey Mayrovitz

CONTENTS

INTRODUCTION

In considering the role of electromagnetic fields (EMFs) as applied to cardiac function, there are several broad roles open for contemplation. One is the use of EMF to provide cardiac monitoring and diagnostic information not easily obtainable via other methods. This aspect relies on the magnetic field produced by the heart being detectible and signals so detected to provide functional information regarding cardiac status. One example is the measuring of fetal heart rate (HR) parameters (Moraes, Murta et al. 2012, Van Hare 2013, Wacker-Gussmann, Paulsen et al. 2014, Batie, Bitant et al. 2018), a procedure not easily done with standard methods. Another example would be the measuring of physiological parameters such as cardiac volume changes via magnetic susceptibility changes (Wikswo 1980, Wikswo, Opfer et al. 1980, Roth and Wikswo 1986).

A second category of potential and possible use of EMF is as a treatment modality for cardiac dysfunction. Such dysfunction may include HR or rhythm disorders or disorders related to cardiac blood flow inadequacy or cardiac muscle dysfunction (Yuan, Wei et al. 2010, Li, Yuan et al. 2015).

A third category of potential applications of EMF is cardiac protection. This relates to how exposure to suitable magnetic fields might reduce risk of future cardiac dysfunction. Examples may include reduction of blood viscosity with pulsed electromagnetic fields (PEMFs)(Tao and Huang 2011, Tao, Wu et al. 2017) as a way to diminish vascular resistance, the use of PEMF to increase cardiac endothelial cell proliferation (Li, Yuan et al. 2015), or the use of bioelectromagnetic actions to reduce blood pressure (Tasic, Djordjevic et al. 2017).

A fourth general category of cardiac–biomagnetic interaction relates to unintended or uncontrolled cardiac effects caused by external sources. Examples may include potential cardiac effects associated with electromagnetic therapy applied to other body parts (Yoshida, Yoshino et al. 2001, Wang, Hensley et al. 2016) or even changes in the geomagnetic field (McCraty, Atkinson et al. 2017, Jarusevicius, Rugelis et al. 2018, Žiubrytė, Jaruševičius et al. 2018a,b).

EMF approaches, devices, and methods that relate to its potential utility in the mentioned categories have been proposed and evaluated and, in some cases, have demonstrated functional utility for detection and treatment. Although, some approaches and methods have theoretically sound foundations, some may not have yet been applied or even evaluated. The main purpose of this chapter is to offer a framework from which past, present, and possible future EMF applications related to cardiac and vascular function may be appreciated in the context of the associated physiological and pathophysiological underpinnings.

MONITORING CARDIAC FUNCTION UTILIZING THE HEART'S MAGNETIC FIELD

PHYSIOLOGICAL PROCESSES

Cardiac transmembrane ionic currents associated with myocardial depolarization and repolarization sweep through the heart causing it to contract and then relax resulting in blood pumping and refilling, respectively. Electrical potential changes (action potentials, APs) associated with these waves of depolarization and repolarization are caused by Na^+, K^+, and Ca^{++} ion movements across cardiac cell membranes through specific ion channels. The current and AP changes are sensed by surface electrodes to generate the well-known electrocardiogram (ECG). A typical ECG pattern has a P-wave associated with atrial contraction, a QRS complex associated with ventricular depolarization, and a T-wave associated with ventricular repolarization (Figure 7.1). The initiation of this electrical activity normally starts at the sinoatrial node (SAN) located in the right atrium. It spontaneously depolarizes and acts as the heart's pacemaker. The ventricles have electrical activity during the QT interval (Figure 7.1) but are electrically quiet during the TQ interval. The AP associated with ventricular myocytes and conduction fibers is divided into phases (ϕ) with ϕ_4 being the resting phase, ϕ_0 a rapid depolarization phase that is mainly associated with a rapid influx of Na^+, a ϕ_2 plateau phase involving Ca^{++} influx and K^+ efflux, and ϕ_3 a repolarization phase in which Ca^{++} influx is diminished and counterbalanced by

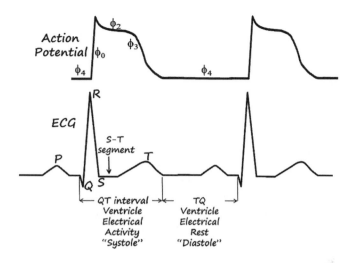

FIGURE 7.1 A cardiac ventricular action potential (AP) and a corresponding electrocardiogram (ECG) showing the waves and time correlation between the AP and the ECG. Ventricular electrical activity occurs during the QT interval. The phases (ϕ) of the AP are indicated. (Courtesy of Dr. HN Mayrovitz.)

K^+ efflux. Depolarization and repolarization currents are quite small but give rise to magnetic fields that can be sensed by suitable devices. The ECG is used to estimate the mean QRS electric vector to help diagnose ventricular hypertrophy and other conduction abnormalities. The ECG vector (mean electrical axis) is reported similar to the direction of the magnetic vector (Barry, Fairbank et al. 1977).

THE CARDIAC MAGNETIC FIELD

Magnetic fields due to cardiac currents are very small, in the neighborhood of 1 nT if measured close to the heart surface(McBride, Roth et al. 2010) and about 0.1 nT at the body surface (Geselowitz 1973). Despite such small magnitudes, Stratbucker and coworkers were able to use a toroidal solenoid containing 17,640 turns to successfully detect cardiac generated fields that were in time synchrony with ECGs obtained from isolated guinea pig hearts (Stratbucker, Hyde et al. 1963). Their estimated value for the magnetic dipole moment of the QRS complex was in good agreement with measured values.

The first measurements on nonisolated hearts appear to have been made by Baule and McFee employing two multiturn coils (2×10^6 turns) over the chest (Baule and McFee 1963). Using this method, it was determined that the voltage induced in the coil (v) depends on the number of turns (n), the permeability of the core (μ), the loop cross-sectional area (A), and the angle (θ) that the loop axis makes with respect to magnetic field strength vector **H**.

The detected voltage is approximately $v = n\mu\cos(\theta)dH/dt$. Since H is a vector, its full simultaneous specification would require the use of three mutually orthogonal coils (Geselowitz 1979). Due to extraneous signals, even in magnetically shielded areas, the signal to noise ratio can be a limitation in obtaining consistently reliable data. Such measurements now are most often done with a superconducting quantum interference device (SQUID) used in adult (Fujino, Sumi et al. 1984) and in fetal (Zhuravlev, Rassi et al. 2002) applications to obtain magnetic counterpart of the ECG called the magnetocardiogram (MCG). An example of the ECG–MCG correspondence is shown in Figure 7.2 that is drawn based on and representative of values of Brockmeier and colleagues (Brockmeier, Schmitz et al. 1997).

Detection of the magnetic field associated with cardiac currents can be a useful adjunct to other diagnostic monitoring methods. Assessment methods using simulations indicate that the magnetic field associated with cardiac repolarization APs is greater than that which would be predicted simply based on considerations of time rates of change of transmembrane potentials (Barach and Wikswo 1994). This observation suggests involvement of other factors. Some such factors examined relate to the extent to which changes in the conductivity of cardiac tissue and blood might affect the measured external field. Further simulation results suggest that changes in blood and myocardium conductivity are importantly involved (Czapski, Ramon et al. 1996). But, given the long history of the ECG as an immensely useful diagnostic tool, the use of MCG might at first glance be questionably redundant. However, arguments have been made suggesting otherwise.

It has been suggested that there is information within the MCG signal that is not available in the ECG signal. One aspect of this difference may relate to the fact that the ECG detects cardiac electrical activity as a flux source but the MCG detects it as a vortex source (Plonsey 1972). Furthermore, some theoretical work supports the concept of basic informational differentials between electric and magnetic sensing (Roth and Wikswo 1986). An example of this in the heart relates to the fact that in certain myocardial regions, because of the anatomical arrangement of muscle fibers, some cardiac currents contribute to the ECG, but some do not. The currents that do

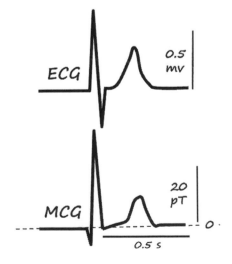

FIGURE 7.2 Illustrating the similarity between the pattern of a measured ECG and magnetocardiogram (MCG). (Courtesy of Dr. HN Mayrovitz.)

not contribute to the ECG would however be detectible via the MCG, thus yielding additional information. Measurements using SQUID in rabbit hearts (McBride, Roth et al. 2010) were felt to be consistent with this concept. The value of such added information in a clinical setting remains to be established.

There are other aspects to be considered when comparing potential utilities of MCG versus ECG. MCG signals do not require skin contact and are much less affected by the electrical conductivity of tissues and fluids that lie between heart and skin surfaces than ECG signals. MCG is also useful for a variety of fetal cardiac evaluations as shown for fetal tachycardia (Abe, Hamada et al. 2005), changes in the fetal QRS complex (Horigome, Takahashi et al. 2000), characterizing accessary pathway features (Kandori, Hosono et al. 2003), fetal third degree block (Hosono, Shinto et al. 2002), fetal atrial fibrillation (Kandori, Hosono et al. 2002), and, more recently, significant advancements in detecting a variety of fetal arrhythmias (Stingl, Paulsen et al. 2013, Yu, Van Veen et al. 2013, Kiefer-Schmidt, Lim et al. 2014, Wacker-Gussmann, Paulsen et al. 2014). In other applications, the potential advantage of MCG versus ECG in detecting the dominant frequency of atrial fibrillation in adults has also been reported (Yoshida, Ogata et al. 2015). Other cardiac conditions for which MCG has demonstrated potential utility above that offered by standard ECG include detecting certain abnormalities associated with ischemic heart disease (Watanabe and Yamada 2008), improved prediction of future major cardiac events in patients with dilated cardiomyopathy (Kawakami, Takaki et al. 2016), and other cardiac conditions (Gapelyuk, Wessel et al. 2007, Schirdewan, Gapelyuk et al. 2007, Van Leeuwen, Hailer et al. 2008, Gapelyuk, Schirdewan et al. 2010).

In addition to potential direct clinical applications of MCG, there appears to be room for its use to study and uncover physiological and pathologic features not easily discoverable via ECG data. A beautiful example is to be found in the pioneering work of Cohen and coworkers who studied the mechanism and underlying aspects of the now well-known shift in the ECG S-T segment that occurs with cases of STEMI myocardial infarction (Loomba and Arora 2009). At the time, details of such shifts were unclear. It was reasoned that one possibility was that the shift was caused by cardiac currents flowing only during the S-T segment interval and that those currents arose due to regional differences in membrane APs. This possibility was termed "primary shift" or "true shift". The other possibility was that the observed S-T segment shift was due to the presence of a continuous steady injury current that was interrupted only during the S-T interval. This was termed "secondary shift" or apparent shift". Part of this process is shown in Figure 7.3 that shows conditions during the ventricular electrical resting phase (during the ECG T-Q interval).

In part A of Figure 7.3, the membrane potentials between normal myocytes are not different in magnitude (negative inside with respect to outside), so the potential difference between them is zero. If some cells become ischemic, as shown in part B, reduced O_2 available to their Na^+-K^+ pumps results in diminished pump function causing partial depolarization of these cells. This gives rise to a potential gradient between normal and ischemic cells that becomes a source of an "injury current".

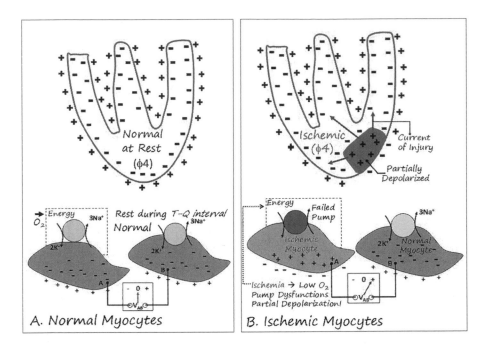

FIGURE 7.3 Illustrating the change in cardiac cellular transmembrane potentials induced by ischemia. In (A) is shown normal healthy cells and in (B) is shown what can happen when one of the cells (or regions) becomes ischemic. The low oxygen alters the cell's sodium-potassium pump function causing a difference in electrical potential between healthy and ischemic cells. Such potential differences can give rise to so-called injury currents. (Courtesy of Dr. HN Mayrovitz.)

The direction of the injury current depends on the phase of the cardiac AP under consideration as illustrated in Figure 7.4.

In Figure 7.4, a region of the ventricular wall near the endocardium is assumed to experience an inadequate blood flow (ischemia). The reduced availability of O_2 causes partial depolarization of the ischemic cell's resting AP that also has a reduced amplitude as shown dotted in the figure. The S-T segment of the ECG occurs during ventricular systole. During this time, current flow is from normal cells to ischemic cells as indicated. However, during ventricular diastole (T-Q interval of Figure 7.1), an injury current would flow from ischemic to healthy cells. The use of MCG to study this process was first determined in dogs (Cohen, Norman et al. 1971, Cohen and Kaufman 1975) followed by measurements in humans (Cohen, Savard et al. 1983, Savard, Cohen et al. 1983) using SQUID measures.

An illustrative example of the evolution of S-T segment shifts with exercise, as investigated using MCG, is shown in Figure 7.5 drawn based on the data of Cohen, Savard et al. (1983). Results from the dog experiments implicated an initial acute ischemia-related induction of a steady "injury current" that was then interrupted via depolarization during the S-T interval. This would correspond to part-labeled "systole" in Figure 7.4. This process gave rise to the visualized S-T segment shift that

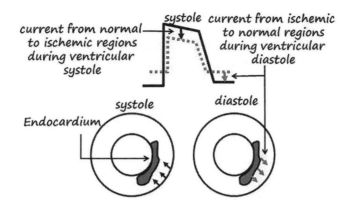

FIGURE 7.4 Illustrating the directional changes in injury currents associated with a region of myocardial ischemia that depends on the time interval of the cardiac action potential. The time interval labeled "systole" is associated with the electrically active QT interval. The interval labeled "diastole" is associated with the quiescent TQ interval. (Courtesy of Dr. HN Mayrovitz.)

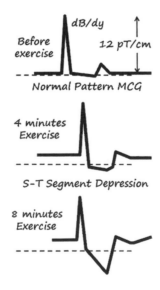

FIGURE 7.5 Illustrating the progressive changes in an MCG associated with evolving ischemia. (Courtesy of Dr. HN Mayrovitz.)

would be considered "secondary" according to the nomenclature of these authors. After this initial change, there seemed to be a transition to a true S-T segment shift.

However, in using MCG to study S-T segment shifts in a person with a left bundle branch block (LBBB) and in three persons with early repolarization (ER) syndrome (Savard, Cohen et al. 1983), the magnetic data showed no evidence of a continuous current baseline shift. In these conditions, a "primary shift" mechanism was assumed to be the main cause. For the three persons with ER, the explanation for the shift would be that abnormally early repolarization in some myocardial regions causes a voltage gradient that drives current between those regions. This current would start during the early part of phase 3 and would closely correspond to the abnormal currents shown in Figures 7.3 and 7.4 but would not require an ischemic basis. The S-T segment shift the authors observed in the patient with LBBB may also be accounted for on the basis of regional differences in repolarization, not due ischemia but instead due to delayed conduction causing a right-to-left voltage gradient during repolarization. Subsequent evaluation of a person with coronary artery disease undergoing stress testing indicated that an ischemia-dependent S-T segment depression (of which Figure 7.5 is representative) is related to both true and apparent mechanisms (Cohen, Savard et al. 1983).

The details herein provided concerning these prior studies helps visualize the potential for MCG measurements. It is noteworthy that most of these and other experimental observations and conclusions would not have been possible without the unique ability of the MCG to measure direct current (DC) fields and their changes, a feature not possible with an ECG. However, the fact that DC measurements are possible means that low-frequency filtering is not applicable, thereby increasing the risk of external interference. In addition, signals arising due to DC fields originating in other organs (e.g., the gastrointestinal system) need to be considered in any analysis and interpretive process of any composite MCG signal.

Some progress has been made in utilizing modalities other than SQUID systems to detect MCG-related information. One example is the use of arrays of magnetoresistive sensors that respond to external cardiac magnetic fields by changing their resistance. A pilot study indicated the possibility of detecting P, QRS, and T waves if suitable averaging methods are used (Shirai, Hirao et al. 2018). Another type of potential advance in the field relates to enhanced processing of acquired MCG signals that offer the possibility for improved diagnosis and characterization of adult ischemic heart disease based on MCG features (Tao, Zhang et al. 2018) and on processing of fetal MCG data (Escalona-Vargas, Wu et al. 2018).

Other advancements in the use of MCG parameters have focused on developing scoring systems with associated threshold values to better characterize and detect coronary artery disease with adequate positive and negative predictive levels (Shin, Park et al. 2018a,b). Partly because standard MCG utilization usually requires enhanced room shielding for proper operation the widespread use of this technology for research and clinically related applications has been limited. Other emerging approaches have evaluated the possibility of MCG use in nonshielded environments using repeatability measures of various MCG-related parameters (Sorbo, Lombardi et al. 2018). However, the use of multichannel MCG remains an evolving methodology as demonstrated by the recent report of its use in detecting patients at high risk for lethal arrhythmias (Kimura, Takaki et al. 2017) and its reported increased discriminating ability compared to ECG to characterize cardiac repolarization features (Smith, Langley et al. 2006).

CARDIAC IMPACTS AND APPLICATIONS OF
APPLIED BIOMAGNETIC FIELDS

When considering impacts of applied PEMF on cardiac parameters, there are two general categories to consider. One relates to the heart's exposure to fields that are not specifically directed toward affecting heart parameters but cause changes as "side effects". This includes potential effects of transcranial magnetic stimulation (TMS) and fields associated with magnetic resonance imaging (MRI) (Chakeres, Kangarlu et al. 2003). The other general category relates to specific targeted applications that are designed to alter cardiac parameters to treat cardiac arrhythmias and other aspects of dysfunction. In each category, an applicable concept is one of an approximate magnetic dipole of current source $\mathbf{I}$ amps in a coil of area A. For such a dipole at some distance (r) from the biological target, $\mathbf{H}(r,\theta) = [\mathbf{m}/4\pi r^3] (1 + 3\cos^2 \theta)^{1/2}$. In this equation, $\mathbf{m}$ is the coil's dipole magnetic moment ($\mathbf{I}$A), and θ is the angle with respect to its axis.

CALCIUM CURRENTS AS A TARGET

The way in which such fields may produce their cardiac effects is unclear. However, it is known that calcium dynamics play multiple roles in relation to cardiac electrical and mechanical properties and function, and there is widespread data demonstrating bioelectromagnetic-Ca^{++} interactions (Smith, McLeod et al. 1987, Blackman, Benane et al. 1988, Walleczek and Budinger 1992, Pilla, Muehsam et al. 1999, Coulton, Barker et al. 2000, Zhao, Yang et al. 2008, Muehsam and Pilla 2009, Lu, Du et al. 2015). Thus, it is likely EMF-Ca^{++} linkages are also involved in cardiac cellular aspects.

Experimental work with the frog heart model has suggested a potential frequency and intensity window that can alter Ca^{++} dynamics (Schwartz, House et al. 1990). The increase in Ca^{++} efflux from hearts exposed to 240 MHz modulated at 16 Hz showed these effects, but the effects were not observed when unmodulated or modulated at 0.5 Hz. Other effects, reported when electrical and mechanical features were recorded during SMF exposure (0.34–1.56 T), included rate irregularities and reductions in contraction force and diastolic relaxation (Reno and Beischer 1966). Whether these changes are related to Ca^{++} dynamics is not verified, but it is clear that all observed physiological changes could be linked to Ca^{++} dynamics.

Although frog hearts do not depend on Ca^{++} release from sarcoplasmic reticulum for contraction as human heart does, reduction in free Ca^{++} during systole and reduced reuptake of cytosol Ca^{++} during diastole could explain the reduced contraction force and reduced diastolic relaxation observed. Further, the electrical rhythm changes might be explained by altered sinus node diastolic depolarization features associated with field-induced Ca^{++} changes. In part, motivated by these and other observations, several innovative ideas have been put forward broadly describing noninvasive ways to impact cardiac function and arrhythmias (Laniado, Kamil et al. 2005, Kassab, Navia et al. 2013, Scheinowitz, Nhaissi et al. 2018).

It was noted earlier that there is some evidence that geomagnetic field variations might be associated with various cardiac-related dysfunctions including myocardial infarction (Jarusevicius, Rugelis et al. 2018), atrial fibrillation (Žiubrytė, Jaruševičius et al. 2018), and angina pectoris (Žiubrytė, Jaruševičius et al. 2018). It is possible to speculate that such correlations are related to the fact that the excitation frequency for cardiac cell Ca^{++} transient currents depends on small variations in a copresent DC magnetic field ranging from about 44 to 49 µT (Fixler, Yitzhaki et al. 2012).

Transcutaneous Magnetic Stimulation (TMS)

Transcutaneous magnetic stimulation (TMS) is a rapidly growing electromagnetic application and, as the name implies, is the use of PEMF to stimulate brain tissue with applicator coils placed on the scalp surface. This allows the magnetic field to pass through the skull to the brain tissue region of interest. Because in some cases there are a series of pulses associated with the treatment, it is further characterized as repetitive TMS or simply rTMS as compared to a single pulse treatment modality (sTMS). One early treatment target for this modality was depression that received FDA clearance in 2008. For this condition, a series of 20–25 treatments over 4–6 weeks has been used by some practitioners. Although reported to be effective for a variety of conditions, there has been concern as to its possible impacts on cardiac function. Some workers have reported associated cardiac changes. These include increased HR and a possible shift in the balance between parasympathetic and sympathetic activity (Cabrerizo, Cabrera et al. 2014) and other heart-rate related effects when using rTMS to treat autistic children (Wang, Hensley et al. 2016). Since it appears that the clinical use of TMS is expanding, it is likely that further cardiac impacts will be discovered.

Magnetic Resonance Imaging (MRI)

A second source of cardiac exposure to bioelectromagnetic fields is associated with the use of MRI for diagnostic purposes. Static magnetic fields (SMFs) associated with such MRI systems now include 1.5 T, 3.0 T, and the increasing use of 7 T devices that permit better resolution than lower-intensity devices(Park, Kang et al. 2018). These higher intensities may be especially useful for the assessment of brain small vessel disease (Geurts, Bhogal et al. 2018, Geurts, Zwanenburg et al. 2018). Future applications of the 7 T field are likely to provide improved resolution in spinal cord (Barry, Vannesjo et al. 2018), breast (Li and Rispoli 2019), peripheral nerves(Yoon, Biswal et al. 2018), quantifying cerebral spinal fluid movement (Markenroth Bloch, Toger et al. 2018), and a variety of other applications (Barisano, Sepehrband et al. 2018). Given this situation, it is natural to consider the possible effect of these high field intensities on cardiac-related parameters. One report has indicated that exposure to an SMF of 2 T reduces resting HR during the first 10 min of exposure (Jehenson, Duboc et al. 1988), but that HR is subsequently normalized. Assessment of HR changes in squirrel monkeys in response to exposure to 7 T fields for about an hour has also shown HR reductions (Beischer and Knepton 1964). Contrastingly, other studies (Chakeres, Kangarlu et al. 2003) have indicated no effect of MRI-generated static fields measured with intensities ranging from 0.8 to 8.0 T. The question of impacts of high-intensity fields on cardiac parameters remains unclear.

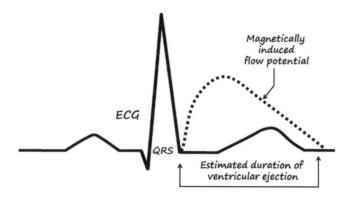

FIGURE 7.6 Illustrating the timing of a magnetically induced flow potential associated with blood flow interacting with an intense static magnetic field. (Courtesy of Dr. HN Mayrovitz.)

MAGNETOHEMODYNAMIC CONSIDERATIONS

An additional consideration with respect to high-intensity applications is the potential impact that occurs due to interactions between applied magnetic fields and blood hemodynamics. Movement of blood within a static field gives rise to a generated electrical potential, a process that forms the basis of electromagnetic blood flow measurement in large blood vessels (Kolin, Assali et al. 1957). Impacts of such externally applied fields have been considered as they affect blood flow in larger arteries (Chen and Saha 1984, Okazaki, Maeda et al. 1987, Sud and Sekhon 1989, Kinouchi, Yamaguchi et al. 1996) and blood and blood flow in small blood vessels (Takeuchi, Mizuno et al. 1995, Higashi, Asjoda et al. 1997, Haik, Pai et al. 2001, Bali and Awasthi 2011). However, the possible role of a magnetically induced voltage affecting cardiac function has not been clarified. Such a generated voltage, termed "flow potential", was initially observed on ECG recordings made when squirrel monkeys were exposed to static magnetic fields ranging from 4 T to 10 T (Beischer and Knepton 1964). This flow potential measured at the skin surface was about 0.5 mv and was estimated to be about 100 mV when extrapolated back to the heart.

An interesting feature of this flow potential was that it appeared to occur maximally during ventricular ejection and thereby might be useable as an index of ascending aortic blood flow as schematically illustrated in Figure 7.6. Whether or not this induced potential would serve as an ectopic impulse to produce cardiac arrhythmias likely depends on its magnitude and timing. As illustrated in Figure 7.6, the flow potential occurs during a time interval when most of the ventricular myocardium is refractory. However, it could have an impact on atrial cells.

HEART RATE VARIABILITY (HRV)

Several studies have examined the potential impact of magnetic fields associated with MRI on HRV parameters. The significance of HRV is in part related to its characterization of cardiac status as determined by the sympathetic–parasympathetic balance controlled by the autonomic nervous system (ANS). This relationship is based on the fact that sympathetic and parasympathetic (vagus) nerve traffic to the sinoatrial node largely determines HR changes measured as variations in ECG R–R intervals. Thus, HRV may be broadly viewed as a neurocardiac process (Shaffer and Ginsberg 2017), and its complexity should be kept in mind (Pagani, Lombardi et al. 1986) when interpreting changes.

Since vagus control can effectuate more rapid changes in HR than sympathetic control, the higher-frequency spectral components of the interbeat time series (R–R intervals) are considered to represent parasympathetic activity and the lower-frequency components viewed as being indices of sympathetic activity. For spectral analysis purposes, the ratio of high-frequency power to low-frequency power is taken as an index of the balance between parasympathetic and sympathetic activity. A higher ratio would be considered good from a cardiovascular perspective. Similarly, for time-domain analyses, a greater HRV as determined by the variance of consecutive R–R intervals is considered good from a cardiovascular perspective. In fact, most studies indicate a reduction in HRV as predictive of negative cardiovascular morbidity and mortality outcomes. However, in some conditions, it appears that an increase in certain HRV temporal parameters are predictive of vascular decline as in the progression of small blood vessel disease in elderly persons (Yamaguchi, Wada et al. 2015). Intracardiac conduction abnormalities can also lead to increased HRV, so increases should not always be interpreted as "good".

Effects of diagnostic MRI (1.5 T) on HRV of 42 patients indicated an increase in HRV as determined by increased R–R interval standard deviations (SDs) (Derkacz, Gawrys et al. 2018). In another study, a 30 min MRI exposure with a 1.5 T field was also associated with increases in time-domain HRV parameters as well as increases in all spectral components (Sert, Akti et al. 2010). However, the potential impacts of MRI procedural anxiety on HRV changes (Pfurtscheller, Schwerdtfeger et al. 2018) should be considered in all such comparisons. A report (Ghadimi-Moghadam, Mortazavi et al. 2018) concerning effects of a relatively low-intensity MRI (1.5 T) on staff workers and test subjects suggests that further research is needed especially in light of the pending increasing use of higher-power units. It is noteworthy that short intervals of head exposure to low-intensity static fields (200 µT) (Koppel, Vilcane et al. 2015) and short- and long-time (12 h) whole body exposure to extremely low-frequency (ELF 50–1,000 Hz) at 100 µT (Kurokawa, Nitta et al. 2003) did not significantly alter any HRV parameter. However, there is now some evidence that environmental magnetic fields associated with Schumann resonance power changes can increase HRV with associated increases in parasympathetic activity (Alabdulgader, McCraty et al. 2018).

POTENTIAL CARDIAC AND VASCULAR TARGETS OF BIOELECTROMAGNETIC THERAPY

CARDIAC ARRHYTHMIAS

Normally, HR is controlled by actions within the sinoatrial node (SAN) located in the right atrium. Its function depends strongly on actions of a so-called ionic "funny" current, I_f, (DiFrancesco, 2010) which if increased causes HR to increase and if decreased causes HR to decrease. Although normal SAN pacemaker activity can operate independently of the ANS, the SAN is richly innervated, and its activity is modulated by nerve traffic from both sympathetic and parasympathetic nervous systems. Increased sympathetic activity acts on β-adrenergic receptors causing I_f to increase thereby increasing HR. Contrastingly, increased parasympathetic activity via the vagus nerve acts on muscarinic receptors and causes I_f and also HR to decrease. Acetylcholine (Ach) is the neurotransmitter involved in the slowing of the HR by the vagus and is rapidly degraded via the action of acetylcholinesterase (AChE), thereby rendering vagus control responsible for handling higher rates of change in HR.

Recognizing the importance of Ach and AChE, Wei Young developed a frog (*Rana pipiens*) vagal heart model that, among other uses, could be exposed to a magnetic field with subsequent effects being observed and described (Young 1969). Exposure to a static field of 0.27 T resulted in an increase in the hydrolysis of ACh and increased AChE activity. These changes caused the duration of vagal inhibition of the SAN to be lessened. However, when the field was increased to 1.0 T, arrhythmias were often detected. Based on multiple observations, it was speculated that the magnetic field was producing a conformational change in enzyme molecules at low intensity but damage at higher intensity. The precise explanation for these HR effects remains absent, but observations points out the complexity involved in these processes. Figure 7.7 illustrates a reported effect of a 10 pT excitation field at 16 Hz obtained from a pig model and herein drawn based on prior data (Laniado, Kamil et al. 2005). As illustrated, the effect was to increase HR and reduce the PR interval measured on the ECG. Although the mechanisms causing these changes are not defined, the fact that such changes can occur rapidly (within 2 min of application) provides experimental data from which to pursue and investigate such issues.

Under some circumstances, the normal pacing actions are lost when the electrical activity of the atrium becomes chaotic with different parts of the atrium depolarizing rapidly and randomly. One form of this chaotic activity is atrial fibrillation (AF) that is the most widely experienced sustained cardiac arrhythmia (Ellervik, Roselli et al. 2019,

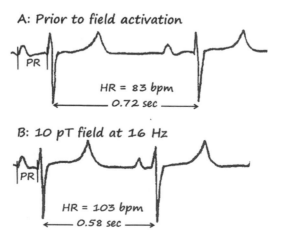

A: Prior to field activation

PR

HR = 83 bpm

0.72 sec

B: 10 pT field at 16 Hz

PR

HR = 103 bpm

0.58 sec

FIGURE 7.7   Illustrating an effect of a 16 Hz field on heart rate and atrial-to-ventricle conduction.

Roberts 2019, Salem, Shoemaker et al. 2019). AF is thought to occur as a consequence of interactions between initiating electrical triggers that impact cardiac tissue that is susceptible to allowing such triggers to cause AF. The resultant AF can be transient and rapidly self-terminating or persistent in need of treatment (medicine or cardioversion or ablation) or a permanent type that is unresponsive to treatment. Ablation treatment, which is invasive, is reported to be 60%–70% effective (Woods and Olgin 2014).

In contrast to invasive interventional procedures, some workers have suggested the use of noninvasive approaches that target the ANS as a way to treat AF and other cardiac arrhythmias (Li, Zhou et al. 2015). The ANS is theorized to be a target for such intervention because of its potential role as a trigger for AF. This being partly related to a β-adrenergic effect that can cause an increase in myocardial automaticity or promote ectopic focal impulses of either the early afterdepolarization (EAD) or delayed afterdepolarization (DAD) types (Chen, Chen et al. 2014). Parasympathetic factors may also be involved (Liu and Nattel 1997).

The neural influences of both sympathetic and parasympathetic nerves on cardiac electrical activity is largely dependent on their interaction with ganglionated plexi that are mostly imbedded in fat pads lying on atrial and ventricular epicardial surfaces (Pauza, Skripka et al. 2000). These plexi contain sympathetic and parasympathetic components with afferent, efferent, and interconnecting neurons (Stavrakis and Po 2017) and are likely involved in initiation and maintenance of AF (Nishida, Datino et al. 2014). Various approaches are there in the medical arsenal to treat this condition, but stoppage of the atrial fibrillation in theory might occur if the atrium, in which these chaotic circulating depolarizations are present, could be rendered transiently refractory, thereby snuffing out the arrhythmia via the application of bio-electromagnetic energy of the proper configuration. An example of an externally applied magnetic field signal, derived from adding together single cycles of two different sinusoidal signals, has been proposed to serve as a leadless magnetic cardiac pacemaker. One form of a signal, shown in Figure 7.8, has been claimed to stabilize

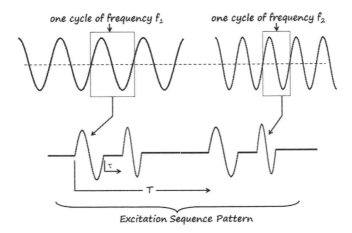

FIGURE 7.8 Illustrating the synthesis of a complex excitation pattern based on the addition of single cycles of differing frequency. (Courtesy of Dr. HN Mayrovitz.)

an erratic heartbeat as assessed in an isolated canine heart (Ramon and Bardy 1992). As drawn herein, the parameters τ and T are subject to variation to best match target tissues. This needs to be done empirically. This pattern is of course but one of many possible signal patterns that may impact biological activity.

CARDIOPROTECTION

The possibility that exposure to low-intensity PEMF could reduce the negative effects of experimental myocardial infarction (MI) was carefully investigated using a rat model (Barzelai, Dayan et al. 2009). Following ligation of the left anterior descending artery, rats were housed either in cages in which the floor was composed of 69 current loops that generated a 16 Hz field of 80–100 nT at their heart level or in nonfield exposed enclosures. Assessment of cardiac function 4 weeks post MI induction showed a greater shortening fraction in rats exposed to the magnetic field and more blood vessels in surviving tissues of the 4-week remodeled myocardium (Barzelai, Dayan et al. 2009). It is unclear if the reported protective effect was related to the increased small blood vessel density.

There is other evidence that exposure to PEMF is more generally associated with angiogenic processes in experimental animals rendered ischemic(Pan, Dong et al. 2013), but it can be stimulatory or inhibitory (Strelczyk, Eichhorn et al. 2009, Wang, Yang et al. 2009, Markov 2010, Delle Monache, Angelucci et al. 2013) possibly dependent on applied field frequency and intensity. For example, a 60 Hz, 2 mT field is reported to diminish new vessel formation potentially through downregulation of the vascular endothelial growth factor (VEGF) pathway (Delle Monache, Angelucci et al. 2013). However, cardioprotective results were also shown for rats in which the left coronary artery had been ligated. Cardiac capillary density 4 weeks post ligation measured in rats treated with 15 Hz fields (6 mT) for 3 h/day was significantly greater than in both rats not treated and those treated at 10 Hz (Yuan, Wei et al. 2010). An accompanying differential elevation in VEGF may have been involved in this process. Furthermore, rats treated with 15 Hz fields had less infarct size and had better cardiac function assessed via maximal rate of change of ventricular pressure. In part based on such findings in experimental animals, various devices have been proposed as ways to promote cardiac angiogenesis, with one approach illustrated in Figure 7.9 based on one of several embodiments in a 2001 patent (March 2001).

In searching for mechanisms, it was hypothesized that myocyte protection may be due to a PEMF-related opening of adenosine-triphosphate-activated potassium channels (K_{ATP}) (Barzelai, Dayan et al. 2009, Fixler, Yitzhaki et al. 2012, Aharonovich, Scheinowitz et al. 2016). These channels are present in myocyte sarcolemma, but most are normally not open (Nichols, Singh et al. 2013). However, reduced blood flow can open these channels causing increased outward K+ flux. This produces a cellular hyperpolarization and a more rapid AP repolarization. This in turn reduces the action potential duration (APD) and also reduces Ca++ entry into the myocyte thereby reducing myocardial contractility. This would conceptually provide a temporary energy conserving process and, if augmented by the application of the PEMF, could help explain the findings.

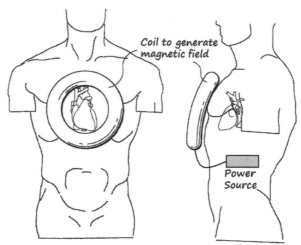

A proposed way to induce cardiac angiogenesis

FIGURE 7.9 Illustrating a proposed method for portable cardiac electromagnetic therapy.

Further insight into a potential linkage between magnetic fields, cardiac K_{ATP} channels, and Ca^{++} currents emerges from studies of rat cardiomyocytes in which cultured cells were exposed for 30 min (16 Hz, 40 nT) and Ca^{++} transients measured every 5 min during exposure (Fixler, Yitzhaki et al. 2012). Exposure was associated with a decrease in peak Ca^{++} transients that, based on K_{ATP} channel blocking agent effects, was concluded to be most likely dependent on field-induced changes in these channels. This mechanistic concept, regarding field-induced alterations in Ca^{++} dynamics, follows elements of general theoretical and applied thoughts and findings of several of the pioneers in the field including Arthur Pilla (Pilla 1974, Pilla and Markov 1994, Pilla 2013), Abraham Liboff and Bruce Mcleod (McLeod and Liboff 1986, Smith, McLeod et al. 1987, Liboff and Parkinson 1991, McLeod, Liboff et al. 1992, Liboff 1997, Vincze, Szasz et al. 2008, Liboff, Poggi et al. 2017), Carl Blackman (Blackman, Benane et al. 1975, 1982, 1988), and others including the esteemed editor of this volume, Marko Markov.

VASCULAR TARGETS

Although there has been little specific work related to coronary vessels, it is likely that information gained from biomagnetic work on peripheral blood vessels will impact future aspects of cardiac applications. There have been many investigations of static magnetic field impacts on biological systems to assess potential impacts of magnets of various designs and magnetic fields of various patterns and intensities on blood cells, vessels, and blood flow. A view held by some is that static magnet field effects are most evident when applied to physiological states that are deviated from normal. Work done by our group using static magnets on healthy persons has for the most part been consistent with that concept. In one example, skin blood flow was measured on the index finger when one hand was exposed to local magnetic fields of about 50 mT at the thenar eminence and about 40 mT along the finger for 36 min with corresponding shams simultaneously applied to the other hand (Mayrovitz, Groseclose et al. 2001). The difference between the sham and magnet flows was not significant at any time point whether measured with laser Doppler flowmetry or laser Doppler imaging. Other experiments did not detect a flow effect (Mayrovitz, Groseclose et al. 2005), whereas experiments using a different magnet intensity showed a clear decrease in skin blood flow (Mayrovitz and Groseclose 2005a). Figures 7.10 and 7.11 illustrate the type of arrangements and data obtained in such experiments. In Figure 7.10 is shown laser Doppler blood flow measurement in the middle finger at two sites with the finger gently resting on a neodymium magnet with a surface field of 0.4 T. The blood flow recordings are shown when the finger is first resting on a sham magnet for 16 min followed by its resting on the magnet for 32 min. Figure 7.11 illustrates simultaneous skin blood flow measurements while one finger (F2) rests on a magnet and another (F4) rests on a sham-like surface. Such protocols have both fingers first resting on the sham, and then one of the shams is replaced with the magnet. The recorded blood flow data shows how similar the skin blood flow in the fingers is and the absence of a clearly observable effect of the magnet in this case.

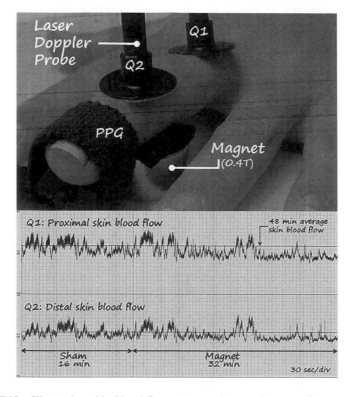

FIGURE 7.10 Illustrating skin blood flow at two points on the same finger exposed to a SMF. (Courtesy of Dr. HN Mayrovitz.)

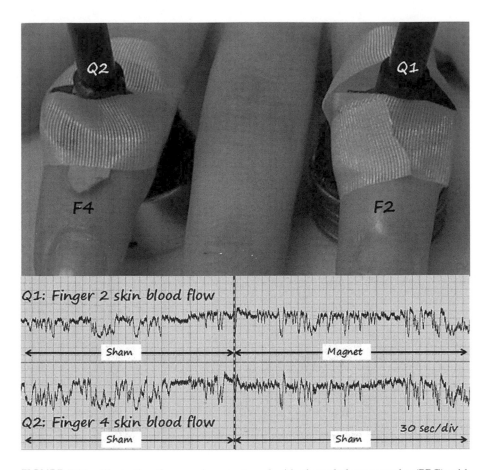

FIGURE 7.11 Illustrating finger pulses measured with photoplethysmography (PPG) with fingers exposed to a vertical SMF or SHAM magnets. The decrease in the PPG signal is due to vasoconstriction attributed to deep inspiration not the magnetic field. (Courtesy of Dr. HN Mayrovitz.)

Figure 7.12 illustrates a fully instrumented method using simultaneous skin blood flow, skin temperature, and photoplethysmography (PPG) using the same finger on opposite hands in which one finger is exposed to a magnet and the other exposed to a sham. From the point of view of set-up and subject convenience, this is a recommended approach. The data shown is the PPG signal demonstrating the similarity of these two hand sides in this parameter. The dramatic decrease in both PPG signals is due to a rapid inhalation taken by the subject. This type of response is due to vasoconstriction caused by the so-called inspiratory gasp (Mayrovitz and Groseclose 2002a,b, 2005b). An example of another configuration is illustrated in Figure 7.13 that depicts skin blood flow recorded in the middle finger before and during exposure to a magnet with a surface field of 0.4 T arranged with the field parallel to the long axis of the finger. The change in the flow pattern over the first 3 min of exposure is noted but not yet explained. Additional descriptions and further details of these types of vascular–magnetic dynamics may be found in several book chapters (Mayrovitz 2004, 2015, 2017).

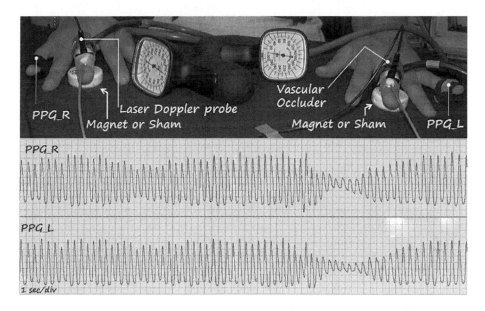

FIGURE 7.12 Illustrating simultaneous finger skin blood flow on two different fingers of the same hand while exposed to a magnet or a sham. (Courtesy of Dr. HN Mayrovitz.)

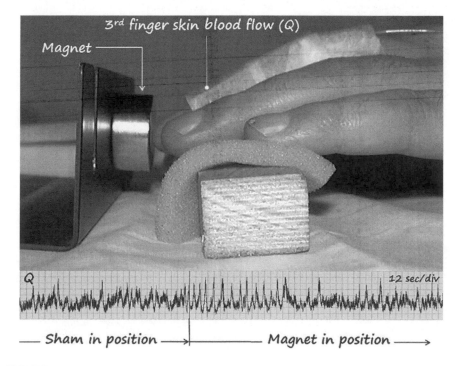

FIGURE 7.13 Illustrating finger dorsum skin blood flow when the finger is exposed to high field intensity acting mainly longitudinally along the finger. (Courtesy of Dr. HN Mayrovitz.)

The impact of time-varying magnetic fields as it relates to possible induced changes in vascular-related processes has also been investigated in our laboratory. One example of this is illustrated in Figure 7.14 that illustrates a method to produce pulsating magnetic fields with varying frequency and intensities at skin target sites. At a fixed rotation speed of the turn table, the number of impulses experienced at the target depends on the number of magnets placed on the surface. The example shows an excitation of 20 magnetic pulses per minute with skin blood flow shown prior to and during the exposure. A slight decrease in skin blood flow of both fingers was noted in this experiment with initiation of pulse exposure.

Another work in our laboratory has focused on the potential utility of bioelectromagnetics for lymphatic vessels especially as it relates to the clinical condition known as lymphedema. One such important application is the use of electromagnetic tissue properties to evaluate the tissue dielectric constant (TDC) to assess the

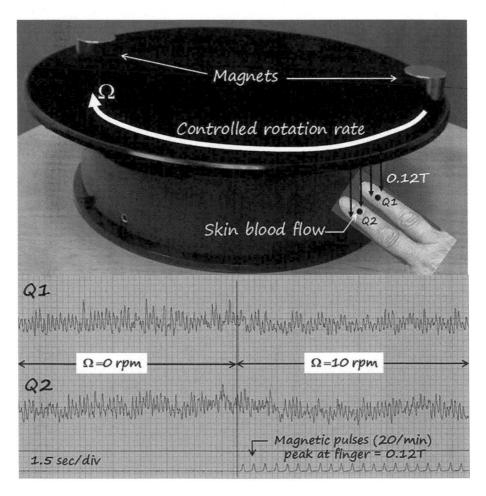

FIGURE 7.14 Illustrating finger skin blood flow while exposed to a time varying magnetic field associated with rotating magnets. (Courtesy of Dr. HN Mayrovitz.)

early presence of lymphedema and to track its natural or treatment-related lymph-edema (Mayrovitz, Weingrad et al. 2014, 2015, Mayrovitz, Mikulka et al. 2018, Mayrovitz and Weingrad 2018). Other clinical applications have shown utility in the assessment of tissue edema in persons with diabetes (Mayrovitz, McClymont et al. 2013, Mayrovitz, Volosko et al. 2017) and as a possible index of the onset of lower-extremity edema in congestive heart failure. Figure 7.15 illustrates the utiliza-tion of this method in which a 300 MHz signal is used to interrogate the tissue in contact with the probe and the reflected electromagnetic energy is used to calculate the TDC as a direct index of tissue water and its change. The measurement principle is based on that of the open-ended coaxial line (Stuchly, Athey et al. 1982, Aimoto and Matsumoto 1996, Gabriel, Lau et al. 1996). Panels A through D illustrate typical measurement sites using two different size probes with the larger probe measuring to a greater depth. The bar graph in Figure 7.15 shows TDC values for different depth measurements on the thigh made 1 month apart in a group of volunteer subjects. It demonstrates that TDC values at certain sites decrease with increasing depth due to decreasing water content but may remain stable over time.

An additional application relevant to lymphedema is the possibility of using PEMF to assist with therapy. Work in this area relates to the potential effects of PEMF on both vascular and lymphatic networks. One study, using PEMF to modulate a

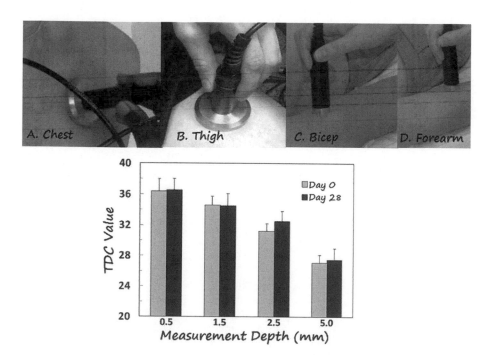

FIGURE 7.15   Illustrating the use of a 300 Hz probe to measure the skin-to-fat tissue dielec-tric constant (TDC) at various tissue depths. This is a unique application of bioelectromag-netic fields to measure tissue properties mainly related to changes in water content especially in cases of edema and lymphedema. (Courtesy of Dr. HN Mayrovitz.)

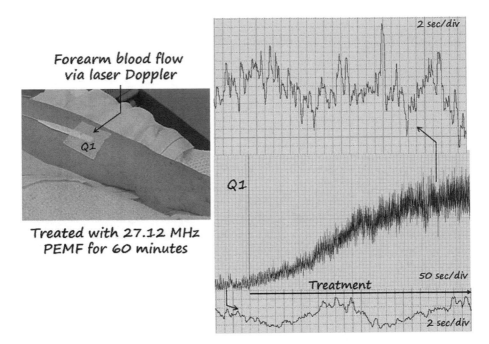

Forearm blood flow via laser Doppler

Treated with 27.12 MHz PEMF for 60 minutes

Q1

2 sec/div

Q1

Treatment

50 sec/div

2 sec/div

FIGURE 7.16 Illustrating the effect of PEMF at 27.12 MHz in patients with arm lymphedema as a result of treatment for breast cancer. The top and bottom parts of the figure show a magnified expanded time scale corresponding to the time points shown by the arrows. The elevated blood flow during treatment is significant. (Courtesy of Dr. HN Mayrovitz.)

27.12 MHz carrier, showed significant improvement in women with breast-cancer-related lymphedema (Mayrovitz, Sims et al. 2002). Figure 7.16 shows a typical blood flow response. Though not measured, it is believed that lymphatic flow may proceed in a similar way. One theory herein offered is that the increased blood pulsations, shown occurring with progressive treatment time, serve as stimuli to promote augmentation of lymphatic emptying thereby accounting for the observed reduction in lymphedema. Further research in this area is needed to verify and further extend this approach.

REFERENCES

Abe, K., H. Hamada, Y. J. Chen, A. Abe, H. Watanabe, Y. Fujiki, H. Yoshikawa, T. Murakami and H. Horigome (2005). "Successful management of supraventricular tachycardia in a fetus using fetal magnetocardiography." *Fetal Diagn Ther* **20**(5): 459–462.

Aharonovich, Y., M. Scheinowitz and S. Zlochiver (2016). "Cardiac KATP channel modulation by 16Hz magnetic fields - A theoretical study." *Conf Proc IEEE Eng Med Biol Soc* **2016**: 161–164.

Aimoto, A. and T. Matsumoto (1996). "Noninvasive method for measuring the electrical properties of deep tissues using an open-ended coaxial probe." *Med Eng Phys* **18**(8): 641–646.

Alabdulgader, A., R. McCraty, M. Atkinson, Y. Dobyns, A. Vainoras, M. Ragulskis and V. Stolc (2018). "Long-term study of heart rate variability responses to changes in the solar and geomagnetic environment." *Sci Rep* **8**(1): 2663.

Bali, R. and U. Awasthi (2011). "Mathematical model of blood flow in small blood vessel in the presence of magnetic field." *Applied Mathematics* **2**: 264–269.

Barach, J. P. and J. P. Wikswo, Jr. (1994). "Magnetic fields from simulated cardiac action currents." *IEEE Trans Biomed Eng* **41**(10): 969–974.

Barisano, G., F. Sepehrband, S. Ma, K. Jann, R. Cabeen, D. J. Wang, A. W. Toga and M. Law (2018). "Clinical 7 T MRI: Are we there yet? A review about magnetic resonance imaging at ultra-high field." *Br J Radiol* **92**(1094): 20180492.

Barry, R. L., S. J. Vannesjo, S. By, J. C. Gore and S. A. Smith (2018). "Spinal cord MRI at 7T." *Neuroimage* **168**: 437–451.

Barry, W. H., W. M. Fairbank, D. C. Harrison, K. L. Lehrman, J. A. Malmivuo and J. P. Wikswo, Jr. (1977). "Measurement of the human magnetic heart vector." *Science* **198**(4322): 1159–1162.

Barzelai, S., A. Dayan, M. S. Feinberg, R. Holbova, S. Laniado and M. Scheinowitz (2009). "Electromagnetic field at 15.95–16 Hz is cardio protective following acute myocardial infarction." *Ann Biomed Eng* **37**(10): 2093–2104.

Batie, M., S. Bitant, J. F. Strasburger, V. Shah, O. Alem and R. T. Wakai (2018). "Detection of fetal arrhythmia using optically-pumped magnetometers." *JACC Clin Electrophysiol* **4**(2): 284–287.

Baule, G. and R. McFee (1963). "Detection of the magnetic field of the heart." *Am Heart J* **66**: 95–96.

Beischer, D. E. and J. C. Knepton, Jr. (1964). "Influence of strong magnetic fields on the electrocardiogram of squirrel monkeys (Saimiri Sciureus)." *Aerosp Med* **35**: 939–944.

Blackman, C. F., S. G. Benane, D. J. Elliott, D. E. House and M. M. Pollock (1988). "Influence of electromagnetic fields on the efflux of calcium ions from brain tissue in vitro: a three-model analysis consistent with the frequency response up to 510 Hz." *Bioelectromagnetics* **9**(3): 215–227.

Blackman, C. F., S. G. Benane, L. S. Kinney, W. T. Joines and D. E. House (1982). "Effects of ELF fields on calcium-ion efflux from brain tissue in vitro." *Radiat Res* **92**(3): 510–520.

Blackman, C. F., S. G. Benane, C. M. Weil and J. S. Ali (1975). "Effects of nonionizing electromagnetic radiation on single-cell biologic systems." *Ann N Y Acad Sci* **247**: 352–366.

Brockmeier, K., L. Schmitz, J. D. Bobadilla Chavez, M. Burghoff, H. Koch, R. Zimmermann and L. Trahms (1997). "Magnetocardiography and 32-lead potential mapping: repolarization in normal subjects during pharmacologically induced stress." *J Cardiovasc Electrophysiol* **8**(6): 615–626.

Cabrerizo, M., A. Cabrera, J. O. Perez, J. de la Rua, N. Rojas, Q. Zhou, A. Pinzon-Ardila, S. M. Gonzalez-Arias and M. Adjouadi (2014). "Induced effects of transcranial magnetic stimulation on the autonomic nervous system and the cardiac rhythm." *Sci World J* **2014**: 349718.

Chakeres, D. W., A. Kangarlu, H. Boudoulas and D. C. Young (2003). "Effect of static magnetic field exposure of up to 8 Tesla on sequential human vital sign measurements." *J Magn Reson Imaging* **18**(3): 346–352.

Chen, I. H. and S. Saha (1984). "Analysis of an intensive magnetic filed on blood flow." *J Bioelectricity* **3**(1 & 2): 293–298.

Chen, P. S., L. S. Chen, M. C. Fishbein, S. F. Lin and S. Nattel (2014). "Role of the autonomic nervous system in atrial fibrillation: Pathophysiology and therapy." *Circ Res* **114**(9): 1500–1515.

Cohen, D. and L. A. Kaufman (1975). "Magnetic determination of the relationship between the S-T segment shift and the injury current produced by coronary artery occlusion." *Circ Res* **36**(3): 414–424.

Cohen, D., J. C. Norman, F. Molokhia and W. Hood, Jr. (1971). "Magnetocardiography of direct currents: S-T segment and baseline shifts during experimental myocardial infarction." *Science* **172**(3990): 1329–1333.

Cohen, D., P. Savard, R. D. Rifkin, E. Lepeschkin and W. E. Strauss (1983). "Magnetic measurement of S-T and T-Q segment shifts in humans. Part II: Exercise-induced S-T segment depression." *Circ Res* **53**(2): 274–279.

Coulton, L. A., A. T. Barker, J. E. Van Lierop and M. P. Walsh (2000). "The effect of static magnetic fields on the rate of calcium/calmodulin-dependent phosphorylation of myosin light chain." *Bioelectromagnetics* **21**(3): 189–196.

Czapski, P., C. Ramon, L. L. Huntsman, G. H. Bardy and Y. Kim (1996). "Effects of tissue conductivity variations on the cardiac magnetic fields simulated with a realistic heart-torso model." *Phys Med Biol* **41**(8): 1247–1263.

Delle Monache, S., A. Angelucci, P. Sanita, R. Iorio, F. Bennato, F. Mancini, G. Gualtieri and R. C. Colonna (2013). "Inhibition of angiogenesis mediated by extremely low-frequency magnetic fields (ELF-MFs)." *PLoS One* **8**(11): e79309.

Derkacz, A., J. Gawrys, K. Gawrys, M. Podgorski, A. Magott-Derkacz, R. Poreba and A. Doroszko (2018). "Effect of electromagnetic field accompanying the magnetic resonance imaging on human heart rate variability - a pilot study." *Int J Inj Contr Saf Promot* **25**(2): 229–231.

DiFrancesco, D. (2010). "The role of the funny current in pacemaker activity." *Circ Res* **106**(3): 434–446.

Ellervik, C., C. Roselli, I. E. Christophersen, A. Alonso, M. Pietzner, C. M. Sitlani, S. Trompet, D. E. Arking, B. Geelhoed, X. Guo, M. E. Kleber, H. J. Lin, H. Lin, P. MacFarlane, E. Selvin, C. Shaffer, A. V. Smith, N. Verweij, S. Weiss, A. R. Cappola, M. Dorr, V. Gudnason, S. Heckbert, S. Mooijaart, W. Marz, B. M. Psaty, P. M. Ridker, D. Roden, D. J. Stott, H. Volzke, E. J. Benjamin, G. Delgado, P. Ellinor, G. Homuth, A. Kottgen, J. W. Jukema, S. A. Lubitz, S. Mora, M. Rienstra, J. I. Rotter, M. B. Shoemaker, N. Sotoodehnia, K. D. Taylor, P. van der Harst, C. M. Albert and D. I. Chasman (2019). "Assessment of the relationship between genetic determinants of thyroid function and atrial fibrillation: A Mendelian randomization study." *JAMA Cardiol* **4**(2): 144–152.

Escalona-Vargas, D., H. T. Wu, M. G. Frasch and H. Eswaran (2018). "A comparison of five algorithms for fetal magnetocardiography signal extraction." *Cardiovasc Eng Technol* **9**(3): 483–487.

Fixler, D., S. Yitzhaki, A. Axelrod, T. Zinman and A. Shainberg (2012). "Correlation of magnetic AC field on cardiac myocyte Ca(2+) transients at different magnetic DC levels." *Bioelectromagnetics* **33**(8): 634–640.

Fujino, K., M. Sumi, K. Saito, M. Murakami, T. Higuchi, Y. Nakaya and H. Mori (1984). "Magnetocardiograms of patients with left ventricular overloading recorded with a second-derivative SQUID gradiometer." *J Electrocardiol* **17**(3): 219–228.

Gabriel, S., R. W. Lau and C. Gabriel (1996). "The dielectric properties of biological tissues: II. Measurements in the frequency range 10 Hz to 20 GHz." *Phys Med Biol* **41**(11): 2251–2269.

Gapelyuk, A., A. Schirdewan, R. Fischer and N. Wessel (2010). "Cardiac magnetic field mapping quantified by Kullback-Leibler entropy detects patients with coronary artery disease." *Physiol Meas* **31**(10): 1345–1354.

Gapelyuk, A., N. Wessel, R. Fischer, U. Zacharzowsky, L. Koch, D. Selbig, H. Schutt, B. Sawitzki, F. C. Luft, R. Dietz and A. Schirdewan (2007). "Detection of patients with coronary artery disease using cardiac magnetic field mapping at rest." *J Electrocardiol* **40**(5): 401–407.

Geselowitz, D. B. (1973). "Electric and magnetic field of the heart." *Annu Rev Biophys Bioeng* **2**: 37–64.

Geselowitz, D. B. (1979). "Magnetocardiography: An overview." *IEEE Trans Biomed Eng* **26**(9): 497–504.

Geurts, L. J., A. A. Bhogal, J. C. W. Siero, P. R. Luijten, G. J. Biessels and J. J. M. Zwanenburg (2018). "Vascular reactivity in small cerebral perforating arteries with 7T phase contrast MRI - A proof of concept study." *Neuroimage* **172**: 470–477.

Geurts, L. J., J. J. M. Zwanenburg, C. J. M. Klijn, P. R. Luijten and G. J. Biessels (2018). "Higher pulsatility in cerebral perforating arteries in patients with small vessel disease related stroke, a 7T MRI study." *Stroke*: STROKEAHA118022516. doi: 10.1161/STROKEAHA.118.022516.

Ghadimi-Moghadam, A., S. M. J. Mortazavi, A. Hosseini-Moghadam, M. Haghani, S. Taeb, M. A. Hosseini, N. Rastegariyan, F. Arian, L. Sanipour, S. Aghajari, S. A. R. Mortazavi, A. Soofi and M. R. Dizavandi (2018). "Does exposure to static magnetic fields generated by magnetic resonance imaging scanners raise safety problems for personnel?" *J Biomed Phys Eng* **8**(3): 333–336.

Haik, Y., V. Pai and C.-J. C. Chen (2001). "Apparent viscosity of human blood in a high static magnetic field." *J Magn Magn Mater* **225**: 180–186.

Higashi, T., N. Asjoda and T. Takeuchi (1997). "Orientation of blood cells in static magnetic field." *Physica B* **237–238**: 616–620.

Horigome, H., M. I. Takahashi, M. Asaka, S. Shigemitsu, A. Kandori and K. Tsukada (2000). "Magnetocardiographic determination of the developmental changes in PQ, QRS and QT intervals in the foetus." *Acta Paediatr* **89**(1): 64–67.

Hosono, T., M. Shinto, Y. Chiba, A. Kandori and K. Tsukada (2002). "Prenatal diagnosis of fetal complete atrioventricular block with QT prolongation and alternating ventricular pacemakers using multi-channel magnetocardiography and current-arrow maps." *Fetal Diagn Ther* **17**(3): 173–176.

Jarusevicius, G., T. Rugelis, R. McCraty, M. Landauskas, K. Berskiene and A. Vainoras (2018). "Correlation between changes in local earth's magnetic field and cases of acute myocardial infarction." *Int J Environ Res Public Health* **15**(3). doi: 10.3390/ijerph15030399.

Jehenson, P., D. Duboc, T. Lavergne, L. Guize, F. Guerin, M. Degeorges and A. Syrota (1988). "Change in human cardiac rhythm induced by a 2-T static magnetic field." *Radiology* **166**(1 Pt 1): 227–230.

Kandori, A., T. Hosono, Y. Chiba, M. Shinto, S. Miyashita, M. Murakami, T. Miyashita, K. Ogata and K. Tsukada (2003). "Classifying cases of fetal Wolff-Parkinson-White syndrome by estimating the accessory pathway from fetal magnetocardiograms." *Med Biol Eng Comput* **41**(1): 33–39.

Kandori, A., T. Hosono, T. Kanagawa, S. Miyashita, Y. Chiba, M. Murakami, T. Miyashita and K. Tsukada (2002). "Detection of atrial-flutter and atrial-fibrillation waveforms by fetal magnetocardiogram." *Med Biol Eng Comput* **40**(2): 213–217.

Kassab, G. S., J. A. Navia and Y. Huo (2013). Devices, systems and methods for pacing, resynchonization and defibrillation therapy. US Patent 8,396,566. USA, CVDevices. 8396566.

Kawakami, S., H. Takaki, S. Hashimoto, Y. Kimura, T. Nakashima, T. Aiba, K. F. Kusano, S. Kamakura, S. Yasuda and M. Sugimachi (2016). "Utility of high-resolution magnetocardiography to predict later cardiac events in nonischemic cardiomyopathy patients with normal QRS duration." *Circ J* **81**(1): 44–51.

Kiefer-Schmidt, I., M. Lim, H. Preissl, R. Draganova, M. Weiss, H. Abele, K. O. Kagan and J. Henes (2014). "Fetal magnetocardiography (fMCG) to monitor cardiac time intervals in fetuses at risk for isoimmune AV block." *Lupus* **23**(9): 919–925.

Kimura, Y., H. Takaki, Y. Y. Inoue, Y. Oguchi, T. Nagayama, T. Nakashima, S. Kawakami, S. Nagase, T. Noda, T. Aiba, W. Shimizu, S. Kamakura, M. Sugimachi, S. Yasuda, H. Shimokawa and K. Kusano (2017). "Isolated late activation detected by magnetocardiography predicts future lethal ventricular arrhythmic events in patients with arrhythmogenic right ventricular cardiomyopathy." *Circ J* **82**(1): 78–86.

Kinouchi, Y., H. Yamaguchi and T. S. Tenforde (1996). "Theoretical analysis of magnetic field interactions with aortic blood flow." *Bioelectromagnetics* **17**(1): 21–32.

Kolin, A., N. Assali, G. Herrold and R. Jensen (1957). "Electromagnetic determination of regional blood flow in unanesthetized animals." *Proc Natl Acad Sci USA* **43**(6): 527–540.

Koppel, T., I. Vilcane, M. Carlberg, P. Tint, R. Priiman, K. Riisik, H. Haldre and L. Visnapuu (2015). "The effect of static magnetic field on heart rate variability – an experimental study." *Agron Res* **13**(3): 765–774.

Kurokawa, Y., H. Nitta, H. Imai and M. Kabuto (2003). "Can extremely low frequency alternating magnetic fields modulate heart rate or its variability in humans?" *Auton Neurosci* **105**(1): 53–61.

Laniado, S., Z. Kamil and E. Nhaissi (2005). Method and apparatus for non-invasive therapy of cardiovascular ailments using weak pulsed electromagnetic radiation. US Patent 02,22,625 A1.

Li, F., Y. Yuan, Y. Guo, N. Liu, D. Jing, H. Wang and W. Guo (2015). "Pulsed magnetic field accelerate proliferation and migration of cardiac microvascular endothelial cells." *Bioelectromagnetics* **36**(1): 1–9.

Li, S., X. Zhou, L. Yu and H. Jiang (2015). "Low level non-invasive vagus nerve stimulation: a novel feasible therapeutic approach for atrial fibrillation." *Int J Cardiol* **182**: 189–190.

Li, X. and J. V. Rispoli (2019). "Toward 7T breast MRI clinical study: Safety assessment using simulation of heterogeneous breast models in RF exposure." *Magn Reson Med* **81**(2): 1307–1321.

Liboff, A. R. (1997). "Electric-field ion cyclotron resonance." *Bioelectromagnetics* **18**(1): 85–87.

Liboff, A. R. and W. C. Parkinson (1991). "Search for ion-cyclotron resonance in an Na(+)-transport system." *Bioelectromagnetics* **12**(2): 77–83.

Liboff, A. R., C. Poggi and P. Pratesi (2017). "Weak low-frequency electromagnetic oscillations in water." *Electromagn Biol Med* **36**(2): 154–157.

Liu, L. and S. Nattel (1997). "Differing sympathetic and vagal effects on atrial fibrillation in dogs: Role of refractoriness heterogeneity." *Am J Physiol* **273**(2 Pt 2): H805–816.

Loomba, R. S. and R. Arora (2009). "ST elevation myocardial infarction guidelines today: A systematic review exploring updated ACC/AHA STEMI guidelines and their applications." *Am J Ther* **16**(5): e7–e13.

Lu, X. W., L. Du, L. Kou, N. Song, Y. J. Zhang, M. K. Wu and J. F. Shen (2015). "Effects of moderate static magnetic fields on the voltage-gated sodium and calcium channel currents in trigeminal ganglion neurons." *Electromagn Biol Med* **34**(4): 285–292.

March, K. L. (2001). Methods of treating cardiovascular disease by angiogenesis. U.S. Patent 6,200,259.

Markenroth Bloch, K., J. Toger and F. Stahlberg (2018). "Investigation of cerebrospinal fluid flow in the cerebral aqueduct using high-resolution phase contrast measurements at 7T MRI." *Acta Radiol* **59**(8): 988–996.

Markov, M. S. (2010). "Angiogenesis, magnetic fields and 'window effects'." *Cardiology* **117**(1): 54–56.

Mayrovitz, H. N. (2004). Electromagentic linkages in soft tissue wound healing. In P. J. Rosch and M. Markov, *Bioelectric Medicine*. New York, Marcel Dekker: pp. 461–483.

Mayrovitz, H. N. (2015). Electromagnetic fields for soft tissue wound healing. In M. S. Markov, *Electromagnetic Fields in Biology and Medicine*. Boca Raton, FL, CRC Press: pp. 231–251.

Mayrovitz, H. N. (2017). Blood and vascular targets for magnetic dosing. In M. S. Markov, *Dosimetry in Bioelectromagnetics*. Boca Raton, FL, CRC Press: pp. 285–313.

Mayrovitz, H. N. and E. E. Groseclose (2002a). "Inspiration-induced vascular responses in finger dorsum skin." *Microvasc Res* **63**(2): 227–232.

Mayrovitz, H. N. and E. E. Groseclose (2002b). "Neurovascular responses to sequential deep inspirations assessed via laser-Doppler perfusion changes in dorsal finger skin." *Clin Physiol Funct Imaging* **22**(1): 49–54.

Mayrovitz, H. N. and E. E. Groseclose (2005a). "Effects of a static magnetic field of either polarity on skin microcirculation." *Microvasc Res* **69**(1–2): 24–27.

Mayrovitz, H. N. and E. E. Groseclose (2005b). "Inspiration-induced vasoconstrictive responses in dominant versus non-dominant hands." *Clin Physiol Funct Imaging* **25**(2): 69–74.

Mayrovitz, H. N., E. E. Groseclose and D. King (2005). "No effect of 85 mT permanent magnets on laser-Doppler measured blood flow response to inspiratory gasps." *Bioelectromagnetics* **26**(4): 331–335.

Mayrovitz, H. N., E. E. Groseclose, M. Markov and A. A. Pilla (2001). "Effects of permanent magnets on resting skin blood perfusion in healthy persons assessed by laser Doppler flowmetry and imaging." *Bioelectromagnetics* **22**(7): 494–502.

Mayrovitz, H. N., A. McClymont and N. Pandya (2013). "Skin tissue water assessed via tissue dielectric constant measurements in persons with and without diabetes mellitus." *Diabetes Technol Ther* **15**(1): 60–65.

Mayrovitz, H. N., A. Mikulka and D. Woody (2018). "Minimum detectable changes associated with tissue dielectric constant measurements as applicable to assessing lymphedema status." *Lymphat Res Biol* **17**(3): 322–328.

Mayrovitz, H. N., N. Sims and J. M. Macdonald (2002). "Effects of pulsed radio frequency diathermy on postmastectomy arm lymphedema and skin blood flow: A pilot investigation." *Lymphology* **35** (suppl): 353–356.

Mayrovitz, H. N., I. Volosko, B. Sarkar and N. Pandya (2017). "Arm, leg, and foot skin water in persons with diabetes mellitus (DM) in relation to HbA1c assessed by tissue dielectric constant (TDC) technology measured at 300 MHz." *J Diabetes Sci Technol* **11**(3): 584–589.

Mayrovitz, H. N. and D. N. Weingrad (2018). "Tissue dielectric constant ratios as a method to characterize truncal lymphedema." *Lymphology* **51**(3): 125–131.

Mayrovitz, H. N., D. N. Weingrad and S. Davey (2014). "Tissue dielectric constant (TDC) measurements as a means of characterizing localized tissue water in arms of women with and without breast cancer treatment related lymphedema." *Lymphology* **47**(3): 142–150.

Mayrovitz, H. N., D. N. Weingrad and L. Lopez (2015). "Patterns of temporal changes in tissue dielectric constant as indices of localized skin water changes in women treated for breast cancer: a pilot study." *Lymphat Res Biol* **13**(1): 20–32.

McBride, K. K., B. J. Roth, V. Y. Sidorov, J. P. Wikswo and F. J. Baudenbacher (2010). "Measurements of transmembrane potential and magnetic field at the apex of the heart." *Biophys J* **99**(10): 3113–3118.

McCraty, R., M. Atkinson, V. Stolc, A. A. Alabdulgader, A. Vainoras and M. Ragulskis (2017). "Synchronization of human autonomic nervous system rhythms with geomagnetic activity in human subjects." *Int J Environ Res Public Health* **14**(7). doi: 10.3390/ijerph14070770.

McLeod, B. R. and A. R. Liboff (1986). "Dynamic characteristics of membrane ions in multifield configurations of low-frequency electromagnetic radiation." *Bioelectromagnetics* **7**(2): 177–189.

McLeod, B. R., A. R. Liboff and S. D. Smith (1992). "Electromagnetic gating in ion channels." *J Theor Biol* **158**(1): 15–31.

Moraes, E. R., L. O. Murta, O. Baffa, R. T. Wakai and S. Comani (2012). "Linear and nonlinear measures of fetal heart rate patterns evaluated on very short fetal magnetocardiograms." *Physiol Meas* **33**(10): 1563–1583.

Muehsam, D. J. and A. A. Pilla (2009). "A Lorentz model for weak magnetic field bioeffects: part II--secondary transduction mechanisms and measures of reactivity." *Bioelectromagnetics* **30**(6): 476–488.

Nichols, C. G., G. K. Singh and D. K. Grange (2013). "KATP channels and cardiovascular disease: suddenly a syndrome." *Circ Res* **112**(7): 1059–1072.

Nishida, K., T. Datino, L. Macle and S. Nattel (2014). "Atrial fibrillation ablation: Translating basic mechanistic insights to the patient." *J Am Coll Cardiol* **64**(8): 823–831.

Okazaki, M., N. Maeda and T. Shiga (1987). "Effects of an inhomogeneous magnetic field on flowing erythrocytes." *Eur Biophys J* **14**(3): 139–145.

Pagani, M., F. Lombardi, S. Guzzetti, O. Rimoldi, R. Furlan, P. Pizzinelli, G. Sandrone, G. Malfatto, S. Dell'Orto, E. Piccaluga and et al. (1986). "Power spectral analysis of heart rate and arterial pressure variabilities as a marker of sympatho-vagal interaction in man and conscious dog." *Circ Res* **59**(2): 178–193.

Pan, Y., Y. Dong, W. Hou, Z. Ji, K. Zhi, Z. Yin, H. Wen and Y. Chen (2013). "Effects of PEMF on microcirculation and angiogenesis in a model of acute hindlimb ischemia in diabetic rats." *Bioelectromagnetics* **34**(3): 180–188.

Park, C. A., C. K. Kang, Y. B. Kim and Z. H. Cho (2018). "Advances in MR angiography with 7T MRI: From microvascular imaging to functional angiography." *Neuroimage* **168**: 269–278.

Pauza, D. H., V. Skripka, N. Pauziene and R. Stropus (2000). "Morphology, distribution, and variability of the epicardiac neural ganglionated subplexuses in the human heart." *Anat Rec* **259**(4): 353–382.

Pfurtscheller, G., A. Schwerdtfeger, D. Fink, C. Brunner, C. S. Aigner, J. Brito and A. Andrade (2018). "MRI-related anxiety in healthy individuals, intrinsic BOLD oscillations at 0.1 Hz in precentral gyrus and insula, and heart rate variability in low frequency bands." *PLoS One* **13**(11): e0206675.

Pilla, A. A. (1974). "Electrochemical information transfer at living cell membranes." *Ann N Y Acad Sci* **238**: 149–170.

Pilla, A. A. (2013). "Nonthermal electromagnetic fields: From first messenger to therapeutic applications." *Electromagn Biol Med* **32**(2): 123–136.

Pilla, A. A. and M. S. Markov (1994). "Bioeffects of weak electromagnetic fields." *Rev Environ Health* **10**(3–4): 155–169.

Pilla, A. A., D. J. Muehsam, M. S. Markov and B. F. Sisken (1999). "EMF signals and ion/ligand binding kinetics: Prediction of bioeffective waveform parameters." *Bioelectrochem Bioenerg* **48**(1): 27–34.

Plonsey, R. (1972). "Capability and limitations of electrocardiography and magnetocardiography." *IEEE Trans Biomed Eng* **19**(3): 239–244.

Ramon, C. and G. H. Bardy (1992). Leadless Magnetic Cardiac Pacemaker. U.S. Patent 5,170,784.

Reno, V. R. and D. E. Beischer (1966). "Cardiac excitability in high magnetic fields." *Aerosp Med* **37**(12): 1229–1232.

Roberts, J. D. (2019). "Thyroid function and the risk of atrial fibrillation: Exploring potentially causal relationships through Mendelian randomization." *JAMA Cardiol* **4**(2): 97–99.

Roth, B. J. and J. P. Wikswo, Jr. (1986). "Electrically silent magnetic fields." *Biophys J* **50**(4): 739–745.

Salem, J. E., M. B. Shoemaker, L. Bastarache, C. M. Shaffer, A. M. Glazer, B. Kroncke, Q. S. Wells, M. Shi, P. Straub, G. P. Jarvik, E. B. Larson, D. R. Velez Edwards, T. L. Edwards, L. K. Davis, H. Hakonarson, C. Weng, D. Fasel, B. C. Knollmann, T. J. Wang,

J. C. Denny, P. T. Ellinor, D. M. Roden and J. D. Mosley (2019). "Association of thyroid function genetic predictors with atrial fibrillation: A phenome-wide association study and inverse-variance weighted average meta-analysis." *JAMA Cardiol* **4**(2): 136–143.

Savard, P., D. Cohen, E. Lepeschkin, B. N. Cuffin and J. E. Madias (1983). "Magnetic measurement of S-T and T-Q segment shifts in humans. Part I: Early repolarization and left bundle branch block." *Circ Res* **53**(2): 264–273.

Scheinowitz, M., E. Nhaissi, E. Levine and D. Giler (2018). Apparatus for non-invasive therapy of biological tissue using directed magnetic beams. US Patent 10092769 B2.

Schirdewan, A., A. Gapelyuk, R. Fischer, L. Koch, H. Schutt, U. Zacharzowsky, R. Dietz, L. Thierfelder and N. Wessel (2007). "Cardiac magnetic field map topology quantified by Kullback-Leibler entropy identifies patients with hypertrophic cardiomyopathy." *Chaos* **17**(1): 015118.

Schwartz, J. L., D. E. House and G. A. Mealing (1990). "Exposure of frog hearts to CW or amplitude-modulated VHF fields: Selective efflux of calcium ions at 16 Hz." *Bioelectromagnetics* **11**(4): 349–358.

Sert, C., Z. Akti, O. Sirmatel and R. Yilmaz (2010). "An investigation of the heart rate, heart rate variability, cardiac ions, troponin-I and CK-MB in men exposed to 1.5 T constant magnetic fields." *Gen Physiol Biophys* **29**(3): 282–287.

Shaffer, F. and J. P. Ginsberg (2017). "An overview of heart rate variability metrics and norms." *Front Public Health* **5**: 258.

Shin, E. S., J. W. Park and D. S. Lim (2018a). "Magnetocardiography for the diagnosis of non-obstructive coronary artery disease1." *Clin Hemorheol Microcirc* **69**(1–2): 9–11.

Shin, E. S., S. G. Park, A. Saleh, Y. Y. Lam, J. Bhak, F. Jung, S. Morita and J. Brachmann (2018b). "Magnetocardiography scoring system to predict the presence of obstructive coronary artery disease." *Clin Hemorheol Microcirc* **70**(4): 365–373.

Shirai, Y., K. Hirao, T. Shibuya, S. Okawa, Y. Hasegawa, Y. Adachi, K. Sekihara and S. Kawabata (2018). "Magnetocardiography using a magnetoresistive sensor array." *Int Heart J* **60**(1): 50–54.

Smith, F. E., P. Langley, P. van Leeuwen, B. Hailer, L. Trahms, U. Steinhoff, J. P. Bourke and A. Murray (2006). "Comparison of magnetocardiography and electrocardiography: a study of automatic measurement of dispersion of ventricular repolarization." *Europace* **8**(10): 887–893.

Smith, S. D., B. R. McLeod, A. R. Liboff and K. Cooksey (1987). "Calcium cyclotron resonance and diatom mobility." *Bioelectromagnetics* **8**(3): 215–227.

Sorbo, A. R., G. Lombardi, L. La Brocca, G. Guida, R. Fenici and D. Brisinda (2018). "Unshielded magnetocardiography: Repeatability and reproducibility of automatically estimated ventricular repolarization parameters in 204 healthy subjects." *Ann Noninvasive Electrocardiol* **23**(3): e12526.

Stavrakis, S. and S. Po (2017). "Ganglionated plexi ablation: Physiology and clinical applications." *Arrhythm Electrophysiol Rev* **6**(4): 186–190.

Stingl, K., H. Paulsen, M. Weiss, H. Preissl, H. Abele, R. Goelz and A. Wacker-Gussmann (2013). "Development and application of an automated extraction algorithm for fetal magnetocardiography - normal data and arrhythmia detection." *J Perinat Med* **41**(6): 725–734.

Stratbucker, R. A., C. M. Hyde and S. E. Wixson (1963). "The magnetocardiogram--A new approach to the fields surrounding the heart." *IEEE Trans Biomed Eng* **10**: 145–149.

Strelczyk, D., M. E. Eichhorn, S. Luedemann, G. Brix, M. Dellian, A. Berghaus and S. Strieth (2009). "Static magnetic fields impair angiogenesis and growth of solid tumors in vivo." *Cancer Biol Ther* **8**(18): 1756–1762.

Stuchly, M. A., T. W. Athey, G. M. Samaras and G. E. Taylor (1982). "Measurement of radio frequency permittivity of biological tissues with an open-ended coaxial line: Part II - Experimental results." *IEEE Trans Microwave Therory Tech* **30**(1): 87–92.

Sud, V. K. and G. S. Sekhon (1989). "Blood flow through the human arterial system in the presence of a steady magnetic field." *Phys Med Biol* **34**(7): 795–805.

Takeuchi, T., T. Mizuno, T. Higashi, A. Yamagishi and M. Date (1995). "Orientation of red blood cells in high magnetic filed." *J Magn Magn Mater* **140–144**: 1462–1463.

Tao, P., X. Wu, L. Zhao, F. Wang, Q. Xia, Y. Peng and B. Gao (2017). "[Effect of high-intensity alternating magnetic field on viscosity of sheep blood]." *Zhong Nan Da Xue Xue Bao Yi Xue Ban* **42**(12): 1395–1400.

Tao, R. and K. Huang (2011). "Reducing blood viscosity with magnetic fields." *Phys Rev E Stat Nonlin Soft Matter Phys* **84**(1 Pt 1): 011905.

Tao, R., S. Zhang, X. Huang, M. Tao, J. Ma, S. Ma, C. Zhang, T. Zhang, F. Tang, J. Lu, C. Shen and X. Xie (2018). "Magnetocardiography based ischemic heart disease detection and localization using machine learning methods." *IEEE Trans Biomed Eng* **66**(6): 1658–1667.

Tasic, T., D. M. Djordjevic, S. R. De Luka, A. M. Trbovich and N. Japundzic-Zigon (2017). "Static magnetic field reduces blood pressure short-term variability and enhances baroreceptor reflex sensitivity in spontaneously hypertensive rats." *Int J Radiat Biol* **93**(5): 527–534.

Van Hare, G. F. (2013). "Magnetocardiography in the diagnosis of fetal arrhythmias." *Heart Rhythm* **10**(8): 1199–1200.

Van Leeuwen, P., B. Hailer, S. Lange, A. Klein, D. Geue, K. Seybold, C. Poplutz and D. Gronemeyer (2008). "Quantification of cardiac magnetic field orientation during ventricular de- and repolarization." *Phys Med Biol* **53**(9): 2291–2301.

Vincze, G., A. Szasz and A. R. Liboff (2008). "New theoretical treatment of ion resonance phenomena." *Bioelectromagnetics* **29**(5): 380–386.

Wacker-Gussmann, A., H. Paulsen, K. Stingl, J. Braendle, R. Goelz and J. Henes (2014). "Atrioventricular conduction delay in the second trimester measured by fetal magnetocardiography." *J Immunol Res* **2014**: 753953.

Walleczek, J. and T. F. Budinger (1992). "Pulsed magnetic field effects on calcium signaling in lymphocytes: dependence on cell status and field intensity." *FEBS Lett* **314**(3): 351–355.

Wang, Y., M. K. Hensley, A. Tasman, L. Sears, M. F. Casanova and E. M. Sokhadze (2016). "Heart rate variability and skin conductance during repetitive TMS course in children with autism." *Appl Psychophysiol Biofeedback* **41**(1): 47–60.

Wang, Z., P. Yang, H. Xu, A. Qian, L. Hu and P. Shang (2009). "Inhibitory effects of a gradient static magnetic field on normal angiogenesis." *Bioelectromagnetics* **30**(6): 446–453.

Watanabe, S. and S. Yamada (2008). "Magnetocardiography in early detection of electromagnetic abnormality in ischemic heart disease." *J Arrhythmia* **24**: 4–17.

Wikswo, J. P., Jr. (1980). "Noninvasive magnetic detection of cardiac mechanical activity: Theory." *Med Phys* **7**(4): 297–306.

Wikswo, J. P., Jr., J. E. Opfer and W. M. Fairbank (1980). "Noninvasive magnetic detection of cardiac mechanical activity: Experiments." *Med Phys* **7**(4): 307–314.

Woods, C. E. and J. Olgin (2014). "Atrial fibrillation therapy now and in the future: drugs, biologicals, and ablation." *Circ Res* **114**(9): 1532–1546.

Yamaguchi, Y., M. Wada, H. Sato, H. Nagasawa, S. Koyama, Y. Takahashi, T. Kawanami and T. Kato (2015). "Impact of nocturnal heart rate variability on cerebral small-vessel disease progression: a longitudinal study in community-dwelling elderly Japanese." *Hypertens Res* **38**(8): 564–569.

Yoon, D., S. Biswal, B. Rutt, A. Lutz and B. Hargreaves (2018). "Feasibility of 7T MRI for imaging fascicular structures of peripheral nerves." *Muscle Nerve* **57**(3): 494–498.

Yoshida, K., K. Ogata, T. Inaba, Y. Nakazawa, Y. Ito, I. Yamaguchi, A. Kandori and K. Aonuma (2015). "Ability of magnetocardiography to detect regional dominant frequencies of atrial fibrillation." *J Arrhythm* **31**(6): 345–351.

Yoshida, T., A. Yoshino, Y. Kobayashi, M. Inoue, K. Kamakura and S. Nomura (2001). "Effects of slow repetitive transcranial magnetic stimulation on heart rate variability according to power spectrum analysis." *J Neurol Sci* **184**(1): 77–80.

Young, W. (1969). Magnetic field and in situ acetylcholinesterase in the vagal heart system. In M. Barnothy, *Biological Effects of Magnetic Fields*, 2. New York, Plenum Press: pp. 79–102.

Yu, S., B. D. Van Veen and R. T. Wakai (2013). "Detection of T-wave alternans in fetal magnetocardiography using the generalized likelihood ratio test." *IEEE Trans Biomed Eng* **60**(9): 2393–2400.

Yuan, Y., L. Wei, F. Li, W. Guo, W. Li, R. Luan, A. Lv and H. Wang (2010). "Pulsed magnetic field induces angiogenesis and improves cardiac function of surgically induced infarcted myocardium in Sprague-Dawley rats." *Cardiology* **117**(1): 57–63.

Zhao, Y. L., J. C. Yang and Y. H. Zhang (2008). "Effects of magnetic fields on intracellular calcium oscillations." *Conf Proc IEEE Eng Med Biol Soc* **2008**: 2124–2127.

Zhuravlev, Y. E., D. Rassi, A. A. Mishin and S. J. Emery (2002). "Dynamic analysis of beat-to-beat fetal heart rate variability recorded by SQUID magnetometer: quantification of sympatho-vagal balance." *Early Hum Dev* **66**(1): 1–10.

Žiubrytė, G., G. Jaruševičius, J. Jurjonaite, M. Landauskas, R. McCraty and A. Vainoras (2018a). "Correlations between acute atrial fibrillation and local earth magnetic field strength." *J Complexity Health Sci* **1**(2): 31–41.

Žiubrytė, G., G. Jaruševičius, M. Landauskas, R. McCraty and A. Vainoras (2018b). "The local earth magnetic field changes impact on weekly hospitalization due to unstable angina pectoris." *J Complexity Health Sci* **1**(1): 16–25.

8 Toward ELF Magnetic Fields for the Treatment of Cancer

Leonardo Makinistian and Igor Belyaev

CONTENTS

ELF-MF: CARCINOGENIC, INNOCUOUS, AND/OR ANTINEOPLASTIC?

The possibility of a link between childhood leukemia and other cancers and extremely low-frequency magnetic fields (ELF-MFs) has given rise to several epidemiological studies [1–5].

In 2002, ELF-MFs were classified as a possible human carcinogen, Group 2B, by the International Agency for Research on Cancer (IARC) at the World Health Organization (WHO) [6]. ELF fields higher than 0.3 μT have consistently been shown to correlate with increased risks of childhood leukemia in studies in Europe, the U.S. and Japan [2,7]. The increased risk of childhood leukemia with daily average exposure above 0.3–0.4 µT is supported by recent meta-analyses [8,9]. The ELF exposure assessment in epidemiological studies was analyzed by Magne et al. [10]. A multiple correspondence analysis showed the difficulty to build a statistical model predicting child exposure. The distribution of child personal exposure was significantly different from the distribution of exposure during sleep, questioning the exposure assessment in some epidemiological studies. Prenatal exposure has not been considered in most studies although first key events in many childhood leukemias originate in hematopoietic stem cells *in utero* [11]. In addition, socioeconomic status, which may be connected to the leukemia risks, was not assessed in most studies.

Investigations on the association of ELF-MFs with the susceptibility to other types of cancer including brain cancer are sparse, were often based on small number of cases, and provided contradictory results [12–17]. Given that the results on ELF-MFs exposure and glioma risk were hampered by small studies, different methods, and lack of results on subtypes of glioma including exposure in specific time windows, Carlberg et al. assessed lifetime occupations in case-control studies during 1997–2003 and 2007–2009 for all gliomas (1,346 cases) and separately for different glioma types [18]. An ELF-MFs Job-Exposure Matrix was used for associating occupations with ELF exposure (µT). Cumulative exposure (µT-years), average exposure (µT), and maximum exposed job (µT) were calculated. While no risks were found for all gliomas and astrocytoma, grades I, II and III, for astrocytoma grade IV (glioblastoma multiforme), cumulative exposure in the time window 1–14 years resulted in odds ratio (OR) = 1.9, 95% confidence interval (CI) = 1.4–2.6, P linear trend <0.001, and, in the time window 15+ years, it resulted in OR = 0.9, 95% CI = 0.6–1.3, P linear trend = 0.44 in the highest exposure categories 2.75+ and 6.59+ µT-years, respectively. The data indicated an increased risk in late stage (promotion/progression) of astrocytoma grade IV for occupational ELF-MFs exposure.

In contrast, ELF-MFs (and also static magnetic fields, SMFs [19]) have been successfully used for decades for the treatment of bone and joint repair, spinal fusions, leg ulcers, nerve lesions, and pain in general, among others [20–22]. Given their confirmed effectiveness for some pathologies, the further application of these fields for the treatment of others, including cancers, has been of the utmost interest and has received increasing attention in the last years [23,24].

CANCER TREATMENT WITH ELF-MFS: FROM FUNDAMENTAL PHYSICS TO CELLS AND ANIMALS AND TOWARD THE CLINIC

POSSIBLE TARGETS FOR THE TREATMENT OF CANCER

Magnetite

Magnetite (Fe_3O_4), found in magnetosomes that are present in the human body, including brain tissue, is a strong absorber of radiofrequency (RF) radiation between 500 MHz and 10 GHz [25]. Low-frequency magnetic fields might also produce biological effects if they induce ferromagnetic resonance in tissues that contain high concentrations of magnetite [26]. A thorough calculation for the microwave absorption by magnetic nanoparticles in organisms, based on the Landau–Lifshitz equation, corroborates this conclusion [27].

Chromatin

Matronchik and Belyaev have considered the effects of ELF and microwaves (MWs) in the framework of the same model in their attempts to account for similar peculiarities of ELF and MW effects on conformation of chromatin in *E. coli* and human cells (Matronchik and Belyaev 2008). In particular, frequency dependency (of resonance type) of the ELF/MW effects and their dependence on SMF have been considered in their model of slow nonuniform rotation of the charged DNA-domain/ nucleoid under combined effects of ELF/MW and SMF. This model suggested that the combined action of ELF/MW and SMF results in slow nonuniform rotation of the nucleoid with angular speed that depends on Larmor frequency. The main assumption of this model was natural oscillations in chromatin. This assumption is confirmed by the emerging notion of the dynamic oscillations in charged chromatin (see for Ref. [28–30]) including natural rotation of entire nucleus in living cells including neurons [31–35]. Rapid asymmetric oscillations of <1 s are the basis for translational movements and configurational changes in chromatin previously detected over time spans of minutes–hours and are the result of both the stochastic collisions of macromolecules and specific molecular interactions [36]. Recent data indicate that chromatin is a fractal globule in nonequilibrium state [37]. Coherent chromatin movement for several seconds was recently revealed across large regions (4–5 μm) of nuclei using the method of displacement correlation spectroscopy [38]. Regions of coherent motion extended beyond the boundaries of single-chromosome territories, suggesting elastic coupling of motion over length scales much larger than those of genes. These large-scale, coupled motions were ATP dependent and unidirectional for several seconds, perhaps accounting for ATP-dependent directed movement of single genes. Perturbation of major nuclear ATPases such as DNA polymerase, RNA polymerase II, and topoisomerase II eliminated micron-scale coherence, while causing rapid, local movement to increase; i.e., local motions accelerated but became uncoupled from their neighbors. Similar trends were observed in chromatin dynamics upon inducing a direct DNA damage.

Ions and Ionic Channels

Even though there is no full consensus about which are the physical phenomena at the core of the transduction of an ELF-MFs signal by living organisms (i.e., at the very first stage of sensing of the fields, prior to any subsequent cascade of events), there is one family of targets which are shared by most of the interaction models: ions. Indeed, be it either classical models based on direct action of the Lorentz force [39,40] or quantum ones based on the splitting of atomic levels (ion parametric resonance [41,42]) or quantum interference of bound ions [43], all have ions as their primary target. A theoretical model by Panagopoulos et al. has considered effects of ELF and ELF-modulated RF on collective behavior of intracellular ions which affects membrane gating [44–46]. Despite physical differences in and incompleteness of these mechanisms, all of them relate ELF effects with ion cyclotron resonance frequencies and their harmonics/ subharmonics, which were validated by specially designed experiments [47–49].

Hence, it is interesting to retrieve some experimental results from the literature. For instance, amplitude and frequency windows in the efflux of Ca^{+2} through cell membranes in a spinach vesicles model were observed after a 30 min exposure (f between 7 and 72 Hz, DC field between 27 and 37 μT, AC between 13 and 114 $μT_{peak}$) [50]. While that work was not carried out in the context of cancer research, it is a strikingly clear example of how MFs can affect the dynamics of an ion of paramount biological relevance such as calcium. Given that cancer has been proposed to be considered a "channelopathy" [51], and that ionic channels of both the cytoplasmic and the inner mitochondrial membranes are being posed as candidate targets for the development of cancer treatments [52–55], the action of ELF-MFs on ions could, in principle, be customized to affect specific channels. In fact, Buckner et al. [56] reported a Ca^{2+} uptake increase in malignant cells (by 3.2-fold in B16-BL6, 4.2-fold in HeLa, 2.8-fold in MCF-7, and 4.3-fold in MDA-MB-231), while it was not altered in the nonmalignant lines HBL-100, HEK293, or HSG), after 1 h exposure to the forward Thomas-EMF pattern [57]; although the authors reported fields of 5–10 μT, it is worth noting that their fields were highly inhomogeneous, so a full spatial mapping would have probably been advantageous, instead of only reporting a range of amplitudes. They performed inhibitor studies that showed that T-type voltage-dependent Ca^{2+} channels mediated the observed results. Further studies by the same team [58] confirmed and expanded their results, to conclude that the exact spatial and temporal features of the exposure are critical and that departure from the optimal parameters can, indeed, make the anti-proliferative effect disappear. They also found [59] that the observed effects of inhibition of the growth of malignant cells (but not of normal cells) involves contributions from the adenosine 3′,5′-cyclic monophosphate (cAMP) and mitogen-activated protein kinase (MAPK) pathways. Several reviews have recently focused on the possibility of targeting mitochondrial ionic channels for the treatment of cancer [52,53,55].

Free Radicals and the Redox Homeostasis

Apart from ions, free radicals (ionic or not) are another family of targets that have been proposed to be responsible for SMF and ELF-MF sensing, in the context of the radical pair mechanism of magnetoreception and of carcinogenesis [60–62]. The most favored mechanism for the effects from low-level, long-term ELF exposures involves radicals, such as superoxide $•O_2^-$, $•NO$, and the nonradical H_2O_2, which belongs in the family of the reactive oxygen species (ROS) and is readily converted into the radical $•OH$; highly reactive molecules with unpaired electron spins [63]. ROS often mediate the effect of chemical and physical treatments on cancer cells: genotoxic damage is accomplished through severe redox imbalance and overproduction of ROS [64]. The basis of metabolism correction strategies for cancer therapy involves the manipulation of redox homeostasis in cancer cells [65]: they need an optimal level of ROS that supports cancer-promoting but does not lead to irreversible oxidative damage that results in cell death or senescence [66]. In other words, ROS can have not only proapoptotic but also antiapoptotic effects [67]. Therefore, it would be desirable to administer MFs that would increase production of ROS in cancer cells. Mattsson and Simkó [68] reviewed 41 *in vitro* studies and concluded that MFs of 1 mT and above consistently affected oxidative status, while fields below that value also showed the effect but not consistently. In their review, Wang and Zhang [69] found that, in most cases, MFs increased ROS levels in humans, mice, rat cells, and tissues. However, the authors also found studies showing that ROS levels were decreased or not affected by MFs. The activity of redox-responsive mediators could be controlled by coexposure to ELF-MFs and ROS-generating agents in cancer cells [70]. Mitochondria, being the main producers of ROS, have been posed as a target of ELF-MFs to alter the redox homeostasis of the cell [71]. The right manipulation of this balance using ELF-MFs could lead to the development of novel treatments. For instance, Mannerling et al. [72] reported on the influence of MF exposure (50 Hz sine wave, 1 h, 0.025–0.10 mT, vertical or horizontal) on proliferation, cell cycle distribution, superoxide radical anion, and HSP70 protein levels in the human leukemia cell line K562. While the field (0.10 mT, 1 h) did not affect either cell cycle kinetics or proliferation, both vertical and horizontal orientations for 1 h caused a significant and transient increase in HSP70 levels (>twofold, at several field intensities). This exposure also increased (30%–40%) the levels of the superoxide radical anion. Interestingly, the addition of free radical scavengers (melatonin or 1,10-phenanthroline) inhibited the MF-induced increase in HSP70. The authors concluded that an early response to ELF-MFs in K562 cells seemed to be an increased amount of oxygen radicals, leading to HSP70 induction. It is worth noting that the same effect could be detrimental. For instance, from experiments of exposure of rats to 0.01–0.5 mT, 60 Hz, Lai and Singh [73] hypothesized that exposure to a 60 Hz MF initiates an iron-mediated process (e.g., the Fenton reaction) that increases free radical formation in brain cells, leading to DNA strand breaks and cell death that, in turn, could have important implications in public and occupational environment health hazards assessment. Interestingly, and in contrast, Di Loreto et al. [74] found that ELF-MFs (0.1 and 1 mT, 50 Hz) affected positively the cell viability and concomitantly reduced the levels of apoptotic death in rat neuronal primary cultures, with no significant effects on the main antioxidative defenses.

Besides ions and free radicals, there is also the possibility of direct interaction with the magnetic moment of noncharged atoms or rotating molecules [75–77]. This mechanism encompasses the possibility of targeting virtually any molecule of the cell; nevertheless – as with all others – it is not as yet ready to provide a rationale to design specific (SMF plus) ELF-MFs exposures "tuned" to specific molecules of interest for the treatment of cancer (or, for that matter, any other purpose).

Regardless of the exact underlying mechanism of interaction at the detection/transduction stage and of the details of the immediate downstream cascade of events (action on magnetite and/or chromatin; alteration of ion fluxes in ionic channels, free radicals and the cell redox homeostasis, or rotating molecules such as DNA undergoing replication or transcription), when considering a possible development of cancer treatments, there is a natural interest in exposures that produce a decrease of proliferation and/or an increase of cell death in cancer cell cultures *in vitro* and/or an inhibition of tumor growth *in vivo*.

In-Vitro Proliferation Studies

Olsson et al. exposed SPD8 cells, derived from V79 Chinese hamster cells, to a specific combination of static and ELF-MFs, which has been proven to have effects on chromatin conformation in several cell types [78]. The genotoxic agent camptothecin (CPT) was used either as a positive control or simultaneously with the ELF-MFs. Although there was no effect on cell survival in response to ELF-MF exposure, inhibition of cell growth was observed. On the other hand, ELF-MF exposure partly counteracted the growth inhibition seen with CPT. The data suggested that ELF-MF exposure may stimulate or inhibit cell growth depending on the state of the cells. Although ELF-MFs did not induce recombination, a weak but statistically significant DNA fragmentation comparable with CPT-induced fragmentation was observed with PFGE 48 h after exposure to ELF-MFs.

Recent studies provided further evidence that ELF-MFs can affect proliferation of various cell types including stem cells under specific conditions of exposure [79–81]. Bai et al. investigated ELF-MF effects on proliferation of epidermal stem cells (ESCs) [82]. The ESCs obtained from human foreskin were grafted into type-I three-dimensional collagen sponge scaffolds and then were exposed to MFs (50 Hz, 5 mT) for 14 days, 30 min daily. The effects of MFs on growth and proliferation of ESCs were analyzed with staining of hematoxylin and eosin (H&E) and 4',6-diamidino-2-phenylindole (DAPI) under microscope or scanning electron microscope. The data of DAPI staining for 2, 7, 10, and 14 days were collected, respectively, to investigate the cells' proliferation. The MFs promoted ESCs proliferation compared with control.

Cid et al. verified the hypothesis that the ELF-MF effect on cancer progression could be mediated by MF-induced effects on the cellular response to melatonin (MEL), a potentially oncostatic neurohormone [83]. HepG2 cells were exposed to intermittent 50 Hz, 10 µT MF, in the presence or absence of MEL at physiological (10 nM) or pharmacological doses (1 µM). The results indicated that the MF exerts significant cytoproliferative and dedifferentiating effects that can be prevented by 10 nM MEL. Conversely, MEL exerts cytostatic and differentiating effects on HepG2 that are abolished by simultaneous exposure to MF.

Sarimov et al. [47] found that ELF-MFs under conditions of exposure tuned to Zn^{2+} (according to the ion parametric resonance (IPR) model of Blanchard and Blackman: 20 Hz, 43 µT DC, 38.7 $µT_{peak}$ AC for 72 h and 96 h) inhibited the growth (by ~20%–30%) of cancer (cervix cancer and osteosarcoma) and normal cells (normal human fibroblasts) but not lung carcinoma cells. Vincenzi et al. [84] observed an increase of cytotoxicity by lactate dehydrogenase (LDH) release and apoptosis by caspase-3 activation in PC12 (adrenal pheochromocytoma) and U87MG (human glioblastoma) cells but not in cortical neurons after 24 h exposure with pulsed electromagnetic fields (PEMFs, 1.5 mT_{peak} pulses, 10% duty cycle, 75 Hz). Filipovic et al. [85] studied the effect of 50 Hz 10 mT_{rms} fields on breast cancer (MDA-MB-231) and colon carcinoma cells (SW480 and HCT-116) for 24 and 72 h, finding a ~2%–11% inhibition for the colon carcinoma cells and one inhibition >50% for the MDA-MB-231 cells. Akbarnejad et al. [86] tested several combinations of amplitudes (5 and 10 mT) and frequencies (10, 50, and 100 Hz) for 24 h in glioblastoma multiforme (GBM) cell line U87; they found the three possible outcomes: no-effect, enhanced (~25%), and inhibited (~10%–25%) proliferation. Buckner et al. [56] reported growth inhibition (~25%–45%) of the malignant cell lines B16-BL6, MDA-MB-231, MCF-7, and HeLa cells but no-effect on the growth of nonmalignant cells (HBL-100, HEK293T, and HSG cell lines) upon 1 h/day exposure to the so-called forward Thomas-EMF pattern [57].

Bae et al. exposed cultured human cancer cells, DU145 and Jurkat, to ELF-MFs (1 mT, 60 Hz) for 3 days and measured cell proliferation using a 2-(4-Iodophenyl)-3-(4-nitrophenyl)-5-(2,4-disulfophenyl)-2H-tetrazolium (WST-1) assay and a trypan blue exclusion assay [80]. This exposure was repeated for three similar CO_2 incubators at the same location. A periodogram analysis was performed to investigate periodic patterns in the MF and sham effects on cell growth. An MF effect on growth in both cell types was found, which was either promotive or suppressive in a period-dependent manner. This pattern of the MF-induced changes in cell proliferation was consistent in all incubators, with only small variation. The authors linked the observed variation in the ELF effect with the pattern of solar activity and quasiperiodical change in geomagnetic field. The data indicated that the period-dependent converse effect of MFs on cell growth may limit reproducibility.

Jeong et al. investigated the effect of chronic exposure to ELF-MF (1 mT, 50 Hz, 12 days) on astrocytic differentiation of mesenchymal stem cells (MSCs) from adult bone marrow (BM) [87]. ELF-MF exposure reduced the rate of proliferation and enhanced astrocytic differentiation. The ELF-MF-treated cells showed increased levels of the glial fibrillary acidic protein (GFAP, an astrocyte marker), while those of the early neuronal marker (Nestin) and stemness marker (OCT3/4) were down-regulated. The ROS level was significantly elevated by ELF-MF exposure, which strengthens the modulatory role of SIRT1 and SIRT1 downstream molecules (TLE1, HES1, and MASH1) during astrocytic differentiation. After nicotinamide (5 mM) mediated inhibition of SIRT1, levels of TLE1, HES1, and MASH1 were examined; TLE1 was significantly upregulated, and MASH1 was downregulated. These results suggest that ELF-MFs induce astrocytic differentiation through induction of ROS and activation of SIRT1 and SIRT1 downstream molecules.

Crocetti et al. [88] found that DNA damage, apoptosis, and the number of dead MCF7 (human breast cancer) cells increased with respect to control upon exposure to 20 Hz, 3 mT fields for 60 min/day for 3 days. Many others have reported DNA damage upon exposure to ELF-MFs [73,89–92] which if severe enough could lead to cell death.

While the desired effect is to inhibit proliferation, there are, indeed, studies in which no effect of ELF-MFs was observed on the proliferation of cancer cells [93]. This stresses the fact that ELF-MF effects depend on multiple physical and biological variables which, in turn, prompt the convenience (if not need) of screening (see "The Importance of Screening" sectionbelow). This is clearly exemplified by the work of Kwee and Raskmark [94], where an increase of proliferation was shown to be highly dependent on the parameters of the exposure and the cultures.

IN-VIVO TUMOR GROWTH STUDIES

Cameron et al. [95] found a decrease in growth of breast cancer tumors and vascularization for a 120 Hz semi-sine 10–20 mT wave applied for 3–80 min/day for 12 days in syngeneic C3H mice. Zhang et al. [96] reported a significant decrease in sarcoma tumor volume and weight in a Kunming mice model upon 15 min/day for 20 days exposure to pulses of 0.6 to 2.0 T, with a gradient of 10–100 T/m, pulse width of 20–200 ms, and frequency of 0.16 to 1.34 Hz. They also reported an increase of immune cells infiltration and a clearly defined encapsulation of the tumor. Novikov et al. [97] exposed Bagg Albino/c (BALB/c) male mice for 1 h/day for 12 days to a diversity of pure and combined frequencies and different AC intensities for a fixed 42 µT DC MF. They observed, for selected combinations of the parameters, an increase in the percentage of surviving animals by the end of the experiment (day 25). For the optimal parameters, survival percentage was 85% at day 25, while all control animals were dead by day 16. In control animals without tumors, no toxic effect was detected. Buckner et al. [56] found a significant decrease in tumor diameter (35%–50%) in a B16-BL6 melanoma model on C57b mice treated 3 h/day for 18–21 days or 15–16 days. Tatarov et al. [98] used Swiss outbred female nude mice injected with the metastatic mouse breast tumor cell line EpH4-MEK-Bcl2 as breast cancer models. Mice were exposed to 100 mT, 1 Hz half-sine-wave unipolar MFs, 360 min/day (60 min/day and 180 min/day were found not to be effective) for as long as 4 weeks. Their results showed that exposure suppressed tumor growth. More reports can be found in Ref. [99], where the authors point out the relevant difference between tumor growth inhibition and tumor shrinkage, the more common effect being the former.

With regards to *in-vivo* studies such as the ones reviewed above, it is worth noting that when exposing animals bearing tumors, a whole organism is being exposed. In particular, if cells from the immune system were to be stimulated by the exposure, then tumor growth progression could be affected, not directly but indirectly, by enhancement of the organism's inflammatory reaction. Indeed, Rosado et al. [100] envisage the modulation of the immune response by ELF-MFs, which is in line with experiments by some of them, Frahm et al. [101,102], where they found that a 45 min exposure (50 Hz 1 mT) significantly elevated phagocytic activity, free radical release, and interleukin-1b (IL-1b) production, suggesting that ELF-MFs can enhance the physiological function of macrophages, without any genotoxic effect. In their review, Ross and Harrison [103] further elaborate on the possibility of enhancement of the immune response by ELF-MFs with their implications on inflammatory response to infection and tumor control.

TOWARD THE CLINIC

ELF as Coadjutant of Conventional Treatments

Berg et al. [104] found a synergy of 15–20 mT 50 Hz MFs with hyperthermia (41.5°C) and/or bleomycin (a cancerostatic agent) in both *in-vitro* (human erythroleukemia K-562 cells and healthy donor lymphocytes) and *in-vivo* experiments (severe combined immunodeficiency (SCID) mice inoculated with human breast adenocarcinoma MX-1 cells, 3 h/day for 8 days). Interestingly, they found that normal lymphocytes were significantly less responsive to ELF-MF treatment than cancer cells, which represents a clear advantage for minimizing side effects. However, MFs can have the opposite effect: Blackman et al. [105] reported that a 1.2 μT_{rms} 60 Hz MF reduced the inhibitory effect of tamoxifen on the proliferation of MCF-7 human breast cancer cells. Falone et al. [70] provide a thorough review with dozens of examples of ELF-MFs working synergistically with differentiating, cytostatic, or cytotoxic agents (melatonin, tamoxifen, dimethyl sulfoxide, methotrexate, mitomycin C, and cisplatin, among others) and with DNA-damage promoters (γ-rays, benzene, hydrogen peroxide, and menadione, among others). However, they also provide examples of studies reporting the opposite: an antagonistic effect between the chemical agent and the ELF-MFs.

Clinical Trials

Sun et al. [106] carried out a pilot study on patients with advanced non-small-cell lung cancer (NSCLC). They evaluated effects on survival and palliation of general symptoms after a 2h treatment, five times a week, for 6–10 weeks. The MFs were of 0.4 T generated by permanent magnets fixed to an iron plate rotating at 420 revolutions per minute (7 Hz). They concluded that the treatment was effective, well tolerated and safe, and that ELF-MFs may prolong survival and improve general symptoms of advanced NSCLC patients. Ronchetto et al. [107] conducted a pilot study on patients with advanced neoplasm to assess safety and acute toxicity of a treatment with a 5.5 mT 50 Hz MF following a complex modulation that had been proven to be effective in mice by the authors [108]. Toxicity was assessed according to the WHO criteria, and the authors concluded that the MF exposures under study were safe to be administered. It is worth noting that the same modulation (with almost exactly the same average intensity: 5.1 mT instead of 5.5 mT) was used by Yuan et al. [109] in *in-vitro* and *in-vivo* (nude mice) models of nephroblastoma and neuroblastoma, finding that the exposure had an antiproliferation and an antitumor effect. Rick et al. [110] conducted a clinical trial aiming to elucidate whether MFs could be of help for the treatment of chemotherapy-induced peripheral neuropathy. While this is not an example of treatment of cancer, it illustrates how MFs are readily being considered for the clinical practice in oncology.

THE IMPORTANCE OF SCREENING

The existence of amplitude and frequency windows [50] prompts the need of screening through the different parameters of exposure for two reasons: (i) optimization and (ii) avoidance of undesired effects. Indeed, Akbarnejad et al. [86] reported both, inhibition and stimulation of cancer cell proliferation upon different exposures. The work by Cameron et al. [95] is a clear example were screening was used for optimization of results: the authors tested seven different combinations of field intensities and exposure durations, finding reductions of tumor volume to between 25% and 40% of mean tumor volume of controls. Novikov et al. [97] tested 45 different combinations of AC intensities (in the 40–500 nT_{rms} range, collinear with a 42 μT DC field) at pure and combined frequencies (0.5, 1, 4.4, and 16.5 Hz) for the treatment of a mouse Ehrlich ascites carcinoma model, finding outcomes from no-effect to a complete absence of tumors in the treated group. After testing six different combinations of frequency/intensity/duration, Kumar et al. [111] found that a sinusoidal MF of 5 Hz, 4 μT, and 90 min was optimal in lowering the paw edema volume and decreasing the activity of lysosomal enzymes in adjuvant induced arthritis (AIA) in rats (a model for rheumatoid arthritis), while an MF of 6 Hz, 3.5 μT, and 80 min was significantly less effective. Crocetti et al. [88] observed a discrete window of vulnerability of MCF7 (human breast cancer) cells to PEMFs (square, 6 ms width) at 20 Hz, 3 mT, and exposure duration of 60 min/day for 3 days, while their normal counterparts, MCF10, were not affected by the fields. They were able to reach these findings after testing exposures of 30, 60, and 90 min; field intensities of 2, 3, and

5 mT; and also a frequency of 50 Hz. László et al. [112] optimized an arrangement of permanent magnets to maximize the analgesic effect of MFs on mice. They tested 16 different configurations (with SMFs of tens to hundreds of millitesla) and found effects between ~80% and no-effect with respect to control.

Besides the financial aspect, researchers might feel discouraged to undertake the task of screening because it is time consuming, although results might be worth it [30]. A way around this pitfall is the use of inhomogeneous fields like the ones generated by the system proposed by Makinistian [113], the one used by Aarholt with bacteria [114,115], or, simply, any permanent magnet or system of coils, as long as the exposed samples are located away from the region with homogeneous fields [116–119]. Alternatively many identical homogeneous-field exposure systems can be built and used simultaneously, as in the work by Dhiman and Galland [120], who made an exhaustive sweeping of the DC intensity effect on *Arabidopsis Thaliana* using 24 Helmholtz coils at the same time.

HOMOGENEOUS VERSUS INHOMOGENEOUS FIELDS

If a treatment for cancer by exposure to MFs is in mind, then it is pertinent to consider the issue of how the fields will be administered. More precisely, we call here attention to the spatial distribution of the fields: will they be homogeneous or inhomogeneous? For the purpose of illustration, Figure 8.1 shows the hypothetical case of a

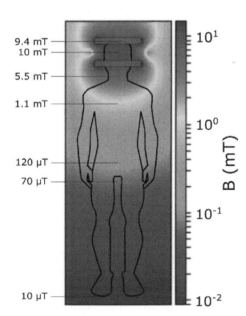

FIGURE 8.1 Simulation of the MF generated by a pair of 300 mm diameter Helmholtz coils (gray rectangles) generating 10 mT at their center. The silhouette of a person 1.70 m tall helps in visualizing the distribution of the field throughout the body, spanning three orders of magnitude. A custom-written Python code was used for solving the Biot–Savart law; spatial resolution: 5 mm; 100 nodes discretization for each of the coils.

clinical trial in which a pair of Helmholtz coils are used to treat a brain tumor with a 10 mT (100 Gauss) field. The simulation shows that the MFs span down to 10 µT (three orders of magnitude) at the feet of the 1.70 m tall patient's silhouette. Since there are studies reporting effects for the µT and the mT ranges (and even for the nT range), there is no *a priori* justification for assuming that the fields outside the zone of interest (the tumor) will be safe for the surrounding, healthy tissues. But the treatment would have not reached the patient had its safety and (apparent) effectiveness not been tested in preclinical *in-vitro* and *in-vivo* studies beforehand. The issue that we want to raise here is that (i) if cells and small animals in the preclinical tests were exposed to homogeneous fields (as it is usually the case), then parts of the patients' body will be exposed to fields not tested in the preclinical studies; and (ii) if inhomogeneous fields are intended to be used on patients (e.g., administered by relatively portable devices), then, in rigor, the same inhomogeneities should have been tested in the preclinical studies. But the very meaning of "the same inhomogeneities" is not so clear when a change of scale is to be implemented and, moreover, when they have to be reproduced for *in-vitro* studies and for small animals. In principle, the most purist, safest way of going from experiments with cellular cultures to experiments with small animals and then finally to clinical trials with humans would be to use homogeneous fields in these three stages. In this way, dosimetry would be much easier (since no spatial mapping of the fields would be needed), but the drawback would be substantial: for the treatment of humans, much bigger facilities [121–123] should be used, in order to house devices big enough to generate homogeneous fields in a volume equal to or greater than a human being. Besides the increment of costs and the loss of portability, the generation of extensive homogeneous fields implies the generation of inhomogeneous ones spreading outward from the volume for treatment into the surrounding area, which could be a hazard for the personnel administrating the treatment [124]. Table 8.1 condensates advantages and disadvantages of the use of homogeneous and inhomogeneous MFs for the purposes of *in-vitro* screening, exposure of small animals, and treatment of humans.

ADHERENCE TO METHODOLOGICAL RECOMMENDATIONS

A full description of the exposure parameters and adherence to other methodological recommendations is highly desirable for the progress of magnetobiology in general and in particular for the development of medical technology for administration of MFs [125–128]. Consideration of temperature as a confounding variable is still not a standard. Falone et al. [70] reported that among a set of more than 50 articles surveyed, 61% did not make a clear check of temperature as a possible confounder, and they prompted the need for stronger efforts to set uniform standards for the experimental community. Another key issue is the one of controls: the fact that samples (or animals) will always be subjected to some MFs (even if it is near zero inside a mu-metal box) renders the impossibility of having a control in the usual sense of the word (i.e., a sample that is not exposed to the agent under study). Hence, not only the MFs affecting the exposed, but also the ones affecting the control sample ought to be reported in order to allow replication attempts. In this regard, a work by Prato et al. [129] is remarkable, since, in their experiments of magnetoreception of mice,

TABLE 8.1

Advantages and Disadvantages of the Use of Homogeneous and Inhomogeneous MFs for Research and Therapeutics

Purpose	Homogeneous MFs		Inhomogeneous MFs	
	Advantages	Disadvantages	Advantages	Disadvantages
In-vitro screening	• Easier dosimetry • Gradients are zero: one less variable to analyze	• Time consuming • More expensive • Different batches of samples for the different parameters	• Several values of the parameter(s) can be tested simultaneously • Less expensive • Same batch of samples for the different parameters	• Difficult dosimetry • Gradients are present: more complex analysis
Exposure of small animals	• Easier dosimetry • Direct up-scaling of the exposure system for humans	• Bigger exposure systems needed • Animals' movements could generate "apparent fields"	• Less expensive • Smaller exposure systems (e.g., small permanent magnet or solenoid)	• Difficult dosimetry • Difficult (if not impossible) to implement if animals are not restrained
Treatment of humans	• Easier dosimetry • Direct up-scaling of the exposure system used with animals • Exposure only to fields tested on animals	• Whole facilities needed • Substantially more expensive	• Less expensive • Portable devices are possible	• Difficult dosimetry • Almost unavoidable exposure to fields not tested on animals

they used three different controls: a mu-metal box (shielding electric and magnetic fields from the environment), a stainless steel box (shielding only the electric fields), and a fiberglass sham box (not affecting the environmental fields). Besides finding that mice can detect 30 Hz MFs as weak as 33 nT, the authors found significant differences between the different controls. This is an undisputable example of the relevance of reporting the control/sham-exposure conditions.

LOOKING AHEAD

More than 50 years have passed since the pioneering studies in magnetobiology [130], and it is undeniable that a massive amount of experimental and theoretical results have been produced. In spite of this, it is fair to note that, in terms of theory, there is still an incomplete comprehension of the interaction of SMFs and ELF-MFs with living matter, and, in terms of experiments, there is still a rather small (if not minute) reproducibility. As if by a parallel lane, the development of magnetotherapy technology has advanced solidly, and it has established itself for the treatment of some diseases but still not for cancers. This step is definitely underway, with many promising *in-vitro* and *in-vivo* experiments and even some clinical trials pointing in the desired direction.

We humbly propose the following pointers that we consider should help magnetobiology accelerate toward the development of MFs for the treatment of cancer:

- Keep looking for a better comprehension of mechanisms of transduction and downstream cascades.
- Keep looking for models that would take as input cancer cell features (e.g., Lucia et al. [131]) or readings of physiological variables from cancer patients (e.g. Costa et al. [132]) and have as output candidate signals for effective treatment.
- Keep looking for devices and/or experimental protocols with greater throughput so that more signals can be tested per unit time.
- Attempt replication of successful reports already available in the literature.
- Adhere to methodological recommendations to increase the chances of reproducibility.
- If a future clinical trial is in mind, the advantages and disadvantages of using homogeneous or inhomogeneous fields should be taken into consideration.

ACKNOWLEDGMENTS

LM thanks financial support from the Universidad Nacional de San Luis, Argentina (grant PROICO 02-0518). IB was supported by the Structural Funds of EU (Protonbeam, ITMS: 26220220129), the Slovak Research and Development Agency (APVV 0669-10, APVV-15-0250), and the Vedecká grantová agentúra (VEGA) Grant Agency (2/0089/18) of the Slovak Republic.

REFERENCES

1. A. Ahlbom, N. Day, M. Feychting, E. Roman, J. Skinner, J. Dockerty, M. Linet, M. McBride, J. Michaelis, J. H. Olsen, T. Tynes, and P. K. Verkasalo, A pooled analysis of magnetic fields and childhood leukaemia, *Br J Cancer*, vol. 83, pp. 692–8, Sep 2000.
2. M. Kabuto, H. Nitta, S. Yamamoto, N. Yamaguchi, S. Akiba, Y. Honda, J. Hagihara, K. Isaka, T. Saito, T. Ojima, Y. Nakamura, T. Mizoue, S. Ito, A. Eboshida, S. Yamazaki, S. Sokejima, Y. Kurokawa, and O. Kubo, Childhood leukemia and magnetic fields in Japan: A case-control study of childhood leukemia and residential power-frequency magnetic fields in Japan, *Int J Cancer*, vol. 119, pp. 643–50, 2006.
3. J. Schuz, Exposure to extremely low-frequency magnetic fields and the risk of childhood cancer: Update of the epidemiological evidence, *Prog Biophys Mol Biol*, vol. 107, pp. 339–42, 2011.
4. J. Schuz, C. Dasenbrock, P. Ravazzani, M. Roosli, P. Schar, P. L. Bounds, F. Erdmann, A. Borkhardt, C. Cobaleda, M. Fedrowitz, Y. Hamnerius, I. Sanchez-Garcia, R. Seger, K. Schmiegelow, G. Ziegelberger, M. Capstick, M. Manser, M. Muller, C. D. Schmid,

D. Schurmann, B. Struchen, and N. Kuster, Extremely low-frequency magnetic fields and risk of childhood leukemia: A risk assessment by the ARIMMORA consortium, *Bioelectromagnetics*, vol. 37, pp. 183–9, 2016.

5. J. Schuz, K. Grell, S. Kinsey, M. S. Linet, M. P. Link, G. Mezei, B. H. Pollock, E. Roman, Y. Zhang, M. L. McBride, C. Johansen, C. Spix, J. Hagihara, A. M. Saito, J. Simpson, L. L. Robison, J. D. Dockerty, M. Feychting, L. Kheifets, and K. Frederiksen, Extremely low-frequency magnetic fields and survival from childhood acute lymphoblastic leukemia: An international follow-up study, *Blood Cancer J*, vol. 2, p. e98, 2012.

6. IARC, Non-ionizing radiation, part 1: Static and extremely low-frequency (ELF) electric and magnetic fields. In: *IARC Monographs on the Evaluation of Carcinogenic Risks to Humans*. Lyon, France: IARC Press, vol. 80, pp. 1–395, 2002.

7. Y. Yang, X. Jin, C. Yan, Y. Tian, J. Tang, and X. Shen, Case-only study of interactions between DNA repair genes (hMLH1, APEX1, MGMT, XRCC1 and XPD) and low-frequency electromagnetic fields in childhood acute leukemia, *Leuk Lymphoma*, vol. 49, pp. 2344–50, 2008.

8. J. C. Teepen and J. A. A. M. van Dijck, Impact of high electromagnetic field levels on childhood leukemia incidence, *Int J Cancer*, vol. 131, pp. 769–78, 2012.

9. L. Zhao, X. Liu, C. Wang, K. Yan, X. Lin, S. Li, and H. Bao, Magnetic fields exposure and childhood leukemia risk: A meta-analysis based on 11,699 cases and 13,194 controls, *Leuk Res*, vol. 38, pp. 269–74, 2014.

10. I. Magne, M. Souques, I. Bureau, A. Duburcq, E. Remy, and J. Lambrozo, Exposure of children to extremely low frequency magnetic fields in France: Results of the EXPERS study, *J Expo Sci Environ Epidemiol*, vol. 27(5), pp. 505–512, 2017.

11. P. Kosik, M. Skorvaga, M. Durdik, L. Jakl, E. Nikitina, E. Markova, K. Kozics, E. Horvathova, and I. Belyaev, Low numbers of pre-leukemic fusion genes are frequently present in umbilical cord blood without affecting DNA damage response, *Oncotarget*, vol. 8, pp. 35824–34, 2017.

12. I. Baldi, G. Coureau, A. Jaffre, A. Gruber, S. Ducamp, D. Provost, P. Lebailly, A. Vital, H. Loiseau, and R. Salamon, Occupational and residential exposure to electromagnetic fields and risk of brain tumors in adults: A case-control study in Gironde, France, *Int J Cancer*, vol. 129, pp. 1477–84, 2011.

13. P. Elliott, G. Shaddick, M. Douglass, K. de Hoogh, D. J. Briggs, and M. B. Toledano, Adult cancers near high-voltage overhead power lines, *Epidemiology*, vol. 24, pp. 184–90, 2013.

14. M. C. Turner, G. Benke, J. D. Bowman, J. Figuerola, S. Fleming, M. Hours, L. Kincl, D. Krewski, D. McLean, M. E. Parent, L. Richardson, S. Sadetzki, K. Schlaefer, B. Schlehofer, J. Schuz, J. Siemiatycki, M. van Tongeren, and E. Cardis, Occupational exposure to extremely low-frequency magnetic fields and brain tumor risks in the INTEROCC study, *Cancer Epidemiol Biomarkers Prev*, vol. 23, pp. 1863–72, 2014.

15. T. Koeman, P. A. van den Brandt, P. Slottje, L. J. Schouten, R. A. Goldbohm, H. Kromhout, and R. Vermeulen, Occupational extremely low-frequency magnetic field exposure and selected cancer outcomes in a prospective Dutch cohort, *Cancer Causes Control*, vol. 25, pp. 203–14, 2014.

16. T. Sorahan, Magnetic fields and brain tumour risks in UK electricity supply workers, *Occup Med (London)*, vol. 64, pp. 157–65, 2014.

17. Y. Zhang, J. Lai, G. Ruan, C. Chen, and D. W. Wang, Meta-analysis of extremely low frequency electromagnetic fields and cancer risk: A pooled analysis of epidemiologic studies, *Environ Int*, vol. 88, pp. 36–43, 2016.

18. M. Carlberg, T. Koppel, M. Ahonen, and L. Hardell, Case-control study on occupational exposure to extremely low-frequency electromagnetic fields and glioma risk, *Am J Ind Med*, vol. 10, p. 22707, 2017.

19. X. Zhang, K. Yarema, and A. Xu, *Biological Effects of Static Magnetic Fields*. Singapore: Springer, 2017.
20. M. S. Markov, Magnetic field therapy: A review, *Electromagn Biol Med*, vol. 26, pp. 1–23, 2007.
21. N. M. Shupak, F. S. Prato, and A. W. Thomas, Therapeutic uses of pulsed magnetic-field exposure: A review, *Radio Sci Bull*, vol. 307, pp. 9–32, 2003.
22. M. S. E. Markov (Ed.), *Electromagnetic Fields in Biology and Medicine*. Boca Raton, FL: CRC Press, 2015.
23. M. Vadala, J. C. Morales-Medina, A. Vallelunga, B. Palmieri, C. Laurino, and T. Iannitti, Mechanisms and therapeutic effectiveness of pulsed electromagnetic field therapy in oncology, *Cancer Med*, vol. 5, pp. 3128–39, 2016.
24. S. Sengupta and V. K. Balla, A review on the use of magnetic fields and ultrasound for non-invasive cancer treatment, *J Adv Res*, vol. 14, pp. 97–111, 2018.
25. J. L. Kirschvink, Microwave absorption by magnetite: A possible mechanism for coupling nonthermal levels of radiation to biological systems, *Bioelectromagnetics*, vol. 17, pp. 187–94, 1996.
26. L. J. Challis, Mechanisms for Interaction Between RF Fields and Biological Tissue, *Bioelectromagnetics Supplement*, vol. 7, pp. S98–S106, 2005.
27. V. N. Binhi, Microwave absorption by magnetic nanoparticles in organisms, *Biophysics*, vol. 56, pp. 1096–8, 2011.
28. I. Y. Belyaev, Radiation-induced DNA repair foci: Spatio-temporal aspects of formation, application for assessment of radiosensitivity and biological dosimetry, *Mutat Res*, vol. 704, pp. 132–41, 2010.
29. A. Y. Matronchik and I. Y. Belyaev, Mechanism for combined action of microwaves and static magnetic field: Slow non uniform rotation of charged nucleoid, *Electromagn Biol Med*, vol. 27, pp. 340–54, 2008.
30. V. N. Binhi, Y. D. Alipov, and I. Y. Belyaev, Effect of static magnetic field on E. coli cells and individual rotations of ion-protein complexes, *Bioelectromagnetics*, vol. 22, pp. 79–86, 2001.
31. C. Lang, S. Grava, T. van den Hoorn, R. Trimble, P. Philippsen, and S. L. Jaspersen, Mobility, microtubule nucleation and structure of microtubule-organizing centers in multinucleated hyphae of Ashbya gossypii, *Mol Biol Cell*, vol. 21, pp. 18–28, 2010.
32. M. Brosig, J. Ferralli, L. Gelman, M. Chiquet, and R. Chiquet-Ehrismann, Interfering with the connection between the nucleus and the cytoskeleton affects nuclear rotation, mechanotransduction and myogenesis, *Int J Biochem Cell Biol*, vol. 42, pp. 1717–28, 2010.
33. J. R. Levy and E. L. Holzbaur, Dynein drives nuclear rotation during forward progression of motile fibroblasts, *J Cell Sci*, vol. 121, pp. 3187–95, 2008.
34. J. Y. Ji, R. T. Lee, L. Vergnes, L. G. Fong, C. L. Stewart, K. Reue, S. G. Young, Q. Zhang, C. M. Shanahan, and J. Lammerding, Cell nuclei spin in the absence of lamin bl, *J Biol Chem*, vol. 282, pp. 20015–26, 2007.
35. P. C. Park and U. De Boni, Dynamics of nucleolar fusion in neuronal interphase nuclei in vitro: Association with nuclear rotation, *Exp Cell Res*, vol. 197, pp. 213–21, 1991.
36. A. Pliss, K. S. Malyavantham, S. Bhattacharya, and R. Berezney, Chromatin dynamics in living cells: Identification of oscillatory motion, *J Cell Physiol*, vol. 228, pp. 609–16, 2013.
37. G. Fudenberg and L. A. Mirny, Higher-order chromatin structure: Bridging physics and biology, *Curr Opin Genet Dev*, vol. 22, pp. 115–24, 2012.
38. A. Zidovska, D. A. Weitz, and T. J. Mitchison, Micron-scale coherence in interphase chromatin dynamics, *Proc Natl Acad Sci U S A*, vol. 110, pp. 15555–60, 2013.
39. D. J. Muehsam and A. A. Pilla, A Lorentz model for weak magnetic field bioeffects: Part I--thermal noise is an essential component of AC/DC effects on bound ion trajectory, *Bioelectromagnetics*, vol. 30, pp. 462–75, 2009.

40. D. J. Muehsam and A. A. Pilla, A Lorentz model for weak magnetic field bioeffects: Part II--secondary transduction mechanisms and measures of reactivity, *Bioelectromagnetics*, vol. 30, pp. 476–88, 2009.

41. V. V. Lednev, Possible mechanism for the influence of weak magnetic fields on biological systems, *Bioelectromagnetics*, vol. 12, pp. 71–5, 1991.

42. S. Engstrom, Dynamic properties of Lednev's parametric resonance mechanism, *Bioelectromagnetics*, vol. 17, pp. 58–70, 1996.

43. V. N. Binhi, *Magnetobiology: Underlying Physical Problems*. San Diego, CA: Academic Press, 2002.

44. D. J. Panagopoulos, A. Karabarbounis, and L. H. Margaritis, Mechanism for action of electromagnetic fields on cells, *Biochem Biophys Res Commun*, vol. 298, pp. 95–102, 2002.

45. D. J. Panagopoulos, N. Messini, A. Karabarbounis, A. L. Philippetis, and L. H. Margaritis, A mechanism for action of oscillating electric fields on cells, *Biochem Biophys Res Commun*, vol. 272, pp. 634–40, 2000.

46. D. J. Panagopoulos and L. H. Margaritis, The identification of an intensity 'window' on the bioeffects of mobile telephony radiation, *Int J Radiat Biol*, vol. 86, pp. 358–66, 2010.

47. R. Sarimov, E. Markova, F. Johansson, D. Jenssen, and I. Belyaev, Exposure to ELF magnetic field tuned to Zn inhibits growth of cancer cells, *Bioelectromagnetics*, vol. 26, pp. 631–8, 2005.

48. I. Y. Belyaev and E. D. Alipov, Frequency-dependent effects of ELF magnetic field on chromatin conformation in Escherichia coli cells and human lymphocytes, *Biochim Biophys Acta*, vol. 1526, pp. 269–76, 2001.

49. B. Poniedzialek, P. Rzymski, H. Nawrocka-Bogusz, F. Jaroszyk, and K. Wiktorowicz, The effect of electromagnetic field on reactive oxygen species production in human neutrophils in vitro, *Biol Med*, vol. 32, pp. 333–41, 2013.

50. C. L. Baureus Koch, M. Sommarin, B. R. Persson, L. G. Salford, and J. L. Eberhardt, Interaction between weak low frequency magnetic fields and cell membranes, *Bioelectromagnetics*, vol. 24, pp. 395–402, 2003.

51. A. Litan and S. A. Langhans, Cancer as a channelopathy: Ion channels and pumps in tumor development and progression, *Front Cell Neurosci*, vol. 9, p. 86, 2015.

52. M. Bachmann, R. Costa, R. Peruzzo, E. Prosdocimi, V. Checchetto, and L. Leanza, Targeting mitochondrial ion channels to fight cancer, *Int J Mol Sci*, vol. 19, p. 2060, 2018.

53. L. Leanza, V. Checchetto, L. Biasutto, A. Rossa, R. Costa, M. Bachmann, M. Zoratti, and I. Szabo, Pharmacological modulation of mitochondrial ion channels, *Br J Pharmacol*, vol. 176, pp. 4258–4283, 2018.

54. L. Leanza, A. Manago, M. Zoratti, E. Gulbins, and I. Szabo, Pharmacological targeting of ion channels for cancer therapy: In vivo evidences, *Biochim Biophys Acta*, vol. 1863, pp. 1385–97, 2016.

55. R. Peruzzo, L. Biasutto, I. Szabo, and L. Leanza, Impact of intracellular ion channels on cancer development and progression, *Eur Biophys J*, vol. 45, pp. 685–707, 2016.

56. C. A. Buckner, A. L. Buckner, S. A. Koren, M. A. Persinger, and R. M. Lafrenie, Inhibition of cancer cell growth by exposure to a specific time-varying electromagnetic field involves T-type calcium channels, *PLoS One*, vol. 10, p. e0124136, 2015.

57. A. W. Thomas, M. Kavaliers, F. S. Prato, and K. P. Ossenkopp, Antinociceptive effects of a pulsed magnetic field in the land snail, Cepaea nemoralis, *Neurosci Lett*, vol. 222, pp. 107–10, 1997.

58. C. A. Buckner, A. L. Buckner, S. A. Koren, M. A. Persinger, and R. M. Lafrenie, The effects of electromagnetic fields on B16-BL6 cells are dependent on their spatial and temporal character, *Bioelectromagnetics*, vol. 38, pp. 165–74, 2017.

59. C. A. Buckner, A. L. Buckner, S. A. Koren, M. A. Persinger, and R. M. Lafrenie, Exposure to a specific time-varying electromagnetic field inhibits cell proliferation via cAMP and ERK signaling in cancer cells, *Bioelectromagnetics*, vol. 39, pp. 217–30, 2018.

60. P. J. Hore and H. Mouritsen, The radical-pair mechanism of magnetoreception, *Annu Rev Biophys*, vol. 45, pp. 299–344, 2016.

61. J. Juutilainen, M. Herrala, J. Luukkonen, J. Naarala, and P. J. Hore, Magnetocarcinogenesis: Is there a mechanism for carcinogenic effects of weak magnetic fields? *Proc Biol Sci*, vol. 285, p. 20180590, 2018.

62. M. Simko, Cell type specific redox status is responsible for diverse electromagnetic field effects, *Curr Med Chem*, vol. 14, pp. 1141–52, 2007.

63. F. Barnes and B. Greenenbaum, Some effects of weak magnetic fields on biological systems: RF fields can change radical concentrations and cancer cell growth rates, *IEEE Power Electron Mag*, vol. 3, pp. 60–8, 2016.

64. B. Marengo, M. Nitti, A. L. Furfaro, R. Colla, C. D. Ciucis, U. M. Marinari, M. A. Pronzato, N. Traverso, and C. Domenicotti, Redox homeostasis and cellular antioxidant systems: Crucial players in cancer growth and therapy, *Oxid Med Cell Longev*, vol. 2016, p. 6235641, 2016.

65. J. Kim, J. Kim, and J. S. Bae, ROS homeostasis and metabolism: A critical liaison for cancer therapy, *Exp Mol Med*, vol. 48, p. e269, 2016.

66. A. Schulze and A. L. Harris, How cancer metabolism is tuned for proliferation and vulnerable to disruption, *Nature*, vol. 491, pp. 364–73, 2012.

67. H. U. Simon, A. Haj-Yehia, and F. Levi-Schaffer, Role of reactive oxygen species (ROS) in apoptosis induction, *Apoptosis*, vol. 5, pp. 415–8, 2000.

68. M. O. Mattsson and M. Simko, Grouping of experimental conditions as an approach to evaluate effects of extremely low-frequency magnetic fields on oxidative response in in vitro studies, *Front Public Health*, vol. 2, p. 132, 2014.

69. H. Wang and X. Zhang, Magnetic fields and reactive oxygen species, *Int J Mol Sci*, vol. 18, 2017.

70. S. Falone, S. Santini, Jr., V. Cordone, G. Di Emidio, C. Tatone, M. Cacchio, and F. Amicarelli, Extremely low-frequency magnetic fields and redox-responsive pathways linked to cancer drug resistance: Insights from co-exposure-based in vitro studies, *Front Public Health*, vol. 6, p. 33, 2018.

71. S. J. Santini, V. Cordone, S. Falone, M. Mijit, C. Tatone, F. Amicarelli, and G. Di Emidio, Role of mitochondria in the oxidative stress induced by electromagnetic fields: Focus on reproductive systems, *Oxid Med Cell Longevity*, vol. 2018, p. 18, 2018.

72. A. C. Mannerling, M. Simko, K. H. Mild, and M. O. Mattsson, Effects of 50-Hz magnetic field exposure on superoxide radical anion formation and HSP70 induction in human K562 cells, *Radiat Environ Biophys*, vol. 49, pp. 731–41, 2010.

73. H. Lai and N. P. Singh, Magnetic-field-induced DNA strand breaks in brain cells of the rat, *Environ Health Perspect*, vol. 112, pp. 687–94, 2004.

74. S. Di Loreto, S. Falone, V. Caracciolo, P. Sebastiani, A. D'Alessandro, A. Mirabilio, V. Zimmitti, and F. Amicarelli, Fifty hertz extremely low-frequency magnetic field exposure elicits redox and trophic response in rat-cortical neurons, *J Cell Physiol*, vol. 219, pp. 334–43, 2009.

75. V. N. Binhi and F. S. Prato, A physical mechanism of magnetoreception: Extension and analysis, *Bioelectromagnetics*, vol. 38, pp. 41–52, 2017.

76. V. N. Binhi and F. S. Prato, Rotations of macromolecules affect nonspecific biological responses to magnetic fields, *Sci Rep*, vol. 8, p. 13495, 2018.

77. V. Binhi, A primary physical mechanism of the biological effects of weak magnetic fields, *Biophysics*, vol. 61, pp. 170–6, 2016.

78. G. Olsson, I. Y. Belyaev, T. Helleday, and M. Harms-Ringdahl, ELF magnetic field affects proliferation of SPD8/V79 Chinese hamster cells but does not interact with intrachromosomal recombination, *Mutat Res Genet Toxicol Environ Mutagen*, vol. 493, pp. 55–66, 2001.

79. M. Jadidi, M. Safari, and A. Baghian, Effects of extremely low frequency electromagnetic fields on cell proliferation, *Koomesh*, vol. 15, pp. 1–10, 2013.

80. J. E. Bae, J. Y. Do, S. H. Kwon, S. D. Lee, Y. W. Jung, S. C. Kim, and K. S. Chae, Electromagnetic field-induced converse cell growth during a long-term observation, *Int J Radiat Biol*, vol. 89, pp. 1035–44, 2013.

81. B. Segatore, D. Setacci, F. Bennato, R. Cardigno, G. Amicosante, and R. Iorio, Evaluations of the effects of extremely low-frequency electromagnetic fields on growth and antibiotic susceptibility of escherichia coli and pseudomonas aeruginosa, *Int J Microbiol*, vol. 2012, p. 587293, 2012.

82. W.-F. Bai, M.-S. Zhang, H. Huang, H.-X. Zhu, and W.-C. Xu, Effects of 50 Hz electromagnetic fields on human epidermal stem cells cultured on collagen sponge scaffolds, *Int J Radiat Biol*, vol. 88, pp. 523–30, 2012.

83. M. A. Cid, A. Ubeda, M. L. Hernandez-Bule, M. A. Martinez, and M. A. Trillo, Antagonistic effects of a 50 Hz magnetic field and melatonin in the proliferation and differentiation of hepatocarcinoma cells, *Cell Physiol Biochem*, vol. 30, pp. 1502–16, 2012.

84. F. Vincenzi, M. Targa, C. Corciulo, S. Gessi, S. Merighi, S. Setti, R. Cadossi, P. A. Borea, and K. Varani, The anti-tumor effect of A3 adenosine receptors is potentiated by pulsed electromagnetic fields in cultured neural cancer cells, *PLoS One*, vol. 7, p. e39317, 2012.

85. N. Filipovic, T. Djukic, M. Radovic, D. Cvetkovic, M. Curcic, S. Markovic, A. Peulic, and B. Jeremic, Electromagnetic field investigation on different cancer cell lines, *Cancer Cell Int*, vol. 14, p. 84, 2014.

86. Z. Akbarnejad, H. Eskandary, C. Vergallo, S. N. Nematollahi-Mahani, L. Dini, F. Darvishzadeh-Mahani, and M. Ahmadi, Effects of extremely low-frequency pulsed electromagnetic fields (ELF-PEMFs) on glioblastoma cells (U87), *Electromagn Biol Med*, vol. 36, pp. 238–47, 2017.

87. W. Y. Jeong, J. B. Kim, H. J. Kim, and C. W. Kim, Extremely low-frequency electromagnetic field promotes astrocytic differentiation of human bone marrow mesenchymal stem cells by modulating SIRT1 expression, *Biosci Biotechnol Biochem*, vol. 29, pp. 1–7, 2017.

88. S. Crocetti, C. Beyer, G. Schade, M. Egli, J. Frohlich, and A. Franco-Obregon, Low intensity and frequency pulsed electromagnetic fields selectively impair breast cancer cell viability, *PLoS One*, vol. 8, p. e72944, 2013.

89. S. Ivancsits, E. Diem, A. Pilger, H. W. Rudiger, and O. Jahn, Induction of DNA strand breaks by intermittent exposure to extremely-low-frequency electromagnetic fields in human diploid fibroblasts, *Mutat Res*, vol. 519, pp. 1–13, 2002.

90. F. Focke, D. Schuermann, N. Kuster, and P. Schar, DNA fragmentation in human fibroblasts under extremely low frequency electromagnetic field exposure, *Mutat Res*, vol. 683, pp. 74–83, 2010.

91. J. L. Phillips, N. P. Singh, and H. Lai, Electromagnetic fields and DNA damage, *Pathophysiology*, vol. 16, pp. 79–88, 2009.

92. M. J. Ruiz-Gomez and M. Martinez-Morillo, Electromagnetic fields and the induction of DNA strand breaks, *Electromagn Biol Med*, vol. 28, pp. 201–14, 2009.

93. C. Consales, M. Panatta, A. Butera, G. Filomeni, C. Merla, M. T. Carri, C. Marino, and B. Benassi, 50-Hz magnetic field impairs the expression of iron-related genes in the in vitro SOD1(G93A) model of amyotrophic lateral sclerosis, *Int J Radiat Biol*, pp. 1–28, 2018.

94. S. Kwee and P. Raskmark, Changes in cell proliferation due to environmental non-ionizing radiation 1. ELF electromagnetic fields, *Bioelectrochem Bioenerg*, vol. 36, pp. 109–14, 1995.

95. I. L. Cameron, M. S. Markov, and W. E. Hardman, Optimization of a therapeutic electromagnetic field (EMF) to retard breast cancer tumor growth and vascularity, *Cancer Cell Int*, vol. 14, p. 125, 2014.

96. X. Zhang, H. Zhang, C. Zheng, C. Li, and W. Xiong, Extremely low frequency (ELF) pulsed-gradient magnetic fields inhibit malignant tumour growth at different biological levels, *Cell Biol Int*, vol. 26, pp. 599–603, 2002.

97. V. V. Novikov, G. V. Novikov, and E. E. Fesenko, Effect of weak combined static and extremely low-frequency alternating magnetic fields on tumor growth in mice inoculated with the Ehrlich ascites carcinoma, *Bioelectromagnetics*, vol. 30, pp. 343–51, 2009.

98. I. Tatarov, A. Panda, D. Petkov, K. Kolappaswamy, K. Thompson, A. Kavirayani, M. M. Lipsky, E. Elson, C. C. Davis, S. S. Martin, and L. J. DeTolla, Effect of magnetic fields on tumor growth and viability, *Comp Med*, vol. 61, pp. 339–45, 2011.

99. I. L. Cameron, M. S. Markov, and W. E. Hardman, Daily exposure to a pulsed electromagnetic field for inhibition of cancer growth therapeutic implications. In: *Electromagnetic Fields in Biology and Medicine*, S. M. Markov, (Ed.), First edn. Boca Ratón, FL: CRC Press, 2015, pp. 311–7.

100. M. M. Rosado, M. Simko, M. O. Mattsson, and C. Pioli, Immune-modulating perspectives for low frequency electromagnetic fields in innate immunity, *Front Public Health*, vol. 6, p. 85, 2018.

101. J. Frahm, M. O. Mattsson, and M. Simko, Exposure to ELF magnetic fields modulate redox related protein expression in mouse macrophages, *Toxicol Lett*, vol. 192, pp. 330–6, 2010.

102. J. Frahm, M. Lantow, M. Lupke, D. G. Weiss, and M. Simko, Alteration in cellular functions in mouse macrophages after exposure to 50 Hz magnetic fields, *J Cell Biochem*, vol. 99, pp. 168–77, 2006.

103. C. Ross and B. S. Harrison, An introduction to electromagnetic field therapy and immune function: A brief history and current status, *J Sci Appl Biomed*, vol. 3, pp. 18–29, 2015.

104. H. Berg, B. Gunther, I. Hilger, M. Radeva, N. Traitcheva, and L. Wollweber, Bioelectromagnetic field effects on cancer cells and mice tumors, *Electromagn Biol Med*, vol. 29, pp. 132–43, 2010.

105. C. F. Blackman, S. G. Benane, and D. E. House, The influence of 1.2 microT, 60 Hz magnetic fields on melatonin- and tamoxifen-induced inhibition of MCF-7 cell growth, *Bioelectromagnetics*, vol. 22, pp. 122–8, 2001.

106. C. Sun, H. Yu, X. Wang, and J. Han, A pilot study of extremely low-frequency magnetic fields in advanced non-small cell lung cancer: Effects on survival and palliation of general symptoms, *Oncol Lett*, vol. 4, pp. 1130–4, 2012.

107. F. Ronchetto, D. Barone, M. Cintorino, M. Berardelli, S. Lissolo, R. Orlassino, P. Ossola, and S. Tofani, Extremely low frequency-modulated static magnetic fields to treat cancer: A pilot study on patients with advanced neoplasm to assess safety and acute toxicity, *Bioelectromagnetics*, vol. 25, pp. 563–71, 2004.

108. S. Tofani, M. Cintorino, D. Barone, M. Berardelli, M. M. De Santi, A. Ferrara, R. Orlassino, P. Ossola, K. Rolfo, F. Ronchetto, S. A. Tripodi, and P. Tosi, Increased mouse survival, tumor growth inhibition and decreased immunoreactive p53 after exposure to magnetic fields, *Bioelectromagnetics*, vol. 23, pp. 230–8, 2002.

109. L. Q. Yuan, C. Wang, K. Zhu, H. M. Li, W. Z. Gu, D. M. Zhou, J. Q. Lai, D. Zhou, Y. Lv, S. Tofani, and X. Chen, The antitumor effect of static and extremely low frequency magnetic fields against nephroblastoma and neuroblastoma, *Bioelectromagnetics*, vol. 39, pp. 375–85, 2018.

110. O. Rick, U. von Hehn, E. Mikus, H. Dertinger, and G. Geiger, Magnetic field therapy in patients with cytostatics-induced polyneuropathy: A prospective randomized placebo-controlled phase-III study, *Bioelectromagnetics*, vol. 38, pp. 85–94, 2017.

111. V. S. Kumar, D. A. Kumar, K. Kalaivani, A. C. Gangadharan, K. V. Raju, P. Thejomoorthy, B. M. Manohar, and R. Puvanakrishnan, Optimization of pulsed electromagnetic field therapy for management of arthritis in rats, *Bioelectromagnetics*, vol. 26, pp. 431–9, 2005.

112. J. Laszlo, J. Reiczigel, L. Szekely, A. Gasparics, I. Bogar, L. Bors, B. Racz, and K. Gyires, Optimization of static magnetic field parameters improves analgesic effect in mice, *Bioelectromagnetics*, vol. 28, pp. 615–27, 2007.

113. L. Makinistian, A novel system of coils for magnetobiology research, *Rev Sci Instrum*, vol. 87, p. 114304, 2016.

114. E. Aarholt, E. A. Flinn, and C. W. Smith, Effects of low-frequency magnetic fields on bacterial growth rate, *Phys Med Biol*, vol. 26, pp. 613–21, 1981.

115. E. Aarholt, E. A. Flinn, and C. W. Smith, Magnetic fields affect the lac operon system, *Phys Med Biol*, vol. 27, pp. 606–10, 1982.

116. V. N. Binhi and C. F. Blackman, Analysis of the structure of magnetic fields that induced inhibition of stimulated neurite outgrowth, *Bioelectromagnetics*, vol. 26, pp. 684–9, 2005.

117. C. F. Blackman, J. P. Blanchard, S. G. Benane, and D. E. House, Empirical test of an ion parametric resonance model for magnetic field interactions with PC-12 cells, *Bioelectromagnetics*, vol. 15, pp. 239–60, 1994.

118. C. F. Blackman, J. P. Blanchard, S. G. Benane, and D. E. House, The ion parametric resonance model predicts magnetic field parameters that affect nerve cells, *FASEB J*, vol. 9, pp. 547–51, 1995.

119. C. F. Blackman, J. P. Blanchard, S. G. Benane, and D. E. House, Effect of ac and dc magnetic field orientation on nerve cells, *Biochem Biophys Res Commun*, vol. 220, pp. 807–11, 1996.

120. S. K. Dhiman and P. Galland, Effects of weak static magnetic fields on the gene expression of seedlings of Arabidopsis thaliana, *J Plant Physiol*, vol. 231, pp. 9–18, 2018.

121. P. Doynov, H. D. Cohen, M. R. Cook, and C. Graham, Test facility for human exposure to AC and DC magnetic fields, *Bioelectromagnetics*, vol. 20, pp. 101–11, 1999.

122. D. H. Nguyen, L. Richard, and J. F. Burchard, Exposure chamber for determining the biological effects of electric and magnetic fields on dairy cows, *Bioelectromagnetics*, vol. 26, pp. 138–44, 2005.

123. A. W. Thomas, D. J. Drost, and F. S. Prato, Magnetic field exposure and behavioral monitoring system, *Bioelectromagnetics*, vol. 22, pp. 401–7, 2001.

124. J. Karpowicz, Environmental and safety aspects of the use of EMF in medical environment. In: *Electromagnetic Fields in Biology and Medicine*, S. M. Markov (Ed.), First edn. Boca Ratón, FL: CRC Press, 2015, pp. 341–62.

125. A. P. Colbert, H. Wahbeh, N. Harling, E. Connelly, H. C. Schiffke, C. Forsten, W. L. Gregory, M. S. Markov, J. J. Souder, P. Elmer, and V. King, Static magnetic field therapy: A critical review of treatment parameters, *Evid Based Complement Alternat Med*, vol. 6, pp. 133–9, 2009.

126. L. Makinistian, D. J. Muehsam, F. Bersani, and I. Belyaev, Some recommendations for experimental work in magnetobiology, revisited, *Bioelectromagnetics*, vol. 39, pp. 556–64, 2018.

127. P. A. Valberg, Designing EMF experiments: What is required to characterize exposure? *Bioelectromagnetics*, vol. 16, pp. 396–401 discussion 402–6, 1995.

128. M. Misakian, A. R. Sheppard, D. Krause, M. E. Frazier, and D. L. Miller, Biological, physical, and electrical parameters for in vitro studies with ELF magnetic and electric fields: A primer, *Bioelectromagnetics*, vol. Suppl 2, pp. 1–73, 1993.

129. F. S. Prato, D. Desjardins-Holmes, L. D. Keenliside, J. M. DeMoor, J. A. Robertson, and A. W. Thomas, Magnetoreception in laboratory mice: Sensitivity to extremely low-frequency fields exceeds 33 nT at 30 Hz, *J R Soc Interface*, vol. 10, p. 20121046, 2013.

130. M. N. Zhadin, Review of russian literature on biological action of DC and low-frequency AC magnetic fields, *Bioelectromagnetics*, vol. 22, pp. 27–45, 2001.

131. U. Lucia, G. Grisolia, A. Ponzetto, and F. Silvagno, An engineering thermodynamic approach to select the electromagnetic wave effective on cell growth, *J Theor Biol*, vol. 429, pp. 181–9, 2017.

132. F. P. Costa, A. C. de Oliveira, R. Meirelles, M. C. Machado, T. Zanesco, R. Surjan, M. C. Chammas, M. de Souza Rocha, D. Morgan, A. Cantor, J. Zimmerman, I. Brezovich, N. Kuster, A. Barbault, and B. Pasche, Treatment of advanced hepatocellular carcinoma with very low levels of amplitude-modulated electromagnetic fields, *Br J Cancer*, vol. 105, pp. 640–8, 2011.

9 Terahertz Electromagnetic Fields in Diagnostic and Therapeutic Settings – Potentials and Challenges

Myrtill Simkó and Mats-Olof Mattsson

CONTENTS

INTRODUCTION

The use of THz (THz = 10^{12} Hz, 4 meV photon energy) frequencies/technologies, specifically THz spectroscopy, is an established technology, which has been used for decades in the submillimeter and far-infrared branches of astronomy. However, this technology has been further developed and has applications in industrial testing procedures, airport security systems as body scanners, and the military sector. Since the THz radiation is not ionizing and has low photon energy, it does not damage living tissues and/or the DNA. Some frequencies within the THz range can penetrate several millimeters of tissue with low water content (e.g., fat tissue) and reflect back, and thus, differences in water content and density of a tissue can be detected. Therefore, THz spectroscopy and imaging is increasingly used also in medical diagnostics, such as the detection of epithelial cancers.

THz radiation resides in the frequency region from 300 GHz (the high-frequency end of the millimeter wave (MMW) band) up to 10 or even 30 THz (the edge of the far-infrared light band; wavelength: 1 mm–15 μm), which also is called submillimeter radiation, THz waves, THz light, T-rays, T-waves, T-light, T-lux, THz gap, or just THz (Sirtori 2002). Beside the use of THz radiation in astronomy ("the submillimeter band"), technologies have been developed for both ends of the THz spectrum (MMW and optical waves).

The penetration depth of THz is typically less than that of microwaves, but it can penetrate nonconducting materials. However, THz fields have limited penetration capacity through fog and clouds and cannot penetrate liquid water or metal. Because of the strong absorption in water, THz waves have a limited penetration depth in tissues, up to a few hundred micrometers (Jae Oh et al. 2013).

Nevertheless, different materials and also living tissues are semitransparent to THz waves and have so-called "THz fingerprints". For example, organic molecules show strong absorption from GHz to THz through rotational and vibration transitions, providing such fingerprints in the THz band. Thus, those molecules can be imaged, identified, and analyzed by means of THz radiation (Romanenko et al. 2017). Therefore, it is even possible to get information about dynamic changes of large molecules in the subpicosecond range without causing any tissue or cell damage, allowing investigation even in *in-vivo* or in "wet lab" conditions (Romanenko et al. 2017).

THz has also received interest since it can be used for making imaging possible in locations not accessible to relevant investigations with conventional techniques and with reasonable resolution. Suggested applications may include dermatological investigations to identify cancerous lesions, wound assessment, and even dental investigations (Humphreys et al. 2004). Clinical trials using THz imaging that are underway worldwide are met with high expectations since THz imaging is a noninvasive technique for early detection of cancer. The case for cancer diagnostics has been recently reviewed by Yu and coworkers (Yu et al. 2012).

In contrast, the therapeutic potential of THz waves is sparingly investigated, although, for example, certain skin conditions could be targets for THz treatment (Titova et al. 2013). The potential therapeutic effects of THz exposure may be due to heating, which would be the case if exposures to high-intensity continuous waves are used (Kristensen et al. 2010). On the other hand, calculations and modeling suggest that if the exposures are to short pulses, any intracellular heating is negligible and the effect would be mediated by nonthermal interactions between the THz irradiation and biomolecules (Kristensen et al. 2010; Cherkasova et al. 2009). Alexandrov et al. (2010) have furthermore argued that the main effect of THz radiation is to influence the stability of the DNA (and thus to influence gene expression). By means of modeling approaches, they found that spatial perturbation of DNA could occur above a certain threshold that is determined by the intensity and the frequency of the THz field and by the exposure duration.

Regarding any therapeutic applications of THz waves, it is mainly the potentials and prospects for cancer treatment that have been highlighted in the sporadic investigations which rely on certain non-thermal cellular effects under THz exposure (see Fedorov

(2017) for a recent overview including studies related to medical use from the former Soviet Union). Sporadic evidence has been provided that THz radiation interacts with cellular components at multiple levels including chromosomes, DNA, genes, and proteins (reviewed in Wang et al. 2013). Moreover, THz radiation has been hypothesized to be a useful, noncontact tool for the selective control of specific genes and cellular processes. As a matter of fact, continuous wave 2.45 THz radiation in human cells has been demonstrated to trigger significant changes in the expression of numerous mRNAs and microRNAs affecting specific intracellular pathways that are not affected in thermally matched cells exposed to bulk heating (Echchgadda et al. 2016).

As THz waves are nonionizing but have specific properties, especially its high sensitivity to soft tissues, there is an increasing interest in their biomedical applications, both *in vivo* and *ex vivo*. Furthermore, due to the technological development, a number of novel imaging technologies have been developed, although they are still in the early stages of development. In this chapter, we aim to give an overview about the state of the art of medical THz applications and to provide an evaluation of the relevance of major breakthroughs within a reasonably near future.

SPECTROSCOPY AND IMAGING

THz Spectroscopy Principles

THz spectroscopy typically uses a single-point measurement of a homogenous sample, and the resulting THz electric field can be recorded as a function of time. THz spectroscopic techniques are often performed as time-domain spectroscopy (TDS) or frequency-domain spectroscopy (FDS), which are the major equipment and sensing tools in THz applications.

The THz time-domain spectroscopy (THz-TDS) is a spectroscopic technique in which the properties of a material are probed with short pulses of THz radiation. Photoconductive switches irradiated by femtosecond lasers can generate pulsed THz radiation. The signal can be Fourier transformed to provide meaningful spectroscopic information due to the broadband nature of pulsed THz radiation in the frequency range 0.1–5 THz (Yu et al. 2012). THz-TDS measures not only the power but also the electric field of a pulse. Thus, THz-TDS measures both the amplitude and phase information of the frequency components it contains. The amplitude and phase are directly related to the absorption coefficient and refractive index of a sample, and thus the complex permittivity is obtained without requiring any further analysis.

In the THz frequency-domain spectroscopy (THz-FDS), the materials are probed with continuous-wave (CW) THz radiation. The radiation is generated by optical high-bandwidth photoconductors and allows us to determine optical properties of materials in a broad frequency band, including the 10 THz range. Both pulsed and continuous THz waves are being applied for noninvasive testing of different materials.

THz spectrometers can be configured in various ways depending upon the sample properties and geometrical limitations. Transmission measurements function very well on samples that are moderately absorbing and are of low dispersion; materials/samples which are highly absorbing, reflective, or dispersive may instead be best measured in a reflection or attenuated total reflection (ATR) geometry (Baxter and Guglietta 2011).

A relevant advantage of THz spectroscopy is that it is nondestructive since it works with long wavelengths and low energy that does not induce effects such as phase changes or photochemical reactions to living organisms. Biochemical molecular modes, such as weak hydrogen bonding and lattice vibration, lie across the 0.1–10 THz region, and the energy of THz waves of several meV is so low that biological samples are rarely damaged when exposed to a THz beam, which obviously is different from exposures to ultraviolet and X-ray beams (Lee et al. 2018).

In recent decades, there has been increasing interest in applying THz spectroscopy to probe and characterize various biomaterials because most low-frequency biomolecular motions (including vibration and rotation of the molecular skeleton) lie in the same frequency range as THz radiation (for a recent review see Yang et al. 2016). Consequently, biomolecules can be effectively characterized and recognized according to their characteristic spectral fingerprints (Yang et al. 2016).

THz Imaging

Since THz radiations do not affect human tissue because of their low energy, THz imaging has the potential to become a new modality for medical imaging as a noninvasive diagnostic tool. THz imaging technology can extract the intrinsic and morphological properties synchronously from amplitude and phase information. Essential tools for THz biomedical imaging are (i) differences in water content, (ii) structural variations in tissues, and (iii) artificial contrast enhancement, as reviewed by Yang et al. (2016). Structural variations are different tissue microenvironments, different cellular morphologies, and different forms of biomolecules, all of which can change the quality and quantity of the image contrast. Artificial contrast enhancement needs contrast-enhancing agents, e.g., nanorods and temperature increase. This induces the absorption of THz radiation by water molecules (Rønne et al. 1997; Oh et al. 2009, 2012) that makes, e. g., certain carcinomas identifiable due to their high water content.

THz pulsed imaging (TPI) can be seen as an extension of the THz-TDS method. 2D images can be gained with THz-TDS by spatial scanning of either the THz beam or the object itself as spectral information. Thus, geometrical images of the sample can be computer generated to obtain 3D structures and get a view of a layered structure.

By using THz imaging, various biomaterials can be detected and identified. These include molecules (such as nucleic acids and proteins), cells, and tissues. This tool is especially interesting for medical applications such as cancer diagnosis and is also an alternative to other diagnostic technologies for wound assessment or teeth structure imaging for caries identification. Examples of these diagnostic tools will be discussed in the next section.

MEDICAL DIAGNOSTICS

Cancer Detection

The main idea of using THz imaging for cancer diagnostics is to map changes in water content of the tissue in question. It seems that cancerous or diseased tissues contain more interstitial water as a result of abundant vascularity or tissue edema than healthy tissue. This fact can be used for THz imaging by THz reflection systems. Water content differences are not only limited to differences between healthy and diseased tissues; it is also seen that fatty and muscle tissues have different THz optical properties, a difference that is dominated by their different water contents (Sun et al. 2017b). Interestingly, water does not exclusively cause the differences observed in the refractive index and absorption coefficient between healthy and, e.g., cirrhotic liver tissues, as shown by Sy et al. (2010). Also structural changes in the cells were responsible in part for the changes of the THz optical properties (Sun et al. 2017a). Thus, the refractive index and absorption coefficient of the tumor tissue are higher than those of a healthy tissue, and such a difference is possible not only due to the higher water content but also due to structural changes of the cancerous tissue. Huang and coworkers used TPI to investigate the properties of several types of healthy organ tissues, including liver, kidney, heart muscle, leg muscle, pancreas, and abdominal fat tissues (Huang et al. 2009). In a later work, they detected a frequency-dependent refractive index and absorption coefficient difference between the tissues (Yu et al. 2012). Thus, fat tissue has a much lower absorption coefficient and refractive index than the kidney or liver tissues. The use of TPI results in differences in tissue water concentration (because of its attenuation) and thus water absorption. In this way, the contrast between muscle and fat tissues and also between healthy skin and basal cell carcinoma in the skin can be visualized. A further aim is to map the exact margins of early-stage tumors by using THz-based methods. The contrast between healthy and cancerous tissue can, for instance, be used during surgery, where cancerous tissue can be detected and removed directly in real-time settings by using THz imaging at high resolution. This would then minimize the cutting off of healthy tissue and exclude the leaving of any part of the carcinoma. Furthermore this reduces the need for a repeat invasive operative intervention(s) and accordingly the risk for the patient.

Skin Cancer

Melanoma and cutaneous squamous cell carcinoma (cSCC) are high-risk skin cancers with the potential to metastasize and cause lethality. Basal cell carcinoma (BCC) is the most common type of skin cancer, which also has the potential to infiltrate and damage surrounding tissues. An early diagnosis and therapy at early stages of tumor growth is necessary to improve treatment success and patient survival.

One of the first THz imaging studies in this area was performed by Woodward et al. (2003a) showing that the TPI technique (in the frequency range 0.1–2.7 THz) is useful to detect BCC *ex vivo*. The authors investigated 21 samples from patients in a blind manner, and results were compared to histology. Cancerous tissue showed increases in THz absorption compared to normal tissue, which was due to either an increase in the interstitial water within the cancerous tissue or a change in the vibrational modes of water molecules with other functional groups. Irrespective of mechanism, the THz imaging technique identified all BCCs in these samples. Since then this technology has been further developed (Woodward et al. 2002, 2003b; Pickwell et al. 2004; Wallace et al. 2004; Chan and Ramer 2018), but it has still not found its full implementation in medical diagnostics. However, early detection and diagnosis of skin cancer is an important issue since it is one of the most common cancers.

Joseph et al. (2014) showed that cross-polarized THz imaging can identify even nonmelanoma skin cancer (NMSC). Fan et al. (2017a) used combined TPI and optical imaging (polarization enhanced reflectance) for NMSC margin delineation. The authors detected that cancerous tissues differ significantly at 0.47 THz from healthy skin tissues and suggested that this technology has a good potential for quick intraoperative delineation of cancers. Other combined technologies such as the use of nanoparticle contrast agents enhances the THz imaging signals by 20–30 times compared with the signals from cancer cells without nanoparticle contrast agents (Oh et al. 2009; Son 2013). This is obtained by increasing the THz reflection signal beam in the cancer cells with nanoparticles upon their irradiation with an infrared laser, due to the temperature rise of water in cancer cells.

Dysplastic nevi are melanoma precursors, and they thus need to be detected and eliminated very early on to escape cancer development. Accordingly, dysplastic and nondysplastic skin nevi were measured by Zaytsev et al. (2015) using *in vivo* THz pulsed spectroscopy. The authors investigated the dielectric characteristics of healthy skin and dysplastic and nondysplastic skin nevi and detected clear differences in dielectric permittivity between dysplastic and nondysplastic skin nevi.

Since it is possible to use THz imaging to differentiate spectral fingerprints of surface proteins that are markers for certain cancers, it may also be possible to identify these cancers at even earlier stages.

WOUND AND BURN INVESTIGATION

Skin burn injuries are common and may in certain cases need additional treatment beyond the natural healing processes that are going on. Then it is critical to know the state of the burn wound, how far the healing is progressing, and how deep the wound is. There are a number of tools available for monitoring burn wound healing (see, e.g., Dutta et al. (2016) for a recent overview), although none of them is considered to be a *panacea* that can replace all the others. In that setting, THz radiation spectroscopy has been investigated regarding its appropriateness as a tool for determining wound depth, severity, and healing progress. This knowledge is needed for optimal treatment to improve healing outcomes.

The key feature of skin, which THz spectroscopy is relying on, is the water content in the superficial layer (primarily the stratum corneum of the epidermis) of the skin. Different states of the skin are represented by different water contents, where, e.g., burn wounds will exhibit different zones with varying water content, which can be observed by investigating THz reflectivity and refraction index (Dutta et al. 2016; Sun et al. 2017b).

There are a number of studies done both *in vivo* and *ex vivo* in several species that have been addressing burn wound diagnosis by means of THz spectroscopy. The possible clinical applicability has been based on studies such as one from (Pickwell et al. 2004), which showed that normal human forearm and palmar skin *in vivo* can be scanned with THz radiation (0.1–3 THz; pulses with a full-width half maximum of 0.3 ps at a power of 100 nW) which provides information about absorption/reflectivity and refractive index. Changes in these parameters were seen to correlate well with the skin's dryness and thus water content. The same imaging approach was also used to demonstrate that the THz radiation can penetrate through wound dressing on the palm of the hand and measure the thickness of the stratum corneum (Huang et al. 2007).

More direct studies of burned tissue by means of THz imaging have also been done *ex vivo*, e.g., on paper (Dougherty et al. 2007). The authors employed THz-TDS (1.5 THz) to investigate tissue samples from chicken breast and from bovines that were sectioned and investigated in the form of 0.4 mm thick tissue slides. The slides were either used as controls or locally heated with a steel rod (350°C) and subsequently investigated. Significant contrast was seen between burned areas, even when covered with gauze, and controls for both species.

One of the key features of a burn wound is the edema (tissue swelling) that occurs. The edema is caused by the accumulation of interstitial water and is thus a sign of increased water content that THz imaging can detect. Several animal studies performed on male Sprague-Dawley rats have elegantly shown the formation of this edema and how it contrasts with the surrounding healthy tissue. Tewari et al. (2012) used a 0.525 THz imaging system to follow development of the edema by taking images every 15–30 min for 7 h after the initiation of burn injuries. It was possible to monitor both the formation and the dissipation of the edema, where the increased local water content was seen as increase in reflectivity. Similar observations were made by Arbab et al. (2011, 2013) who induced second and third degree burns and followed the wound development up to 72 h post burn. The investigations used 0.5–0.7 THz irradiation and found that burn wounds displayed a 30% increase in reflectivity compared to controls. In both studies, the authors also observed that it is possible to detect not only water content but also changes to skin structures such as microvascular capillaries, hair follicles, and sweat glands and their ducts with THz spectroscopy. More severe burns were seen in histological sections to cause more damages to these structures, which was displayed as increased reflectivity in THz spectroscopy. The authors concluded that not only water content but also tissue structure is an endpoint that can be investigated with THz radiation. The combination of these parameters can thus provide information regarding the severity of tissue damage associated with burn damages.

The precision of the THz imaging approach regarding edema formation was determined in a study by Bajwa et al. (2017a) who compared THz imaging *in vivo* with MRI investigations. The experiments were performed on male Sprague-Dawley rats, in which burns were induced on the abdominal skin. Wound development was followed for up to 270 min. Two types of burns were induced (deeper and more superficial, respectively), and in both cases, the results showed a positive correlation between THz and MRI data regarding the extent of edema. The THz imaging used a frequency of 0.525 THz with a 125 GHz bandwidth. The same group (Bajwa et al. 2017b) could also show the usefulness of THz imaging in a study on Lewis rats where skin tissue survival after transplantation surgery was investigated ("tissue flips"). Pieces of skin were transferred from one body part to another and put in place in either a fashion that promoted survival or in a way that caused the transplanted tissue to die. The THz investigation could reveal the prognosis after 24 h, whereas clinical ocular investigation was able to make such a judgment after 48 h.

There are also studies on humans *in vivo* that show the feasibility of THz radiation to demonstrate wound healing. Fan et al. (2017b) used a 1 THz scanning frequency to investigate scar development in four patients up to a few months after injury. Differences in refraction index persisted after this time, whereas contrast in reflectivity compared to healthy skin diminished with time. Importantly, the authors also found that the THz imaging could show if wounds developed so-called hypertrophic scars, which could need additional (surgical) treatment. Application of an occluding structure (such as a plaster) on a skin wound will cause changes in water diffusivity (water passage through a membrane per unit of time) so that more water is retained. This is a positive event since increased hydration is promoting healing by several mechanisms. The degree of hydration under a plaster can be determined by THz imaging, as shown by Sun et al. (2017a) who employed a scanning protocol using 0.3–0.8 THz radiation on human volar forearm.

In summary, the studies performed on burn wounds and other skin conditions suggest that THz imaging has the potential to distinguish between healthy and damaged skin even beneath wound dressing. Most of the diagnostic potential comes from the fact that changes in skin conditions cause changes in tissue hydration, which can be picked up with THz waves. There is also possibly tissue and cell structure changes that can contribute to differences in THz absorption spectra. This is however sparingly investigated. THz imaging may be a promising approach that can replace some other tools, but so far, there are no larger *in-vivo* or human clinical studies that have investigated the realism in such a scenario.

ENDOSCOPIC APPLICATION

The probability to distinguish cancerous tissues from normal ones using THz spectroscopy has also stimulated research and development that aim to develop instruments that are suitable for endoscopy of the gastrointestinal tract. The market already contains instruments based on several technologies that accomplish such a diagnosis. These approaches include MRI, CT, and PET scans; optical coherence tomography; and also conventional endoscopy, which uses white light. The problem with all these technologies is that none of them can provide real-time high-resolution images with microscopic information. Hence, THz spectroscopy could possibly fill this gap if devices could be constructed to function as endoscopes and if the tissue investigation could distinguish between normal and dysplastic tissues with high-enough sensitivity and specificity.

One of the clinical areas where there is some research regarding the potential for THz endoscopy as a diagnostic tool is colorectal cancer. Requirements for a useful instrument were recently outlined (Doradla et al. 2017) and include engineering challenges such as the construction of a single-channel instrument. It is also necessary to collect data regarding absorbance and reflectivity of relevant tissues when exposed to THz irradiation and that the contrast between normal and diseased tissue is large enough to be recognized by the instrument.

The proof of concept has indeed been shown in several studies. Thus, dedicated feasibility studies have been published by, e.g., Reid et al. (2011). The study investigated both healthy and dysplastic tissues from 30 patients. Absorption and refractive index spectra were collected, and models were based on up to 17 different parameters calculated from time- and frequency-domain signals. These models were sufficient to differentiate between healthy and diseased tissues and also to indicate which tissue was not only diseased but also specifically cancerous. The spectra were obtained in the frequency region from 0.3 to 0.8 THz, where specifically frequencies of 0.44–0.60 THz were used for the statistical models. Similar observations were made in several other studies (Wahaia et al. 2011, 2016; Doradla et al. 2014; Chen et al. 2015; Kašalynas et al. 2016), using frequencies most often between 0.5 and 0.6 THz. Importantly several of these studies could distinguish between normal and cancerous tissues in specimens that were fixed in formalin and dehydrated. This would suggest that not only the water content but also structural changes can be detected by the THz irradiation (Kašalynas et al. 2016; Doradla et al. 2014). The studies also revealed that the cancerous tissues absorbed more of the THz waves, with an increase in contrast corresponding to 23% (Wahaia et al. 2016).

A promising study regarding the reliability of this diagnostic approach was published by Chen et al. (2015). They made a quantitative analysis of the absorbance at 0.30–0.335 THz in freshly surgically excised colon tissue with the aim of determining the degrees of sensitivity and specificity. All samples were subjected to normal histopathological microscopic investigations for confirmation of diagnosis. Based on 58 specimens from 31 patients, it was found that the THz investigation found all cases of cancer (sensitivity) and that all patients with healthy tissues were identified as healthy (specificity). The absorbance of the healthy tissue was on average at $9.3–9.8\,mm^{-1}$, where the higher absorbance in the dysplastic tissue was determined to be $11.4–11.8\,mm^{-1}$.

Another cancer form was investigated by Ji et al. (2015) who showed that THz spectroscopy could diagnose early gastric cancer (EGC). They collected reflectivity data from eight patients with EGC. Of these, six were clearly identified on THz images, and they correlated very well with corresponding microscopic images.

In summary, although studies have shown that THz radiation, roughly between 0.3 and 0.8 THz, can identify tumor tissues based on absorption, reflectivity, and refraction index, the technique has seemingly not passed beyond the proof-of-concept stage. There may still be some questions regarding what actually the THz imaging can pick up, since tissue water content cannot be the only indicator of dysplasia, based on the tissue treatment protocols. Finally, practically useful endoscopes are likely not available in the very near future, and it is unknown to what extent development work is in progress.

DENTAL CARE APPLICATION

Another area where THz imaging has been suggested to have potential is in dental medicine. A number of potential specific applications were suggested in a recent review (Asha et al. 2015). The suggestions included detection of caries and teeth erosion, periodontosis diagnosis, detection of oral malignant melanoma, and monitoring of dental restorative materials. There is some support from various studies for all these areas, although any clinical use seems to be far away in time.

Characterization of different dental tissues have been performed by THz imaging, in studies by e.g., Sim et al. (2009), who could distinguish between enamel and dentin. They scanned the different tissues *ex vivo* at 0.1–1 THz and found higher refractive index for enamel than for dentin. Kamburoglu et al. (2014) could distinguish between primary and permanent teeth when doing a THz TDS at 0.5 THz. However, the distinction was not due to differences in refraction index (which was the same) but due to differences in absorption coefficient. This characteristic was significantly higher in primary teeth.

Still other characterizations have been shown by (Crawley et al. 2003) and (Hirmer et al. 2012). The former investigated the enamel thickness in 14 samples and obtained THz-generated values that were within 10 μm of the expected results, based on conventional measurements. Hirmer et al. (2012) compared a number of imaging techniques using wavelengths from visible light down to THz waves in studies of pulp vitality and blood flow, which was feasible also with THz frequencies.

An area with seemingly some practical potential is detection of (early stage) caries (Smith 2004). Several studies have shown results where healthy and carious enamel samples exhibit differences in absorption coefficient and refractory index. Crawley et al. (2003) found that the absorption coefficient in carious teeth was higher than in healthy ones, using a pulsed THz imaging system spanning between 0.1 and 2.7 THz. Similar findings were also reported by Kamburoglu et al. (2014, 2018) in the frequency range of 0.2–0.5 THz.

Still another application that has been investigated in a pilot study is the curing of dental composite materials (Schwerdtfeger et al. 2012). The curing of three composites was investigated with 0.3–1.5 THz pulses, and refractive index and absorption coefficient were calculated. The investigation concluded that the material density increased, which led to a higher refractive index, whereas the absorption coefficient was unaffected.

The few studies available in this niche area suggest that THz imaging can be an alternative to other investigations, primarily regarding caries detection and monitoring of dental filling materials. However, the studies are done *ex vivo*, and it seems that substantial work needs to be done before clinically useful instruments are developed and properly tested.

OTHER APPLICATIONS

In addition to *in vivo* THz imaging, *ex vivo* imaging efforts have also been applied to identify or detect other types of cancers, such as breast cancer (Hassan et al. 2012; Bowman et al. 2014; Grootendorst et al. 2017; Bowman et al. 2018; Ashworth et al. 2009) and colon cancer (Wahaia et al. 2011; Doradla et al. 2013; Eadie et al. 2013), and other types of diseases. For instance, diabetic foot ulcer is a serious complication of diabetes and often initially silent because of attendant neuropathy. Diabetic foot syndrome causes dehydration of the foot skin, increasing the risk of developing severe ulcers, which can eventually necessitate amputation if not treated. Thus, there is a need to detect this condition in its early stages. Hernandez-Cardoso et al. performed a proof-of-concept study to investigate if THz imaging could be a useful technique (Hernandez-Cardoso et al. 2017). The authors applied THz reflection imaging as the screening method for early stages of feet deterioration in diabetic patients and obtained very promising results.

THz imaging of corneal tissue water content (CTWC) is a proposed method for early, accurate detection and study of corneal diseases (Taylor et al. 2015a,b; Sung et al. 2018a). A prototype of noncontact THz imaging of *in vivo* human cornea was recently presented by Sung et al. (2018b) where the authors discuss strategies for optimizing the imaging system design for clinical use.

Hearing impairment in adults is often caused by the late diagnosis and inappropriate treatment of *otitis media* (Kenna 2015). Therefore, a THz otoscope with potential for diagnosing *otitis media* was developed by Ji et al. (2016). The THz otoscope primarily detects the presence of pus on the inner side of the tympanic membrane, which is equivalent to the presence of water.

SAFETY ASPECTS

As already mentioned, THz photons have meV energies, which are very low; thus, they lack sufficient energy to remove tightly bound electrons. Therefore, THz radiation is unlikely to damage tissues and DNA, and so it is considered to be safe to the human body. However, thermal effects can occur leading to tissue damage. Such thermal effects of THz waves were investigated by, e.g., Wilmink et al. (Wilmink et al. 2010). The authors exposed human cells to THz radiation (2.52 THz, $227\,mW/cm^2$, 1–40 min) and found that the majority of Jurkat cells in their experiments underwent apoptosis or necrosis after exposure for 20 min or longer. Exposed Jurkat cells survived for 30 min (60%), and only 20% lived longer. Using human dermal fibroblasts (2.52 THz; $84.8\,mW/cm^2$; durations of 5, 10, 20, 40, and 80 min), they detected a significantly decreased cell viability if the temperature was increased by 3°C. Furthermore the expression of both heat shock proteins and DNA damage markers showed an increasing trend (Wilmink et al. 2011).

Other aspects of safety-related studies are the nonthermal biological effects of THz waves. As shown in Table 9.1, several *in-vitro* studies detected nonthermal effects investigating different biological endpoints (cell proliferation and cell cycle regulation, cell viability, genotoxicity, gene and protein expression, nerve cell function, morphology, oxidative stress, differentiation). As an example, Olshevskaya et al. (Olshevskaya et al. 2008) have observed that THz radiation causes injury to the morphology of neurons in a power- and wavelength-dependent manner *in vitro*. The power densities used in these studies are below the levels that are considered to cause tissue heating; thus the exposure should allow "nonthermal" effects to occur.

A summary analysis of experimental studies investigating possible nonthermal effects of THz radiation was recently provided by Mattsson et al. (2018). They found that there are results in favor of the possibility that nonthermal effects can occur, but such studies are few in number, and the quality of the work is often questionable, so it is not meaningful to draw any further conclusions regarding the presence of such effects and their possible relevance for safety assessments.

However, no effects on genotoxicity and cell proliferation were detected after THz exposure (Zeni et al. 2007; Scarfi et al. 2003; Koyama et al. 2016; Hintzsche et al. 2012, 2013; Williams et al. 2013), morphological changes (Koyama et al. 2016), gene and protein expression (Wilmink and Grundt 2011), cell differentiation (Williams et al. 2013; Clothier and Bourne 2003; Bourne et al. 2008), and also no effects were detected on intracellular stress and antioxidants levels (Bourne et al. 2008). For example, Bourne et al. (2008) exposed primary cultured human keratinocytes and

TABLE 9.1

Selected *In-Vitro* Publications Dealing with Biological Effects (Endpoints) Relevant for Safety Assessment

References	Cell Types	Frequency	Power Density	Exposure Duration	Endpoint/s
Alexandrov et al. (2011)	Mesenchymal mouse stem cells	10 THz, 2.52 THz	1, 3 mW/cm²	2 and 9h	Gene expression
Amicis et al. (2015)	Human primary fibroblasts HFFF2	0.1–0.15 THz			Cell cycle analysis, DNA damage, micronuclei, DNA repair, telomere length
Bock et al. (2010)	Mouse mesenchymal stem cells	10 THz	1 mW/cm²	2 and 6h	Gene expression, proliferation
Bogomazova et al. (2015)	Human embryonic stem cells (hESCs)	2.3 THz	Average power 1.4 W/cm², peak intensity 4 kW/cm²	Not provided	hESC morphology, chromosomal aberrations, DNA double-strand breaks, mitotic index, gene annotation enrichment analysis
Borovkova et al. (2017)	C6 rat glial cell line	0.12–0.18 THz	3.2 mW/cm²	1, 2, 3, 4, and 5 min	Mitochondrial membrane potential (MMP)/apoptosis
Demidova et al. (2013)	*E. coli*/pKatG-gfp	1.5, 2.0, and 2.3 THz	1.4 W/cm²	10 min	Changes in green fluorescent protein (GFP) expression
Demidova et al. (2016)	*E. coli*/pEmrR-GFP, *E. coli*/pCopA-GFP	1–3 THz	1.4 W/cm²	15 min	Oxidative stress
Doria et al. (2004)	Human lymphocytes	0.13 THz	0.16, 0.23, 5.6, 7.8, 11.1 mW/cm²	2 and 60 min	DNA damage
Echchgadda et al. (2014)	Jurkat cells	2.52 THz	636 mW/cm²	40 min	mRNA, miRNA expression
Echchgadda et al. (2016)	Jurkat cells	2.52 THz	637 mW/cm²	41 min	Global gene expression, miRNAs, mRNAs, signaling pathways

(Continued)

TABLE 9.1 (Continued)

Selected *In-Vitro* Publications Dealing with Biological Effects (Endpoints) Relevant for Safety Assessment

References	Cell Types	Frequency	Power Density	Exposure Duration	Endpoint/s
Gallerano et al. (2014)	Human fibroblasts HFFF2	0.10 and 0.15 THz	Few mW		Cytogenetic markers (micronucleus assay, comet assay, chromosome aneuploidy), cell cycle kinetics, protein expression, and ultrastructural observations
Hintzsche et al. (2011)	Human hamster-hybrid cells (AL cells)	0.106 THz	0.043, 0.43, 4.3 mW/cm^2	30 min	Mitosis disturbances, genotoxicity
Korenstein-Ilan et al. (2008)	Human lymphocytes	0.1 THz	0.031 mW/cm^2	1, 2, and 24 h	Genetic effects, proliferation
Olshevskaya et al. (2008)	Isolated neurons of *Lymnaea stagnalis*	0.7–4.3 THz	0.3, 1–10, 30 mW/m^2	Not provided but 1 min in results	Neogenesis of a neural network
Sergeeva et al. (2016)	*Salmonella typhimurium*, TA98, and TA102	2.3 THz	1.4 W/cm^2	5, 10, or 15 min	Mutagenicity test
Titova et al. (2010)	Artificial human skin tissues	0.375 THz	57 mW/cm^2	30 min	H2AX phosphorylation,
Titova et al. (2013)	Artificial human skin tissues	0.2–2.5 THz		10 min	Whole-genome gene expression profiling
Wilmink et al. (2010)	Jurkat cells	2.52 THz	227 mW/cm^2	5, 10, 20, 30, and 40 min	Proliferation, apoptosis
Wilmink et al. (2011)	Human dermal fibroblasts	2.52 THz	227 mW/cm^2	5, 10, 20, 30, and 40 minutes	mRNA gene expression

Furthermore, parameters related to THz EMF exposure have been extracted.

ND 7/23 cell lines characterized by sensory neurons to 0.14 THz radiation for 80 ns, with power densities ranging from 24 to 62 mW/cm^2, but no changes occurred in the glutathione and heat shock protein 70 levels, which are typical indicators of the degree of stress response.

There are some *in-vivo* studies that show temperature-related effects in mice and rats; however, these few studies need further verification. Bondar et al. (2008) investigated behavior effects on mice after exposure to THz waves (3.6 THz, unknown power density, 15 or 30 min). The authors detected maximum effects of THz waves when the laser directly contacted the mice. Furthermore, increased anxiety of experimental animals was observed on the next day after 30 min irradiation.

Using *in-vivo* exposure of mice ear skin (pulsed 2.7 THz, 4 µs pulse width, 61.4 µJ/pulse, 3 Hz repetition rate, with an average power density of 260 mW/cm^2, 30 min), acute inflammatory response was analyzed at the cellular level using laser-scanning confocal microscopy (Hwang et al. 2014). The authors detected a massive recruitment of newly infiltrated neutrophils in the THz-wave-irradiated skin area after 6 hours, which suggests an induction of an acute inflammatory response.

Kim et al. (2013) analyzed whole-genome gene expression of mouse skin after exposure to 3.75 THz to evaluate nonthermal effects of the radiation. They detected mostly wound responses that were predominantly mediated by transforming growth factor-beta (TGF-β) signaling pathways and a delay in the closure of skin punch wounds. The authors suggested that THz radiation initiates a wound-like signal in skin with concomitant increased expression of TGF-β and activation of its downstream target genes, which perturbs the wound healing process *in vivo*.

FUTURE CHALLENGES

There are a number of challenges and opportunities encountered in present-day medicine and healthcare, due to rapid societal changes and scientific progress over the last decades. Presently, more people than ever have access to at least some kind of modern healthcare and also have better opportunities than ever to benefit from appropriate nutrition and education. The population is also reaching advanced ages in numbers never encountered before. These new conditions are changing the spectrum of disease conditions, which necessitates development of novel approaches for diagnosis and therapy. In parallel, there are advances in biology, information technology, and applied chemistry and physics that make it possible to develop novel tools for medical use. These innovations will replace or complement existing technologies because of their appropriateness and cost–benefit profile. From that perspective, the overall question in this chapter is whether approaches built based on MMWs and/or THz waves are realistic.

Clearly, medicine today has outstanding tools for both diagnosis and therapy, although there is room for improvement. Furthermore, certain situations cannot be satisfactorily handled with available techniques. The search for new and improved tools is thus legitimate and a high priority in many national and transnational research programs.

A prerequisite for any stronger impact of THz-wave-based instruments is that devices are further improved regarding practical use in a clinical setting. Until now, studies have been focused on proof of concept regarding the ability for THz imaging to discern relevant changes between different types of biological material and thus distinguish between healthy and diseased tissues in the first place. Many of these studies have been done *ex vivo*, thus paying little or no attention to the use of an instrument in a specific and complex clinical situation.

Requirements for an endoscope based on THz imaging were discussed by Doradla et al. (2017). The authors listed *inter alia* that an endoscope should be built on a single-channel principle, where both emission and reception of the THz waves take place through one single pathway. Furthermore, the device should consist of a flexible THz waveguide instead of using optical fibers.

For THz spectroscopy to enter mainstream medicine, it is also necessary for the technology to demonstrate that "killer apps" exist. Such an application should provide the clinician with an alternative that is safer, cheaper, and at least as reliable and precise as already existing modalities. The studies that are available do not address the costs, but the laser technology that is underpinning the generation of the THz signal has been very expensive so far, although prices are likely to go down in the near future. Safety does not seem to be a major issue when using THz instruments, since the transmitted energy is low and, in most cases, present only for very short times. Aspects of operator safety have nevertheless to be considered (see Mattsson et al. (2018) for further discussions regarding safety aspects). In comparison with certain other technologies, use of THz waves for diagnostics is to be considered less of a problem, since neither ionization nor more general heating occurs. Regarding sensitivity and specificity, too few studies are currently available where THz measurements are directly compared with those of established technologies. Such works are crucial to evaluating the usefulness of THz waves.

A more refined diagnosis based on molecular "fingerprints" in terms of THz absorption and reflection in addition to detecting differences in tissue water content is currently speculative. The studies that have found possible differences between, for example, healthy and diseased cells based on molecular clues are interesting but are too few that any further conclusions cannot be drawn from them.

Any possible therapeutic action of THz waves needs to be based on knowledge about biological effects of THz exposure. A key question here is whether only thermal effects occur after THz exposure or whether there is any possibility for nonthermal effects. The important distinction here is that thermal effects would affect the entire exposed tissue with all its cells. All effects would then be a consequence of increased temperature, which eliminates any "custom-made" treatment of the tissue or the cells. In contrast, if nonthermal effects are possible, understanding their mode of action would allow for specific focus on certain processes or molecules. This issue was discussed in some detail in Mattsson et al. (2018). There may also be the potential for simultaneous use of conventional therapeutic modalities, such as pharmaceuticals and THz waves. An interesting case in point is the work from Yamazaki et al. (2018) who investigated the activation of actin polymerization by THz irradiation and detected the increase of actin polymerization at least in the elongation process. The authors suggested that an explanation for the enhancement of actin polymerization might be the transient increase of temperature due to the absorption of THz irradiation by water molecules. Some actin-binding chemicals act as anticancer drugs because actin dynamics contribute to metastasis of cancer cells. Thus, these results suggest that THz waves could be applied for manipulating biomolecules and cells, since actin dynamics in the cytoplasm play pivotal roles in the proliferation and the motility of cells. Possibly, a potentiation of cytostatic drugs could be accomplished by THz exposure, and at lower drug concentrations, thus causing less side effects for the patients.

CONCLUSION

THz imaging is sparingly but increasingly used in medical diagnostics and considered especially suitable for scanning of epithelial surfaces. This imaging can generate information about the biological materials absorption, reflectivity, and refraction index. The information is related to the water content in the investigated tissue and can also reveal certain cellular structures and molecular information. Based on these findings, a THz investigation can distinguish between healthy and neoplastic tissue with forms of cancer such as skin and intestinal cancer. The health state of the skin can furthermore be analyzed and in that way inform about, e.g., healing rates of skin wounds. It is possible to detect not only the presence of certain disease states but also their extension in space, so that the affected area can be delineated.

Other diagnostic uses are directed toward the eyes (corneal water content) and teeth. In the latter case, reduction in enamel quality, presence of caries, and demineralization are conditions that are amenable for investigation.

The proof-of-concept principle for these and also other diagnostic medical uses has been discussed, and it seems reasonable to state that THz imaging can complement and possibly also partly replace already existing diagnostic tools.

However, any possible widespread clinical use is not realistic at this stage. First of all, there is more or less a complete lack of clinical studies, and even data from well-performed animal studies are scarce. Secondly, the development of proper patient- and user-friendly instruments has not been strong. Finally, production costs for both generators and sensors of THz waves are still high, even though it is likely that costs will drop when demands increase and more producers enter the market.

The case for therapeutic uses is much weaker. Although there are a number of studies that have investigated different biological effects of THz waves, with the present level of knowledge, we cannot say that there are biological effects due to low-energy depositions that do not cause heating. Despite that there are some studies indicating effects on gene expression that could be of clinical interest, the mechanism(s) behind such effects are not clear. It is furthermore unclear if the results of these studies are obtainable in independent replication studies. Basically, this research area is still very open, and substantial efforts are needed before any conclusions can be drawn regarding the therapeutic potential of THz waves.

Overall, the use of THz waves for medical purposes probably has a potential in certain niche, diagnostic, areas. It is likely that THz-based imaging instruments can be used in a fashion complementary to other technologies, when crucial clinical testing and instrument development has advanced further.

REFERENCES

Alexandrov, Boian S., V. Gelev, Alan R. Bishop, Anny Usheva, and Kim Ø. Rasmussen. 2010. "DNA Breathing Dynamics in the Presence of a Terahertz Field." *Physics Letters A* 374 (10): 1214–17. doi:10.1016/j.physleta.2009.12.077.

Alexandrov, Boian S, Kim Ø. Rasmussen, Alan R. Bishop, Anny Usheva, Ludmil B. Alexandrov, Shou Chong, Yossi Dagon, et al. 2011. "Non-Thermal Effects of Terahertz Radiation on Gene Expression in Mouse Stem Cells." *Biomedical Optics Express* 2 (9): 2679–89. doi:10.1364/BOE.2.002679.

De Amicis, Andrea, Stefania De Sanctis, Sara Di Cristofaro, Valeria Franchini, Florigio Lista, Elisa Regalbuto, Emilio Giovenale, et al. 2015. "Biological Effects of in Vitro THz Radiation Exposure in Human Foetal Fibroblasts." *Mutation Research/ Genetic Toxicology and Environmental Mutagenesis* 793: 150–60. doi:10.1016/j. mrgentox.2015.06.003.

Arbab, M. Hassan, Trevor C. Dickey, Dale P. Winebrenner, Antao Chen, Mathew B. Klein, and Pierre D. Mourad. 2011. "Terahertz Reflectometry of Burn Wounds in a Rat Model." *Biomedical Optics Express* 2 (8): 2339. doi:10.1364/BOE.2.002339.

Arbab, M. Hassan, Dale P. Winebrenner, Trevor C. Dickey, Antao Chen, Mathew B. Klein, and Pierre D. Mourad. 2013. "Terahertz Spectroscopy for the Assessment of Burn Injuries in Vivo." *Journal of Biomedical Optics* 18 (7): 077004. doi:10.1117/1.JBO.18.7.077004.

Asha, M. L, Srilakshmi Jasti, H. M. Mahesh Kumar, Basetty Neelakantam Rajarathnam, and Priyanka Basavaraj Lasune. 2015. "Terahertz Pulse Imaging: A Novel Imaging Technique in Dentistry." *World Journal of Pharmaceutical Research.* 4 (10): 2795–2802. www.wjpr.net.

Ashworth, Philip C., Emma Pickwell-MacPherson, Elena Provenzano, Sarah E. Pinder, Anand D. Purushotham, Michael Pepper, and Vincent P. Wallace. 2009. "Terahertz Pulsed Spectroscopy of Freshly Excised Human Breast Cancer." *Optics Express* 17 (15): 12444. doi:10.1364/OE.17.012444.

Bajwa, Neha, Joshua Au, Reza Jarrahy, Shijun Sung, Michael C. Fishbein, David Riopelle, Daniel B. Ennis, et al. 2017a. "Non-Invasive Terahertz Imaging of Tissue Water Content for Flap Viability Assessment." *Biomedical Optics Express* 8 (1): 460. doi:10.1364/BOE.8.000460.

Bajwa, Neha, Shijun Sung, Daniel B. Ennis, Michael C. Fishbein, Bryan N. Nowroozi, Dan Ruan, Ashkan MacCabi, et al. 2017b. "Terahertz Imaging of Cutaneous Edema: Correlation with Magnetic Resonance Imaging in Burn Wounds." *IEEE Transactions on Biomedical Engineering* 64 (11): 2682–94. doi:10.1109/TBME.2017.2658439.

Baxter, Brain J., and William G. Guglietta. 2011. "Terahertz Spectroscopy." *Analytical Chemistry* 83 (12): 4342–68. doi:10.1021/ac200907z.

Bock, Jonathan, Yayoi Fukuyo, Sona Kang, M. Lisa Phipps, Ludmil B. Alexandrov, Kim O. Rasmussen, Alan R. Bishop, et al. 2010. "Mammalian Stem Cells Reprogramming in Response to Terahertz Radiation." *PLoS One* 5 (12). doi:10.1371/journal.pone.0015806.

Bogomazova, A. N., E. M. Vassina, T. N. Goryachkovskaya, V. M. Popik, A. S. Sokolov, N. A. Kolchanov, M. A. Lagarkova, S. L. Kiselev, and S. E. Peltek. 2015. "No DNA Damage Response and Negligible Genome-Wide Transcriptional Changes in Human Embryonic Stem Cells Exposed to Terahertz Radiation." *Scientific Reports* 5: 7749. doi:10.1038/srep07749.

Bondar, N. P., I. L. Kovalenko, D. F. Avgustinovich, A. G. Khamoyan, and N. N. Kudryavtseva. 2008. "Behavioral Effect of Terahertz Waves in Male Mice." *Bulletin of Experimental Biology and Medicine* 145 (4): 401–5. doi:10.1007/s10517-008-0102-x.

Borovkova, Mariia, Maria Serebriakova, Viacheslav Fedorov, Egor Sedykh, Vladimir Vaks, Alexander Lichutin, Alina Salnikova, and Mikhail Khodzitsky. 2017. "Investigation of Terahertz Radiation Influence on Rat Glial Cells." *Biomedical Optics Express* 8 (1): 273. doi:10.1364/BOE.8.000273.

Bourne, Nicola, Richard H. Clothier, Marco D'Arienzo, and Paul Harrison. 2008. "The Effects of Terahertz Radiation on Human Keratinocyte Primary Cultures and Neural Cell Cultures." *Alternatives to Laboratory Animals: ATLA* 36(6): 667–84.

Bowman, Tyler, Magda El-Shenawee, and Shree G. Sharma. 2014. "Terahertz Spectroscopy for the Characterization of Excised Human Breast Tissue." In *2014 IEEE MTT-S International Microwave Symposium (IMS2014)*, pp. 1–4. IEEE. doi:10.1109/MWSYM.2014.6848538.

Bowman, Tyler, Tanny Chavez, Kamrul Khan, Jingxian Wu, Avishek Chakraborty, Narasimhan Rajaram, Keith Bailey, and Magda El-Shenawee. 2018. "Pulsed Terahertz Imaging of Breast Cancer in Freshly Excised Murine Tumors." *Journal of Biomedical Optics* 23 (02): 1. doi:10.1117/1.JBO.23.2.026004.

Chan, King Yuk, and Rodica Ramer. 2018. "Novel Concept of Detecting Basal Cell Carcinoma in Skin Tissue Using a Continuous-Wave Millimeter-Wave Rectangular Glass Filled Probe." *Medical Devices: Evidence and Research* 11 (August): 275–85. doi:10.2147/MDER.S168338.

Chen, Hua, Shihua Ma, Xiumei Wu, Wenxing Yang, and Tian Zhao. 2015. "Diagnose Human Colonic Tissues by Terahertz Near-Field Imaging." *Journal of Biomedical Optics* 20 (3): 036017. doi:10.1117/1.JBO.20.3.036017.

Cherkasova, Olga P., Vyacheslav I. Fedorov, Eugenia F. Nemova, and Alexander S. Pogodin. 2009. "Influence of Terahertz Laser Radiation on the Spectral Characteristics and Functional Properties of Albumin." *Optics and Spectroscopy* 107 (4): 534–37. doi:10.1134/s0030400x09100063.

Clothier, Richard H., and Nicola Bourne. 2003. "Effects of THz Exposure on Human Primary Keratinocyte Differentiation and Viability." *Journal of Biological Physics*, 29:179–85 doi:10.1023/A:1024492725782.

Crawley, David A., Christopher Longbottom, Bryan E. Cole, Craig M. Ciesla, Don Arnone, Vincent P. Wallace, and Michael Pepper. 2003. "Terahertz Pulse Imaging: A Pilot Study of Potential Applications in Dentistry." *Caries Research* 37 (5): 352–59. doi:10.1159/000072167.

Demidova, Elizaveta V., Tatiana N. Goryachkovskaya, Tatiana K. Malup, Svetlana V. Bannikova, Artem I. Semenov, Nikolay A. Vinokurov, Nikolay A. Kolchanov, Vasiliy M. Popik, and Sergey E. Peltek. 2013. "Studying the Non-Thermal Effects of Terahertz Radiation on E. Coli/PKatG-GFP Biosensor Cells." *Bioelectromagnetics* 34 (1): 15–21. doi:10.1002/bem.21736.

Demidova, Elizaveta V., Tatiana N. Goryachkovskaya, Irina A. Mescheryakova, Tatiana K. Malup, Artem I. Semenov, Nikolay A. Vinokurov, Nikolay A. Kolchanov, Vasiliy M. Popik, and Sergey E. Peltek. 2016. "Impact of Terahertz Radiation on Stress-Sensitive Genes of E. Coli Cell." *IEEE Transactions on Terahertz Science and Technology* 6 (3): 435–41. doi:10.1109/TTHZ.2016.2532344.

Doradla, Pallavi, Karim Alavi, Cecil Joseph, and Robert Giles. 2013. "Detection of Colon Cancer by Continuous-Wave Terahertz Polarization Imaging Technique." *Journal of Biomedical Optics* 18 (9): 090504. doi:10.1117/1.JBO.18.9.090504.

Doradla, Pallavi, Karim Alavi, Cecil Joseph, and Robert Giles. 2014. "Single-Channel Prototype Terahertz Endoscopic System." *Journal of Biomedical Optics* 19 (8): 080501. doi:10.1117/1.JBO.19.8.080501.

Doradla, Pallavi, Cecil Joseph, and Robert H. Giles. 2017. "Terahertz Endoscopic Imaging for Colorectal Cancer Detection: Current Status and Future Perspectives." *World Journal of Gastrointestinal Endoscopy* 9 (8): 346. doi:10.4253/wjge.v9.i8.346.

Doria, A., G. P. Gallerano, E. Giovenale, G. Messina, A. Lai, A. Ramundo-Orlando, V. Sposato, et al. 2004. "THz Radiation Studies on Biological Systems at the ENEA FEL Facility." *Infrared Physics and Technology*, 45:339–47. doi:10.1016/j.infrared.2004.01.014.

Dougherty, Joseph P., Gregory D. Jubic, and William L. Kiser Jr. 2007. "Terahertz Imaging of Burned Tissue." Edited by Kurt J. Linden and Laurence P. Sadwick. *Proceedings of SPIE*. doi:10.1117/12.705137.

Dutta, Moumita, Amar S. Bhalla, and Ruyan Guo. 2016. "THz Imaging of Skin Burn: Seeing the Unseen—An Overview." *Advances in Wound Care* 5 (8): 338–48. doi:10.1089/wound.2015.0685.

Eadie, Leila H., Caroline B. Reid, Anthony J. Fitzgerald, and Vincent P. Wallace. 2013. "Optimizing Multi-Dimensional Terahertz Imaging Analysis for Colon Cancer Diagnosis." *Expert Systems with Applications* 40 (6): 2043–50. doi:10.1016/j.eswa.2012.10.019.

Echchgadda, Ibtissam, Jessica E. Grundt, Cesario Z. Cerna, Caleb C. Roth, Bennett L. Ibey, and Gerald J. Wilmink. 2014. "Terahertz Stimulate Specific Signaling Pathways in Human Cells." In *International Conference on Infrared, Millimeter, and Terahertz Waves, IRMMW-THz*. doi:10.1109/IRMMW-THz.2014.6956140.

Echchgadda, Ibtissam, Jessica E. Grundt, Cesario Z. Cerna, Caleb C. Roth, Jason A. Payne, Bennett L. Ibey, and Gerald J. Wilmink. 2016. "Terahertz Radiation: A Non-Contact Tool for the Selective Stimulation of Biological Responses in Human Cells." *IEEE Transactions on Terahertz Science and Technology* 6 (1): 54–68. doi:10.1109/TTHZ.2015.2504782.

Fan, Bo, Victor A. Neel, and Anna N. Yaroslavsky. 2017a. "Multimodal Imaging for Nonmelanoma Skin Cancer Margin Delineation." *Lasers in Surgery and Medicine* 49 (3): 319–26. doi:10.1002/lsm.22552.

Fan, Shuting, Benjamin S. Y. Ung, Edward P. J. Parrott, Vincent P. Wallace, and Emma Pickwell-MacPherson. 2017b. "In Vivo Terahertz Reflection Imaging of Human Scars during and after the Healing Process." *Journal of Biophotonics* 10 (9): 1143–51. doi:10.1002/jbio.201600171.

Fedorov, V. I. 2017. "The Biological Effects of Terahertz Laser Radiation as a Fundamental Premise for Designing Diagnostic and Treatment Methods." *Biophysics* 62 (2): 324–30. doi:10.1134/S0006350917020075.

Gallerano, G. P., E. Giovenale, P. Nenzi, A. De Amicis, S. De Sanctis, S. Di Cristofaro, V. Franchini, et al. 2014. "Effects of THz Radiation on Human Fibroblasts In-Vitro: Exposure Set-up and Biological Endpoints." In *2014 39th International Conference on Infrared, Millimeter, and Terahertz Waves (IRMMW-THz)*, pp. 1–2. IEEE. doi:10.1109/IRMMW-THz.2014.6956056.

Grootendorst, Maarten R., Anthony J. Fitzgerald, Susan G. Brouwer de Koning, Aida Santaolalla, Alessia Portieri, Mieke Van Hemelrijck, Matthew R. Young, et al. 2017. "Use of a Handheld Terahertz Pulsed Imaging Device to Differentiate Benign and Malignant Breast Tissue." *Biomedical Optics Express* 8 (6): 2932. doi:10.1364/BOE.8.002932.

Hassan, Ahmed M., David C. Hufnagle, Magda El-Shenawee, and Gilbert E. Pacey. 2012. "Terahertz Imaging for Margin Assessment of Breast Cancer Tumors." In *2012 IEEE/MTT-S International Microwave Symposium Digest*, pp. 1–3. IEEE. doi:10.1109/MWSYM.2012.6259567.

Hernandez-Cardoso, G. G., S. C. Rojas-Landeros, M. Alfaro-Gomez, A. I. Hernandez-Serrano, I. Salas-Gutierrez, E. Lemus-Bedolla, A. R. Castillo-Guzman, H. L. Lopez-Lemus, and E. Castro-Camus. 2017. "Terahertz Imaging for Early Screening of Diabetic Foot Syndrome: A Proof of Concept." *Scientific Reports* 7 (February): 42124. doi:10.1038/srep42124.

Hintzsche, Henning, Christian Jastrow, Thomas Kleine-Ostmann, Helga Stopper, E. Schmid, and Thorsten Schrader. 2011. "Terahertz Radiation Induces Spindle Disturbances in Human-Hamster Hybrid Cells." *Radiation Research* 175 (5): 569–74. doi:10.1667/RR2406.1.

Hintzsche, Henning, Christian Jastrow, Thomas Kleine-Ostmann, Uwe Kärst, Thorsten Schrader, and Helga Stopper. 2012. "Terahertz Electromagnetic Fields (0.106 THz) Do Not Induce Manifest Genomic Damage In Vitro." *PLoS One* 7 (9). doi:10.1371/journal. pone.0046397.

Hintzsche, Henning, Christian Jastrow, Bernd Heinen, Kai Baaske, Thomas Kleine-Ostmann, Michael Schwerdtfeger, Mohammed Khaled Shakfa, et al. 2013. "Terahertz Radiation at 0.380 THz and 2.520 THz Does Not Lead to DNA Damage in Skin Cells in Vitro." *Radiation Research* 179 (1): 38–45. doi:10.1667/RR3077.1.

Hirmer, Marion, Sergey N. Danilov, Stephan Giglberger, Jürgen Putzger, Andreas Niklas, Andreas Jäger, Karl Anton Hiller, et al. 2012. "Spectroscopic Study of Human Teeth and Blood from Visible to Terahertz Frequencies for Clinical Diagnosis of Dental Pulp Vitality." *Journal of Infrared, Millimeter, and Terahertz Waves* 33 (3): 366–75. doi:10.1007/s10762-012-9872-3.

Huang, S. Y., E. Macpherson, and Y. T. Zhang. 2007. "A Feasibility Study of Burn Wound Depth Assessment Using Terahertz Pulsed Imaging." In *Proceedings of the 4th IEEE-EMBS International Summer School and Symposium on Medical Devices and Biosensors, ISSS-MDBS 2007*, pp. 132–35. IEEE. doi:10.1109/ISSMDBS.2007.4338310.

Huang, S. Y., Y. X. J. Wang, D. K. W. Yeung, A. T. Ahuja, Y. T. Zhang, and E. Pickwell-Macpherson. 2009. "Tissue Characterization Using Terahertz Pulsed Imaging in Reflection Geometry." *Physics in Medicine and Biology*. doi:10.1088/0031-9155/54/1/010.

Humphreys, K., J. P. Loughran, M. Gradziel, W. Lanigan, T. Ward, J. A. Murphy, and C. O'sullivan. 2004. "Medical Applications of Terahertz Imaging: A Review of Current Technology and Potential Applications in Biomedical Engineering." In *Conference Proceedings—Annual International Conference of the IEEE Engineering in Medicine and Biology Society*. IEEE Engineering in Medicine and Biology Society. Conference 2: pp. 1302–5. doi:10.1109/IEMBS.2004.1403410.

Hwang, Yoonha, Jinhyo Ahn, Jungho Mun, Sangyoon Bae, Young Uk Jeong, Nikolay a Vinokurov, and Pilhan Kim. 2014. "In Vivo Analysis of THz Wave Irradiation Induced Acute Inflammatory Response in Skin by Laser-Scanning Confocal Microscopy." *Optics Express* 22 (10): 11465–75. doi:10.1364/OE.22.011465.

Jae Oh, Seung, Sang-Hoon Kim, Kiyoung Jeong, Yeonji Park, Yong-Min Huh, Joo-Hiuk Son, and Jin-suck Suh. 2013. "Measurement Depth Enhancement in Terahertz Imaging of Biological Tissues." *Optics Express* 21: 21299–305. doi:10.1364/OE.21.021299.

Ji, Young Bin, Chan Hyuk Park, Hyunki Kim, Sang Hoon Kim, Gyu Min Lee, Sam Kyu Noh, Tae-In Jeon, et al. 2015. "Feasibility of Terahertz Reflectometry for Discrimination of Human Early Gastric Cancers." *Biomedical Optics Express* 6 (4): 1398. doi:10.1364/BOE.6.001398.

Ji, Young Bin, In-Seok Moon, Hyeon Sang Bark, Sang Hoon Kim, Dong Woo Park, Sam Kyu Noh, Yong-Min Huh, Jin-Seok Suh, Seung Jae Oh, and Tae-In Jeon. 2016. "Terahertz Otoscope and Potential for Diagnosing Otitis Media." *Biomedical Optics Express* 7 (4): 1201–9. doi:10.1364/BOE.7.001201.

Joseph, Cecil S., Rakesh Patel, Victor A. Neel, Robert H. Giles, and Anna N. Yaroslavsky. 2014. "Imaging of Ex Vivo Nonmelanoma Skin Cancers in the Optical and Terahertz Spectral Regions." *Journal of Biophotonics* 7 (5): 295–303. doi:10.1002/jbio.201200111.

Kamburoglu, Kivanc, N. O. Yetimôglu, and H. Altan. 2014. "Characterization of Primary and Permanent Teeth Using Terahertz Spectroscopy." *Dentomaxillofacial Radiology* 43 (6): 20130404. doi:10.1259/dmfr.20130404.

Kamburoglu, Kivanc, Burcu Karagöz, Hakan Altan, and Doğukan Özen. 2018. "An Ex Vivo Comparative Study of Occlusal and Proximal Caries Using Terahertz and X-Ray Imaging." *Dentomaxillofacial Radiology*, 20180250. doi:10.1259/dmfr.20180250.

Kašalynas, Irmantas, Rimvydas Venckevičius, Linas Minkevičius, Aleksander Sešek, Faustino Wahaia, Vincas Tamošiūnas, Bogdan Voisiat, et al. 2016. "Spectroscopic Terahertz Imaging at Room Temperature Employing Microbolometer Terahertz Sensors and Its Application to the Study of Carcinoma Tissues." *Sensors (Switzerland)* 16 (4): 1–15. doi:10.3390/s16040432.

Kenna, Margaret A. 2015. "Acquired Hearing Loss in Children." *Otolaryngologic Clinics of North America.* doi:10.1016/j.otc.2015.07.011.

Kim, Kyu-Tae, Jaehun Park, Sung Jin Jo, Seonghoon Jung, Oh Sang Kwon, Gian Piero Gallerano, Woong-Yang Park, and Gun-Sik Park. 2013. "High-Power Femtosecond-Terahertz Pulse Induces a Wound Response in Mouse Skin." *Scientific Reports* 3 (August): 2296. doi:10.1038/srep02296.

Korenstein-Ilan, Avital, Alexander Barbul, Pini Hasin, Alon Eliran, Avraham Gover, and Rafi Korenstein. 2008. "Terahertz Radiation Increases Genomic Instability in Human Lymphocytes." *Radiation Research* 170 (2): 224–34. doi:10.1667/RR0944.1.

Koyama, Shin, Eijiro Narita, Yoko Shimizu, Takeo Shiina, Masao Taki, Naoki Shinohara, and Junji Miyakoshi. 2016. "Twenty Four-Hour Exposure to a 0.12 THz Electromagnetic Field Does Not Affect the Genotoxicity, Morphological Changes, or Expression of Heat Shock Protein in HCE-T Cells." *International Journal of Environmental Research and Public Health* 13 (8): 793. doi:10.3390/ijerph13080793.

Kristensen, Torben T., Withawat Withayachumnankul, Peter U. Jepsen, and Derek Abbott. 2010. "Modeling Terahertz Heating Effects on Water." *Optics Express* 18 (5): 4727. doi:10.1364/OE.18.004727.

Lee, Kyumin, Kiyong Jeoung, Sang Hoon Kim, Young-Bin Ji, Hyeyoung Son, Yuna Choi, Young-Min Huh, Jin-Suck Suh, and Seung Jae Oh. 2018. "Measuring Water Contents in Animal Organ Tissues Using Terahertz Spectroscopic Imaging." *Biomedical Optics Express* 9 (4): 1582–89. doi:10.1364/BOE.9.001582.

Mattsson, Mats Olof, Olga Zeni, and Myrtill Simkó. 2018. "Is There a Biological Basis for Therapeutic Applications of Millimetre Waves and THz Waves?" *Journal of Infrared, Millimeter, and Terahertz Waves* 39 (9): 863–78. doi:10.1007/s10762-018-0483-5.

Oh, Seung Jae, Jinyoung Kang, Inhee Maeng, Jin-Suck Suh, Yong-Min Huh, Seungjoo Haam, and Joo-Hiuk Son. 2009. "Nanoparticle-Enabled Terahertz Imaging for Cancer Diagnosis." *Optics Express* 17 (5): 3469. doi:10.1364/OE.17.003469.

Oh, Seung Jae, Yong Min Huh, Jin Suck Suh, Jihye Choi, Seungjoo Haam, and Joo Hiuk Son. 2012. "Cancer Diagnosis by Terahertz Molecular Imaging Technique." *Journal of Infrared, Millimeter, and Terahertz Waves* 33 (1): 74–81. doi:10.1007/s10762-011-9847-9.

Olshevskaya, J. S., A. S. Ratushnyak, A. K. Petrov, A. S. Kozlov, and T. A. Zapara. 2008. "Effect of Terahertz Electromagnetic Waves on Neurons Systems." In *Proceedings -2008 IEEE Region 8 International Conference on Computational Technologies in Electrical and Electronics Engineering, SIBIRCON 2008*, 210–11. doi:10.1109/SIBIRCON.2008.4602607.

Pickwell, Emma, Bryan E. Cole, Anthony J. Fitzgerald, Michael Pepper, and Vincent Patrick Wallace. 2004. "In Vivo Study of Human Skin Using Pulsed Terahertz Radiation." *Physics in Medicine and Biology* 49 (9): 1595–1607. doi:10.1088/0031-9155/49/9/001.

Reid, Caroline B., Anthony Fitzgerald, George Reese, Robert Goldin, Paris Tekkis, P. S. O'Kelly, Emma Pickwell-MacPherson, Adam P. Gibson, and Vincent P. Wallace. 2011. "Terahertz Pulsed Imaging of Freshly Excised Human Colonic Tissues." *Physics in Medicine and Biology* 56 (14): 4333–53. doi:10.1088/0031-9155/56/14/008.

Romanenko, Sergii, Ryan Begley, Alan R. Harvey, Livia Hool, and Vincent P. Wallace. 2017. "The Interaction between Electromagnetic Fields at Megahertz, Gigahertz and Terahertz Frequencies with Cells, Tissues and Organisms: Risks and Potential." *Journal of the Royal Society Interface* 14 (137). doi:10.1098/rsif.2017.0585.

Rønne, Cecilie, Lars Thrane, Per-Olof Åstrand, Anders Wallqvist, Kurt Mikkelsen, and Søren R. Keiding. 1997. "Investigation of the Temperature Dependence of Dielectric Relaxation in Liquid Water by THz Reflection Spectroscopy and Molecular Dynamics Simulation." *The Journal of Chemical Physics* 107: 5319–31. doi:10.1063/1.474242.

Scarfì, M. R., M Romano, R Di Pietro, O Zeni, A Doria, G P Gallerano, E Giovenale, G Messina, A Lai, and G Campurra. 2003. "THz Exposure of Whole Blood for the Study of Biological Effects on Human Lymphocytes." *Journal of Biological Physics* 29 (2–3): 171–76.

Schwerdtfeger, Michael, Sina Lippert, Martin Koch, Andreas Berg, Stefan Katletz, and Karin Wiesauer. 2012. "Terahertz Time-Domain Spectroscopy for Monitoring the Curing of Dental Composites." *Biomedical Optics Express* 3 (11): 2842. doi:10.1364/BOE.3.002842.

Sergeeva, Svetlana, Elisaveta Demidova, Olga Sinitsyna, Tatiana Goryachkovskaya, Alla Bryanskaya, Artem Semenov, Irina Meshcheryakova, Grigory Dianov, Vasiliy Popik, and Sergey Peltek. 2016. "2.3THz Radiation: Absence of Genotoxicity/Mutagenicity in Escherichia Coli and Salmonella Typhimurium." *Mutation Research/Genetic Toxicology and Environmental Mutagenesis* 803–804 (June): 34–38. doi:10.1016/j.mrgentox.2016.05.005.

Sim, Yookyeong Carolyn, Inhee Maeng, and Joo-Hiuk Son. 2009. "Frequency-Dependent Characteristics of Terahertz Radiation on the Enamel and Dentin of Human Tooth." *Current Applied Physics* 9 (5): 946–49. doi:10.1016/j.cap.2008.09.008.

Sirtori, Carlo. 2002. "Applied Physics: Bridge for the Terahertz Gap." *Nature.* doi:10.1038/417132b.

Smith, R. Lane. 2004. "Diagnostic Modalities." *Clinical Orthopaedics and Related Research* 427 (Special Issue C): S174. doi:10.1097/01.blo.0000144980.80654.bd.

Son, Joo Hiuk. 2013. "Principle and Applications of Terahertz Molecular Imaging." *Nanotechnology.* doi:10.1088/0957-4484/24/21/214001.

Sun, Qiushuo, Yuezhi He, Kai Liu, Shuting Fan, Edward P. J. Parrott, and Emma Pickwell-MacPherson. 2017a. "Recent Advances in Terahertz Technology for Bio-medical Applications." *Quantitative Imaging in Medicine and Surgery.* doi:10.210 37/qims.2017.06.02.

Sun, Qiushuo, Edward P. J. Parrott, and Emma Pickwell-MacPherson. 2017b. "In Vivo Estimation of the Water Diffusivity in Occluded Human Skin Using Terahertz Rejection Spectroscopy." In *2017 42nd International Conference on Infrared, Millimeter, and Terahertz Waves (IRMMW-THz)*, pp. 1–2. IEEE. doi:10.1109/IRMMW-THz.2017.8066862.

Sung, Shijun, Shahab Dabironezare, Nuria Llombart, Skyler Selvin, Neha Bajwa, Somporn Chantra, Bryan Nowroozi, et al. 2018a. "Optical System Design for Noncontact, Normal Incidence, THz Imaging of in Vivo Human Cornea." *IEEE Transactions on Terahertz Science and Technology* 8 (1): 1–12. doi:10.1109/TTHZ.2017.2771754.

Sung, Shijun, Skyler Selvin, Neha Bajwa, Somporn Chantra, Bryan Nowroozi, James Garritano, Jacob Goell, et al. 2018b. "THz Imaging System for in Vivo Human Cornea." *IEEE Transactions on Terahertz Science and Technology* 8 (1): 27–37. doi:10.1109/TTHZ.2017.2775445.

Sy, Stanley, Shengyang Huang, Yi Xiang J. Wang, Jun Yu, Anil T. Ahuja, Yuan Ting Zhang, and Emma Pickwell-MacPherson. 2010. "Terahertz Spectroscopy of Liver Cirrhosis: Investigating the Origin of Contrast." *Physics in Medicine and Biology* 55 (24): 7587–96. doi:10.1088/0031-9155/55/24/013.

Taylor, Zachary D., James Garritano, Shijun Sung, Neha Bajwa, David B. Bennett, Bryan Nowroozi, Priyamvada Tewari, et al. 2015a. "THz and Mm-Wave Sensing of Corneal Tissue Water Content: Electromagnetic Modeling and Analysis." *IEEE Transactions on Terahertz Science and Technology* 5 (2): 170–83. doi:10.1109/TTHZ.2015.2392619.

Taylor, Zachary D., James Garritano, Shijun Sung, Neha Bajwa, David B. Bennett, Bryan Nowroozi, Priyamvada Tewari. 2015b. "THz and Mm-Wave Sensing of Corneal Tissue Water Content: In Vivo Sensing and Imaging Results." *IEEE Transactions on Terahertz Science and Technology* 5 (2): 184–96. doi:10.1109/TTHZ.2015.2392628.

Tewari, Priyamvada, Colin P. Kealey, David B. Bennett, Neha Bajwa, Kelli S. Barnett, Rahul S. Singh, Martin O. Culjat, Alexander Stojadinovic, Warren S. Grundfest, and Zachary D. Taylor. 2012. "In Vivo Terahertz Imaging of Rat Skin Burns." *Journal of Biomedical Optics* 17 (4): 040503. doi:10.1117/1.JBO.17.4.040503.

Titova, Lyubov V., Ayesheshim K. Ayesheshim, Andrey Golubov, Dawson Fogen, Rocio Rodriguez-Juarez, Frank A. Hegmann, Olga Kovalchuk, et al. 2010. "Intense THz Pulses Cause H2AX Phosphorylation and Activate DNA Damage Response in Human Skin Tissue." *Biomedical Optics Express* 4 (4): 559–568.

Titova, Lyubov V., Ayesheshim K. Ayesheshim, Andrey Golubov, Rocio Rodriguez-Juarez, Rafal Woycicki, Frank A Hegmann, and Olga Kovalchuk. 2013. "Intense THz Pulses Down-Regulate Genes Associated with Skin Cancer and Psoriasis: A New Therapeutic Avenue?" *Scientific Reports* 3: 2363. doi:10.1038/srep02363.

Wahaia, Faustino, Gintaras Valusis, Luis M. Bernardo, Abílio Almeida, Joaquim A. Moreira, Patricia C. Lopes, Jan Macutkevic, et al. 2011. "Detection of Colon Cancer by Terahertz Techniques." *Journal of Molecular Structure* 1006 (1–3): 77–82. doi:10.1016/j.molstruc.2011.05.049.

Wahaia, Faustino, Irmantas Kasalynas, Rimvydas Venckevicius, Dalius Seliuta, Gintaras Valusis, Andrzej Urbanowicz, Gediminas Molis, Fatima Carneiro, Catia D. Carvalho Silva, and Pedro L. Granja. 2016. "Terahertz Absorption and Reflection Imaging of Carcinoma-Affected Colon Tissues Embedded in Paraffin." *Journal of Molecular Structure* 1107 (March): 214–19. doi:10.1016/j.molstruc.2015.11.048.

Wallace, V. P., A. J. Fitzgerald, S. Shankar, N. Flanagan, R. Pye, J. Cluff, and D. D. Arnone. 2004. "Terahertz Pulsed Imaging of Basal Cell Carcinoma Ex Vivo and in Vivo." *British Journal of Dermatology* 151 (2): 424–32. doi:10.1111/j.1365-2133.2004.06129.x.

Wang, Mengxi, Guohui Yang, Wanlu Li, and Qun Wu. 2013. "An Overview of Cancer Treatment by Terahertz Radiation." In *2013 IEEE MTT-S International Microwave Workshop Series on RF and Wireless Technologies for Biomedical and Healthcare Applications, IMWS-BIO 2013-Proceedings.* doi:10.1109/IMWS-BIO.2013.6756170.

Williams, Rachel, Amy Schofield, Gareth Holder, Joan Downes, David Edgar, Paul Harrison, Michele Siggel-king, et al. 2013. "The Influence of High Intensity Terahertz Radiation on Mammalian Cell Adhesion, Proliferation and Differentiation." *Physics in Medicine and Biology* 58 (2): 373–91. doi:10.1088/0031-9155/58/2/373.

Wilmink, Gerald J., and Jessica E. Grundt. 2011. "Invited Review Article: Current State of Research on Biological Effects of Terahertz Radiation." *Journal of Infrared, Millimeter, and Terahertz Waves*, 32:1074–1122. doi:10.1007/s10762-011-9794-5.

Wilmink, Gerald J., Bennett L. Ibey, Caleb L. Roth, Rebecca L. Vincelette, Benjamin D. Rivest, Christopher B. Horn, Joshua Bernhard, Dawnlee Roberson, and William P. Roach. 2010. "Determination of Death Thresholds and Identification of Terahertz (THz)-Specific Gene Expression Signatures." *Proceedings of SPIE* 7562 (December 2014): 75620K. doi:10.1117/12.844917.

Wilmink, Gerald J., Benjamin D. Rivest, Caleb C. Roth, Bennett L. Ibey, Jason A. Payne, Luisiana X. Cundin, Jessica E. Grundt, Xomalin Peralta, Dustin G. Mixon, and William P. Roach. 2011. "In Vitro Investigation of the Biological Effects Associated with Human Dermal Fibroblasts Exposed to 2.52 THz Radiation." *Lasers in Surgery and Medicine* 43 (2): 152–63. doi:10.1002/lsm.20960.

Woodward, Ruth M., Bryan E. Cole, Vincent P. Wallace, Richard J. Pye, Donald D. Arnone, Edmund H. Linfield, and Michael Pepper. 2002. "Terahertz Pulse Imaging in Reflection Geometry of Human Skin Cancer and Skin Tissue." *Physics in Medicine and Biology.* doi:10.1088/0031-9155/47/21/325.

Woodward, R. M., V. P. Wallace, D. D. Arnone, E. H. Linfield, and M. Pepper. 2003a. "Terahertz Pulsed Imaging of Skin Cancer in the Time and Frequency Domain." *Journal of Biological Physics.* doi:10.1023/A:1024409329416.

Woodward, Ruth M., Vincent P. Wallace, Richard J. Pye, Bryan E. Cole, Donald D. Arnone, Edmund H. Linfield, and Michael Pepper. 2003b. "Terahertz Pulse Imaging of Ex Vivo Basal Cell Carcinoma." *Journal of Investigative Dermatology* 120 (1): 72–78. doi:10.1046/j.1523-1747.2003.12013.x.

Yamazaki, Shota, Masahiko Harata, Toshitaka Idehara, Keiji Konagaya, Ginji Yokoyama, Hiromichi Hoshina, and Yuichi Ogawa. 2018. "Actin Polymerization Is Activated by Terahertz Irradiation." *Scientific Reports* 8 (1): 1–7. doi:10.1038/s41598-018-28245-9.

Yang, Xiang, Xiang Zhao, Ke Yang, Yueping Liu, Yu Liu, Weiling Fu, and Yang Luo. 2016. "Biomedical Applications of Terahertz Spectroscopy and Imaging." *Trends in Biotechnology* 34 (10): 810–24. doi:10.1016/j.tibtech.2016.04.008.

Yu, Calvin, Shuting Fan, Yiwen Sun, and Emma Pickwell-Macpherson. 2012. "The Potential of Terahertz Imaging for Cancer Diagnosis: A Review of Investigations to Date." *Quatitative Imaging in Medicine and Surgery* 2 (1): 33–45. doi:10.3978/j. issn.2223-4292.2012.01.04.

Zaytsev, Kirill, Konstantin Kudrin, Valeriy E Karasik, Igor Reshetov, and Stanislav Yurchenko. 2015. "In Vivo Terahertz Spectroscopy of Pigmentary Skin Nevi: Pilot Study of Non-Invasive Early Diagnosis of Dysplasia." *Applied Physics Letters* 106: 53702. doi:10.1063/1.4907350.

Zeni, O., G. P. Gallerano, A. Perrotta, M. Romano, A. Sannino, M. Sarti, M. D'Arienzo, et al. 2007. "Cytogenetic Observations in Human Peripheral Blood Leukocytes Following in Vitro Exposure to THz Radiation: A Pilot Study." *Health Physics* 92 (4): 349–57. doi:10.1097/01.HP.0000251248.23991.35.

10 Nanopulse Stimulation Therapy

A Novel, Nonthermal Method for Clearing Skin Lesions

Richard Nuccitelli

CONTENTS

INTRODUCTION

The cellular responses to pulsed electric fields fall generally into three different pulse duration domains: millisecond, microsecond, and nanosecond. Investigations of the cellular effects of pulsed fields in these time domains have a long history documented by over a thousand publications during the past five decades. The millisecond and microsecond domains have been studied most extensively and are commonly used in therapies such as electrochemotherapy (ECT) (Cemazar and Sersa, 2019), irreversible electroporation (IRE) (Aycock and Davalos, 2019) and gene electrotherapy (GET) (Bulysheva et al., 2019). The study of the nanosecond pulse domain is only about 25 years old, and its distinctly different biological mechanisms and potential clinical uses are still in relatively early stages. Research conducted using pulsed electric fields in the nanosecond time domain demonstrate unique properties that allow nanosecond pulses to penetrate the cell interior with larger amplitudes, unlike microsecond- or millisecond-domain pulses, for which larger amplitudes are not practical due to thermal limitations. The intracellular penetration of nanosecond pulsing can lead to a variety of cellular responses which is why the application of pulses in this domain is often referred to as Nano-Pulse Stimulation™ (NPS™) therapy. Pulse Biosciences (NASDAQ: PLSE) is the first company to begin to commercialize this therapy.

Intracellular Penetration

Ultrashort, nanosecond-domain pulses can exhibit unique intracellular effects because of their unique ability to penetrate into cells and organelles with an amplitude sufficiently large to permeabilize small intracellular organelles. This ability to penetrate beyond the plasma membrane is possible when the pulse rise time reaches full amplitude in the nanosecond range, which is faster than the time required for intracellular and intraorganellar charges to redistribute to neutralize the applied field. This phenomenon is observed in models and experiments studying the membrane charging time of cells (Schoenbach, 2010). For example, when a conductor is placed in an external electric field, the mobile charges in the conductor are moved by the field and redistribute to generate an equal and opposite field so that the net field in the conductor becomes zero after the redistribution has occurred. A cell can be modeled as a conductor surrounded by a membrane bilayer with both resistance and capacitance so that when placed in an applied field, positive ions in the cytoplasm will move toward the negative pole of the field and negative ions will move toward the positive pole until they encounter the plasma membrane and charge the membrane's capacitance (Figure 10.1). The capacitance then generates a field that is equal and opposite to the applied external electric field so that the net field inside the cell will be zero. This charge redistribution takes time that will vary with the

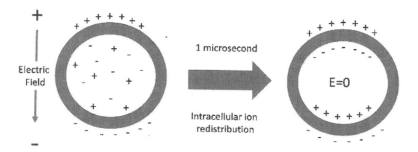

FIGURE 10.1 The redistribution of mobile charges in the cytoplasm in response to an imposed electric field takes about 1 μs and charges the capacitance of the plasma membrane to generate an equal and opposite field so the net field in the cytoplasm is zero.

access resistance to the cells. When cells are tightly packed in a tissue, this polarization will typically take as long as 1 μs (Schoenbach, 2010). Therefore, applied fields with rise times faster than 1 μs will be able to penetrate tissues to exert their force on intracellular organelle membranes. This organelle-penetrating property was demonstrated for nanosecond-domain fields in 2001 (Schoenbach et al., 2001). Since then, many additional examples of small organelles permeabilized by NPS have appeared, including vesicles (Tekle et al., 2005), mitochondria (Vernier, 2011), endoplasmic reticulum (Vernier et al., 2003; White et al., 2004), and nuclei (Thompson et al., 2016).

The penetration of the cytoplasm will occur for longer pulses as well if their rise time is <1 μs. Indeed, 100 μs pulses with fast rise times have been shown to permeabilize large organelles that nearly span the cell, such as the endoplasmic reticulum. For such long organelles, the imposed field does not have to be so large, and the relatively small amplitude microsecond fields with fast rise times can also selectively permeabilize them (Hanna et al., 2017). However, for microsecond pulses to permeabilize smaller organelles like mitochondria, much higher field strengths are required, and such fields would introduce excessive thermal energy and destroy the cell with heat (as discussed below), making any potential changes to organelles impractical.

Large Electric Field

When electric pulses are applied to cells or tissues, they will introduce "Joule heating" which can be calculated based on the current, I, voltage, V, and duration of the pulse, t:

$$Energy = (V)(I)(t) \text{ joules.}$$

Since a joule of energy will heat 1 mL of water by 0.24 °C, it is important to keep the energy delivered by the pulsed electric field in mind to avoid thermal effects. This is very easy for nanosecond-range pulses because they are 1,000 times shorter than those in the microsecond domain. Since the total energy delivered is proportional to the product of the pulse duration, current, and voltage, the shorter pulse duration of a nanopulse makes it possible to increase the applied voltage by this factor of as much as 1,000, so the applied voltage and current can be much larger with minimal temperature change to the cell. This capability to increase the applied voltage is critical for permeabilizing the membranes surrounding small intracellular organelles. The initial targets of pulsed electric fields are the cell membranes because they are the most resistive elements in cells and tissues. The transmembrane voltage difference required to permeabilize a lipid bilayer membrane is 500 mV (Velikonja et al., 2016). In order to generate a voltage gradient of 500 mV across both end membranes perpendicular to the field lines in a 10 μm wide cell, a field of 1 V/10 μm (1 kV/cm) is required, and this is the typical field strength used for ECT mentioned above. However, in order to permeabilize smaller intracellular organelles, higher fields are required. For example, a mitochondrion can be as small as 0.5 μm in diameter and will require 1 V/0.5 μm or 20 kV/cm to permeabilize it. If a 100 ns, 20 kV/cm pulse is applied to a 200 Ω-cm load, the current flowing through the load will be 100 A.

$E = 20$ kV $\times$ 100 A $\times$ 10^{-7} s $= 0.2$ J, and this energy will raise the temperature of 1 mL of the target tissue by <0.05 °C, assuming the heat capacity of water.

In contrast, for a 100 μs pulse having the same voltage and current, $E = 200$ J which would heat the cells exposed by 50°C. Therefore, the ability to apply large voltage gradients across the cytoplasm of cells without significant Joule heating is a very important benefit of nanosecond-domain pulses. Such large voltage gradients are critical for generating transient nanopores in small organelles.

These two unique features of nanosecond-domain pulses, large amplitude and intracellular penetrating ability, make them useful to elicit unique responses of cells and tissues, such as the initiation of vesicle secretion and regulated cell death. This latter response is most commonly used in clinical applications and will be discussed next.

REGULATED CELL DEATH

The first indication that applying pulsed electric fields in the nanosecond domain could influence tumor growth came from the pioneering work of Beebe and Schoenbach (Beebe et al., 2002) who treated subdermal murine fibrosarcoma allografts. Subsequent to their published work, more than 60 papers have been published describing various aspects of tumor ablation using nanosecond pulsed electric fields in murine models. Collectively, these studies suggest that NPS treatment of tumors results in a slower cell death process in treated tissue over a period of days, unlike IRE which generally triggers necrotic death within hours. In a subset of these studies treating malignant tumors, the authors identified that the treatment with NPS initiated immunogenic cell death (ICD), a subset of regulated cell death (RCD) in which tumor-associated antigens are released and presented to the immune system to generate an adaptive immune response. The ICD process involves releasing danger-associated molecular patterns (DAMPs) such as the translocation of calreticulin from the endoplasmic reticulum to the cell surface, as well as the release of both ATP and HMGB1 (Guo et al., 2018; Nuccitelli et al., 2017). This collective body of preclinical research suggests a link between NPS's release of key DAMPs and the subsequent generation of a CD8+-dependent adaptive immune response. This adaptive immune response may prevent growth of rechallenge tumor cells. (Beebe et al., 2018; Chen et al., 2014; Guo et al., 2018; Lassiter et al., 2018; Nuccitelli et al., 2012, 2015; Skeate et al., 2018).

INVESTIGATIONAL STUDIES USING NPS THERAPY

TREATING NORMAL SKIN

In order to characterize the histological effects of a range of energy settings for NPS™ treatment of human skin, a clinical study was initiated treating normal abdominal skin that was scheduled for resection in an abdominoplasty procedure (Kaufman et al., 2019). Six NPS treatment energy levels were evaluated in five patients enrolled in a 60-day longitudinal study in which the patients received NPS treatments 60, 30, 15, 5, and 1 day prior to resection. These treated skin samples were prepared for

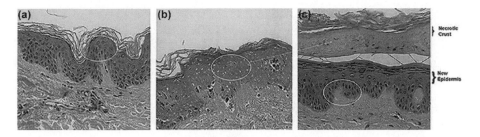

FIGURE 10.2 Hematoxylin- and eosin- stained sections of human skin following NPS therapy. (a) Untreated skin shows strong nucleus labeling in the epidermis (oval). (b) One day after NPS treatment. Oval marks represent nonviable epidermis containing "ghost cells" with intact membranes and hollow nuclei that do not stain. (c) Seven days after NPS treatment. Necrotic epidermis has delaminated forming a crust and a new epidermal layer has regenerated (oval).

histology so that changes in the skin structure over this 60-day period could be detected. One day following NPS treatment, the epidermis is nonviable with evidence of "ghost cells" which appear to have intact cellular membranes but with nuclei that do not stain (Figure 10.2). One week later, a crust appears over the treated region that represents the dead, treated epidermis which has been replaced with a regenerated epidermal layer (Figure 10.2c). A dermatopathologist evaluated the histological sections for indicators of dermal injury, fibroplasia, elastin fiber integrity, and adnexal structure integrity. There was no evidence of thermal injury in the dermis; however, cellular structures such as dermal fibroblasts and sebaceous glands were eliminated in the NPS-treated regions. Recovery of these adnexal structures was observed within 2 weeks.

Once this preliminary dose-response work was completed, lesions residing in or below the epidermis were treated to explore applications of NPS. Thus far, clinical studies of two lesion types have been completed: seborrheic keratosis (SK) and sebaceous gland hyperplasia (SGH).

Seborrheic Keratosis

SK is a benign skin lesion that resides in the epidermis and affects ~83 million Americans (Bickers et al., 2006). It is ubiquitous throughout all populations and is among the top 20 conditions treated by dermatologists (Taylor, 2017). Currently the main methods of treatment to eliminate SK lesions are low-intensity procedures such as curettage, electrodesiccation, cryosurgery, chemical destruction, and laser ablation. However, several common problems such as recurrence, scarring, and pigmentation changes are associated with these approaches. More recently, a topical solution containing high-concentration hydrogen peroxide was cleared by the FDA for use in the United States (Baumann et al., 2018) but it has now been withdrawn from the market.

A recent clinical study explored NPS treatment of SKs (Hruza et al., 2019). This was a prospective, randomized, open-label, multicenter, nonsignificant risk (NSR) study where patients with multiple SK lesions served as their own control. The study protocol was reviewed and approved by an Institutional Review Board (Biomedical Research Institute of America, protocol NP-SK-002) and conformed to the ethical guidelines of the 1975 Declaration of Helsinki.

Treatment Procedures

Fifty-eight patients were enrolled presenting at least four confirmed SK lesions each on the trunk or extremities. Three of the four SK lesions underwent a single treatment with the NPS™ system. The fourth lesion served as an untreated control. The treatment system consists of an electrical pulse console that produces predetermined pulse sequences (a "cycle") of high intensity, ultrashort electrical energy pulses through a hand-piece applicator connected to a sterile, single-patient-use treatment tip, which is applied to the skin on and around the SK lesion. Treatment tips consist of a polymer shell encasing an array of electrically conducting microneedles that penetrate through to the reticular dermis. The two treatment tips used for >90% of the patients treated areas of 2.5 × 2.5 mm and 5.0 × 5.0 mm, respectively (Figure 10.3). Since the larger spot size was limited to 5 mm × 5 mm, larger lesions (69%) required multiple adjacent cycles to cover the full lesion area. Study lesions ranged from 1 to 19 mm in length and width and 0.5–3 mm in height. Prior to SK lesion treatment, a clear acetate "Lesion Map" was created for each lesion location in order to correctly reidentify the location of the original lesion area throughout the 106-day evaluation window. This was needed since complete lesion clearance would reduce visual appearance of the original lesion area (Figure 10.4). After the treatment visit, patients returned at intervals of 7, 30, 60, 90, and 106 days for photographs of the lesion areas and investigator's evaluation of the treated areas as either *clear, mostly clear, partially clear,* or *not clear.*

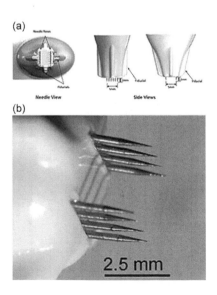

FIGURE 10.3   (a) 5 × 5 mm microneedle tip and (b) 2.5 × 2.5 mm microneedle tip.

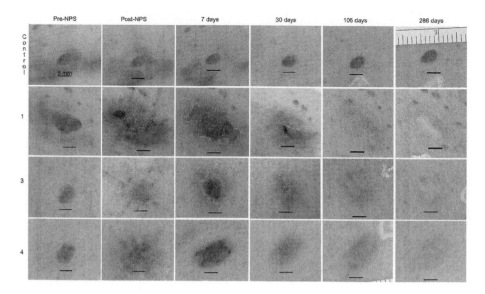

FIGURE 10.4 Pigmented seborrheic keratoses from a single subject shown before (pre) and after (post) a single NPS treatment as well as at four additional time points after treatment. Top row: untreated control on the side of upper body; 1: lesion located on side of upper body; 3,4: lesions located on the back. Black scale bar indicates 5 mm. (reproduced with permission from Hruza, et al. 2019)

Results

Lesion Clearance

The study endpoints for lesion outcomes were based on a combination of investigator assessment, blinded independent assessment of lesion photos, and patient satisfaction at the end of the 106-day study. After an NPS procedure, lesion areas showed initial symptoms of erythema and minor localized edema during the 1-week visit as characterized by a thin crust or, in the case of a larger lesion, a thicker crust. Most treated areas healed within 1 month with at least mild residual hyperpigmentation seen in most lesion areas. During each visit, investigators were asked to observe and rate the lesion skin areas (Figure 10.5). The treated lesion was considered improved when at least partial clearance was observed. Combining all three clearance levels, *clear*, *mostly clear*, and *partially clear*, 88% of the lesions had improved in 1 month with 93% of lesions showing improvement at 106 days post treatment. None of the control SKs demonstrated any lesion clearance. During the optional visit, ~286 days post treatment, 96% of lesions showed improvement.

Hyperpigmentation was present to some degree in 61% of the lesions on day 106 and completely absent in the remainder of lesions. Mild hypopigmentation was noted in 2% of lesions at 106 days. As overlying hyperpigmentation faded, 12% of lesions had mild hypopigmentation noted at 286 days after treatment (Table 10.1). Ongoing studies using lower energies have resulted in a lower rate of hyperpigmentation as well as hypopigmentation so this therapy is yet to be optimized.

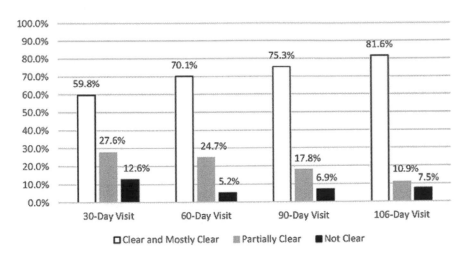

FIGURE 10.5   Percentage of 174 SK lesions scored as *clear, mostly clear, partially clear* and *not clear* at the indicated times post NPS treatment.

TABLE 10.1

SK Treated with NPS™ Technology Can Lead to Hyperpigmentation or Subtle HypoPigmentation that Can Be Usually Resolved with Time

	106 Days		286 Days	
Treated lesions with 286-Day Visits (N = 145)	**N**	**Rate (%)**	**n**	**Rate (%)**
Hyperpigmentation	89	61.38	50	34.48
Subtle hypopigmentation	3	2.07	18	12.24
Patients satisfied or mostly satisfied	174	78	145	87

Discussion

This multicenter study represents the first controlled clinical study to evaluate the role of NPS™ technology in the treatment of SK, a common benign epidermal lesion. In prior studies with normal skin (Kaufman et al., 2019) discussed above, the NPS mechanism was demonstrated to have specificity for cellular structures in both the epidermis and the dermis, with minimal effect on the adjacent acellular dermis. Based on the clinical results from this study showing a high percentage of lesions scored as *clear* or *mostly clear,* the presumed NPS mechanism of nonthermal destruction of keratinocytes resulted in SK lesion clearance with a single NPS treatment. This first clinical study of NPS treatment of a clinical disorder demonstrated its unique capability to target and clear a prototypical cellular lesion that extrudes from the skin surface without damaging dermal tissue at the tissue margin between the cellular lesion and the noncellular components in the dermis.

The NPS procedure demonstrated high effectiveness in removing SKs treated once with 82% of lesions rated as *clear* or *mostly clear* after 106 days. As a comparison, the only FDA-approved drug therapy using a proprietary hydrogen peroxide topical gel formulation applied up to two times in two phase 3 trials resulted in 51% of lesions scored as *clear* or *mostly clear* after 106 days (Baumann et al., 2018).

In conclusion, these results demonstrate that the NPS™ procedure provides a safe and effective treatment for SKs with a low risk of dermal scarring and minimal long-term residual skin effects in most patients. Furthermore, the procedure time is generally less than a minute, avoids surgical excision, and lesion clearance is highly localized with no systemic side effects reported during the study. NPS energy is believed to target cellular structures and has no known effect on fibrous components of the dermis which supports the conclusion that damage is limited to the cellular lesion. Since many superficial cellular lesions have a similar morphology to SK lesions, these promising results likely apply to other common lesions as well.

SEBACEOUS GLAND HYPERPLASIA

Sebaceous Gland Hyperplasia (SGH) is a common condition that appears as white or lightly pigmented papules or bulges on the skin, often with indented centers. They occur when hyperactive sebaceous glands, which are 0.7–1.7 mm below the surface of the skin, produce excess oil (sebum) that is forced out from the skin surface through an existing pore or follicle. There are sebaceous glands all over the body, so the SGH papules can form almost anywhere, though they are more frequently observed and treated when they appear on facial skin. These benign lesions are more likely to occur in middle-aged and older people and are reported to occur in ~1% of the healthy U.S. population. However, the prevalence of SGH has been reported to be as high as 10%–16% in patients receiving long-term immunosuppression (Naldi et al., 2018).

The primary goal of this study was to evaluate the efficacy of a NPS™ device in clearing SGH lesions after one or two NPS treatments (Munavalli et al., 2019).

Methods

This was a prospective, randomized, open-label, multicenter, NSR study where patients with multiple SGH lesions served as their own control. The study protocol was reviewed and approved by an Institutional Review Board (Biomedical Research Institute of America, protocol NP-SH-006) and conformed to the ethical guidelines of the 1975 Declaration of Helsinki.

Patients

Up to 19 patients were recruited in each of five centers with two to five SGHs on their faces for a total of 72 patients.

Treatment Procedures

Principal investigators were identified at five participating study centers. Baseline photographs of each of the two to five selected SGHs were taken prior to local anesthesia or treatment. Local anesthetic lidocaine was injected at the sites of the selected SGHs. One to four lesions were injected with lidocaine; one was treated as a sham and the rest underwent a single NPS treatment. Treatment tips consist of a polymer shell encasing an array of electrically conducting microneedles that penetrate the reticular dermis (Figure 10.6).

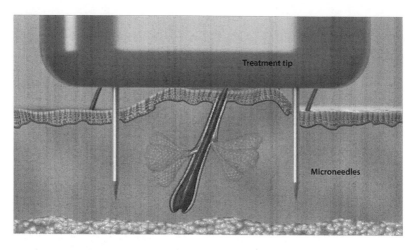

FIGURE 10.6   Treatment tip positioned around SGH.

The NPS treatment of a single SGH lesion entailed selecting the treatment energy level (0.9–4.5 J), applying sterile contact gel, pressing the treatment tip into the skin, and actuating the footswitch to initiate the NPS treatment cycle that lasts <1 min. The two treatment tips used treated areas of 1.5 × 1.5 mm and 2.5 × 2.5 mm, respectively. The patients rated their degree of pain using a standardized pain scale of 0–10. The control lesion underwent an identical process with the exception that no energy was delivered to the lesion.

After the initial treatment visit, patients returned at intervals of 5, 30, 60, and, if applicable, 90 days for photographs of the lesions and investigator evaluation of the treated areas. Photographs were taken of all five lesion areas during each study visit using both a Sony Cyber-shot DSC-RX100 and a dermatoscope (Handyscope, FotoFinder® Systems GmbH, Columbia, MD). Prior to SGH lesion treatment, a clear acetate "Lesion Map" was created for each lesion location in order to correctly reidentify the location of the original lesion area throughout the 60- to 90-day evaluation window. This was needed since complete lesion clearance occurred in most instances and the original lesion area was not visually apparent. A ruler was placed in each photo to enable precise scaling of the images in comparison with the baseline photo of the lesion.

Results

Clinical Outcome

Lesion Clearance   Lesion clearance was assessed by the principal investigators (PIs) during the 30- and 60-day visits for patients that received a single treatment (Figures 10.7 and 10.8). If during the 30-day visit, a subject had one or more lesions that were not rated *clear* or *mostly clear,* the investigator could provide a second NPS™ treatment to those lesions. A total of 18 out of the 226 study lesions on 13 of the 72 patients in the study were treated with a second NPS treatment 30 days after the initial NPS treatment.

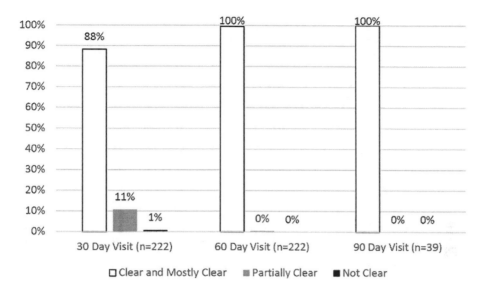

FIGURE 10.7   Percentage of 222 SGH lesions scored as *clear, mostly clear, partially clear,* and *not clear* at indicated times post NPS treatment.

By day 60, most lesion areas required the lesion map in order to identify and photograph the treated area, as neither the lesion nor signs of skin damage were evident to the investigator. None of the control SGHs demonstrated lesion clearance.

Investigator Skin Assessment

Hyperpigmentation   PIs were asked to assess the degree of postinflammatory hyperpigmentation on the 5-, 30-, 60- and, if applicable, 90-day visits. When considering lesions that received both one and two NPS treatments, hyperpigmentation peaked during the 30-day visit, and decreased with time. By the last visit, 55% of lesions were rated *none*, 30% were rated *mild*, 12% were rated *moderate*, 2% were rated *moderately severe*, and 1% were rated *severe*. A total of 39 lesions were assessed during the 90-day visit; 44% and 54% were rated *none* and *mild*, respectively; one lesion (3%) was rated *moderate*; and no lesions were rated *moderately severe* or *severe*.

Erythema and Swelling   Mild erythema was observed in only 3% of lesions prior to treatment. Immediately after the initial NPS treatment, erythema was observed in 91% of lesions: 82% were *mild* and 9% were *moderate* (Figure 10.9a). Erythema was usually present 5 days later, decreased substantially by 30 days, and was only observed in 7% of the treated lesions at 60 days. The typical early response to NPS treatment is shown in Figure 10.10.

Transient swelling was observed immediately following NPS treatment (Figures 10.9b and 10.10a). However, by 5 days, only 37% of lesions showed swelling that decreased to 1% at 30 days, and no swelling was observed at 90 days.

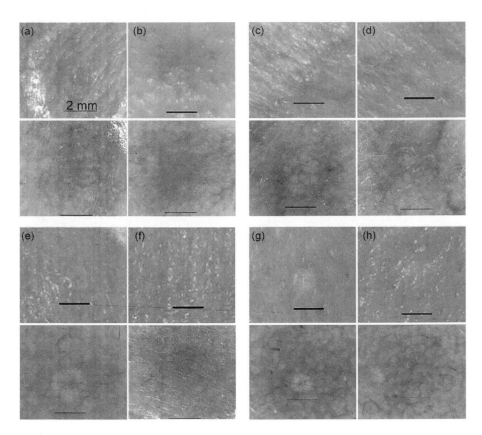

FIGURE 10.8   Pairs of images of four SGHs of four different patients taken before (a, c, e, g) and 60 days after NPS treatment (b, d, f, h). The upper photo in each pair is the reflected light image showing skin surface, and the lower photo is the dermatoscope image of the same region showing sebaceous glands. Scale bar in each image is 2 mm long. (reproduced with permission from Munavalli, et al. 2019)

Volume Loss at the Lesion Site   A natural consequence of the effective destruction of the sebaceous gland is a volume loss at the lesion site. This was reported starting as early as the 30-day visit (see Figures 10.8b and 10.10c). At the last observation for all lesions, 32% of the 222 lesions were noted as having a slight volume loss.

Discussion

This multicenter study represents the first controlled clinical study to evaluate the role of NPS technology in the treatment of SGH, a common benign epidermal lesion. Based on prior studies in normal skin (Hruza et al., 2019; Kaufman et al., 2019), the NPS mechanism demonstrated specificity for cellular structures including the epidermis and minimal effect on the adjacent acellular components in the dermis. Based on the clinical results from this study showing 221 of 222 lesions scored as *clear* or *mostly clear*, the presumed NPS mechanism of nonthermal destruction of

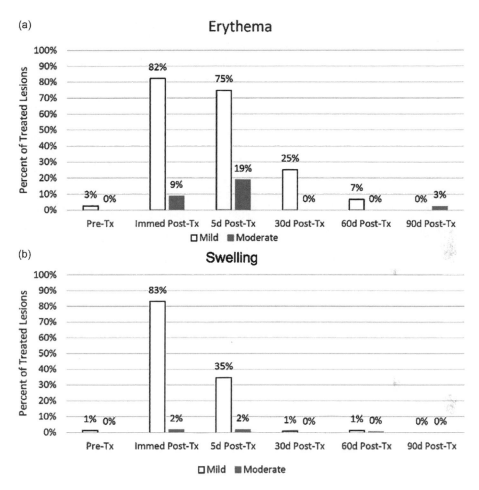

FIGURE 10.9 Percentage of treated SGH lesions with erythema (a) and swelling (b) at the indicated times with respect to NPS treatment (Tx).

sebaceous glands resulted in reliable SGH lesion clearance with a single NPS treatment, with minimal apparent damage to the dermis.

Hyperpigmentation and Skin Depressions

The lesions scored as moderately to severely hyperpigmented increased from 0% at 5 days to 16% at 30 and 60 days. However, by 90 days, only 3% of the lesions were in that category, indicating that the hyperpigmentation may fade over time. Further studies are planned to optimize NPS therapy for SGH which might result in a reduction in the applied energy which may reduce the hyperpigmentation. Another approach would be the use of hydroquinone in darker pigmented patients post procedure as is commonly used to reduce postinflammatory hyperpigmentation caused by other therapies.

Small skin volume losses were evident in some of the treatment sites and 19% of single treatment lesions that were examined at 90 days noted volume loss. This is not

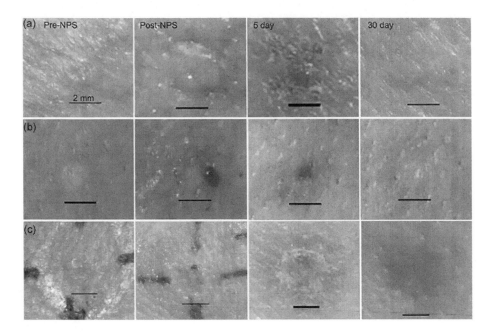

FIGURE 10.10 Images of three separate SGH lesions on different patients responding to NPS treatment. (a–c) Each row has photos of the same lesion taken at the times indicated at the top of row "a". Each scale bar represents 2 mm. (reproduced with permission from Munavalli, et al. 2019)

surprising since the enlarged sebaceous glands were eliminated by the NPS therapy, reducing the local tissue volume. These volume losses appear to improve over time and ongoing dose-ranging studies indicate that a reduction in the treatment energy reduces the amount of volume loss.

Competing Technologies

SGH treatment is challenging due to the need to destroy or excise the entire sebaceous gland. The common destructive modalities including cryosurgery, electrodessication, curettage, shave excision, and topical trichloroacetic acid have a propensity to cause skin discoloration and scarring (Bader and Scarborough, 2000). Another approach used is short burst oral isotretinoin, but it is associated with adverse effects and rapid recurrence upon discontinuation of the medication (Grimalt et al., 1997). Additionally, a patient's hesitance to use oral isotretinoin (due to side effects) for other medical skin conditions, such as acne vulgaris, is widely documented (Bauer et al., 2016). Seven studies using photodynamic therapy and laser therapy have shown some success with small subject numbers (Simmons et al., 2015). One study used a 1,720 nm laser and reported nearly complete clearance of SGH lesions with two treatments on four patients (Winstanley et al., 2012). Another study used a 1,450 nm diode laser to obtain 50%–75% shrinkage of SGH lesions on ten patients (No et al., 2004). Thus, none of the previous studies have reported the success rates and the large numbers of treated patients forming the basis of this work.

Conclusion

These results demonstrate that the NPS procedure provides a safe and effective treatment for SGHs with a low risk of scarring and long-term hyperpigmentation. Furthermore, the treatment time is very short, and the SGH clearance is highly localized with no reported systemic side effects. The mechanism of this localized NPS therapy targets cellular structures within the epidermis and dermis, making it ideal to eliminate sebaceous glands with minimum treatment sessions, a high clearance rate, and high degree of subject satisfaction.

FUTURE APPLICATIONS OF NPS THERAPY

The specificity of NPS™ therapy for cellular targets makes it ideal for treating skin lesions that are cellular in nature. Therefore, clinical studies targeting warts, acne, and basal cell carcinoma are ongoing. Since malignant lesions have morphologies similar to these benign lesions, NPS therapy should be broadly applicable to both. Other applications and delivery methods of NPS therapy are also underway. For example, applicators designed specifically for laparoscopy and endoscopy are under development.

In addition to lesion access, the field strength of about 25–30 kV/cm that must be delivered to all cells in the lesion presents some challenges. Once the field strength reaches 10–15 kV, it can ionize gases in the air to generate arcs between the electrodes. That limits the size of the treatment zone between two microneedle poles to about 5–10 mm. For larger lesions, multiple treatments or larger electrode arrays with multiple poles to be multiplexed may be necessary.

In conclusion, these two completed clinical studies provide strong evidence demonstrating the safety and efficiency of NPS therapy for clearing cellular-based skin lesions with a nonthermal, drug-free modality.

REFERENCES

Aycock, K.N. and R.V. Davalos 2019. Irreversible electroporation: Background, theory and review of recent developments in clinical oncology. *Bioelectricity.* 4:214–234.

Bader, R.S. and D.A. Scarborough. 2000. Surgical pearl: Intralesional electrodesiccation of sebaceous hyperplasia. *Journal of the American Academy of Dermatology.* 42:127–128.

Bauer, L.B., J.N. Ornelas, D.M. Elston, and A. Alikhan. 2016. Isotretinoin: Controversies, facts, and recommendations. *Expert Review of Clinical Pharmacology.* 9:1435–1442.

Baumann, L.S., A. Blauvelt, Z.D. Draelos, S.E. Kempers, M.P. Lupo, J. Schlessinger, S.R. Smith, D.C. Wilson, M. Bradshaw, E. Estes, and S.D. Shanler. 2018. Safety and efficacy of hydrogen peroxide topical solution, 40% (w/w), in patients with seborrheic keratoses: Results from 2 identical, randomized, double-blind, placebo-controlled, phase 3 studies (A-101-SEBK-301/302). *Journal of the American Academy of Dermatology.* 79:869–877.

Beebe, S.J., P. Fox, L.J. Rec, K. Somers, R.H. Stark, and K.H. Schoenbach. 2002. Nanosecond pulsed electric field (nsPEF) effects on cells and tissues: Apoptosis induction and tumor growth inhibition. *IEEE Transactions on Plasma Science.* 30:286–292.

Beebe, S.J., B.P. Lassiter, and S. Guo. 2018. Nanopulse stimulation (NPS) induces tumor ablation and immunity in orthotopic 4T1 mouse breast cancer: A review. *Cancers.* 10:97.

Bickers, D.R., H.W. Lim, D. Margolis, M.A. Weinstock, C. Goodman, E. Faulkner, C. Gould, E. Gemmen, and T. Dall. 2006. The burden of skin diseases: 2004 a joint project of the American Academy of Dermatology Association and the Society for Investigative Dermatology. *Journal of the American Academy of Dermatology.* 55:490–500.

Bulysheva, A., J. Hornef, C. Edelblute, C. Jiang, K. Schoenbach, C. Lundberg, M.A. Malik, and R. Heller. 2019. Coalesced thermal and electrotransfer mediated delivery of plasmid DNA to the skin. *Bioelectrochemistry.* 125:127-133.

Cemazar, M. and G. Sersa 2019. Recent advances in electrochemotherapy. Bioelectricity. 2:204–213.

Chen, R., N.M. Sain, K.T. Harlow, Y.J. Chen, P.K. Shires, R. Heller, and S.J. Beebe. 2014. A protective effect after clearance of orthotopic rat hepatocellular carcinoma by nanosecond pulsed electric fields. *European Journal of Cancer.* 50:2705–2713.

Grimalt, R., J. Ferrando, and J.M. Mascaro. 1997. Premature familial sebaceous hyperplasia: Successful response to oral isotretinoin in three patients. *Journal of the American Academy of Dermatology.* 37:996–998.

Guo, S., Y. Jing, N.I. Burcus, B.P. Lassiter, R. Tanaz, R. Heller, and S.J. Beebe. 2018. Nano-Pulse Stimulation induces potent immune responses, eradicating local breast cancer while reducing distant metastases. *International Journal of Cancer.* 142:629–640.

Hanna, H., A. Denzi, M. Liberti, F.M. Andre, and L.M. Mir. 2017. Electropermeabilization of inner and outer cell membranes with microsecond pulsed electric fields: Quantitative study with calcium ions. *Scientific Reports.* 7:13079.

Hruza, G.J., B.D. Zelickson, M.M. Selim, T.E. Rohrer, J. Newman, H. Park, L. Jauregui, R. Nuccitelli, W.A. Knape, E. Ebbers, and D. Uecker. 2019. Safety and efficacy of Nano-Pulse Stimulation treatment of seborrheic keratoses. *Journal American Academy of Dermatology.* Dec 4. doi:10.1097/DSS.02278. [Epub ahead of print].

Kaufman, D., M. Martinez, L. Jauregui, E. Ebbers, R. Nuccitelli, W.A. Knape, D. Uecker, and D. Mehregan. 2019. A dose-response study of a novel method of selective tissue modification of cellular structures in the skin with Nano-Pulse Stimulation™. *Lasers in Surgery and Medicine.* Aug. 2. Doi.10.1002/lsm.23145.

Lassiter, B.P., S. Guo, and S.J. Beebe. 2018. Nano-pulse stimulation ablates orthotopic rat hepatocellular carcinoma and induces innate and adaptive memory immune mechanisms that prevent recurrence. *Cancers.* 10: 69.

Munavalli, G.S., B.D. Zelickson, M.M. Selim, S.L. Kilmer, T.E. Rohrer, J. Newman, L. Jauregui, W.A. Knape, E. Ebbers, D. Uecker, and R. Nuccitelli. 2019. Safety and efficacy of Nano-Pulse Stimulation treatment of sebaceous gland hyperplasia. *Journal of Dermatologic Surgery.* 2019 Oct 4. doi:10.1097/DSS.02154. [Epub ahead of print].

Naldi, L., A. Venturuzzo, and P. Invernizzi. 2018. Dermatological complications after solid organ transplantation. *Clinical Reviews in Allergy and Immunology.* 54:185–212.

No, D., M. McClaren, V. Chotzen, and S.L. Kilmer. 2004. Sebaceous hyperplasia treated with a 1450-nm diode laser. *Dermatologic Surgery:.* 30:382–384.

Nuccitelli, R., J.C. Berridge, Z. Mallon, M. Kreis, B. Athos, and P. Nuccitelli. 2015. Nanoelectroablation of murine tumors triggers a cd8-dependent inhibition of secondary tumor growth. *PLoS One.* 10:e0134364.

Nuccitelli, R., A. McDaniel, S. Anand, J. Cha, Z. Mallon, J.C. Berridge, and D. Uecker. 2017. Nano-Pulse Stimulation is a physical modality that can trigger immunogenic tumor cell death. *Journal for Immunotherapy of Cancer.* 5:32.

Nuccitelli, R., K. Tran, K. Lui, J. Huynh, B. Athos, M. Kreis, P. Nuccitelli, and E.C. De Fabo. 2012. Non-thermal nanoelectroablation of UV-induced murine melanomas stimulates an immune response. *Pigment Cell Melanoma Res.* 25:618–629.

Schoenbach, K.H. 2010. Bioelectric effect of intense nanosecond pulses. In: *Advanced Electroporation Techniques in Biology and Medicine.* A.G. Pakhomov, D. Miklavcic, and M.S. Markov, editors. Taylor & Francis Group, Boca Raton, FL, 19–50.

Schoenbach, K.H., S.J. Beebe, and E.S. Buescher. 2001. Intracellular effect of ultrashort electrical pulses. *Bioelectromagnetics*. 22:440–448.

Simmons, B.J., R.D. Griffith, L.A. Falto-Aizpurua, F.N. Bray, and K. Nouri. 2015. Light and laser therapies for the treatment of sebaceous gland hyperplasia a review of the literature. *Journal of the European Academy of Dermatology and Venereology: JEADV.* 29:2080–2087.

Skeate, J.G., D.M. Da Silva, E. Chavez-Juan, S. Anand, R. Nuccitelli, and W.M. Kast. 2018. Nano-Pulse Stimulation induces immunogenic cell death in human papillomavirus-transformed tumors and initiates an adaptive immune response. *PLoS One.* 13:e0191311.

Taylor, S. 2017. Advancing the understanding of seborrheic keratosis. *Journal of Drugs in Dermatology.* 16:419–424.

Tekle, E., H. Oubrahim, S.M. Dzekunov, J.F. Kolb, K.H. Schoenbach, and B.P. Chock. 2005. Selective field effects on intracellular vacuoles and vesicle membranes with nanosecond electric pulses. *Biophysical Journal.* 89:274–284.

Thompson, G.L., C.C. Roth, M.A. Kuipers, G.P. Tolstykh, H.T. Beier, and B.L. Ibey. 2016. Permeabilization of the nuclear envelope following nanosecond pulsed electric field exposure. *Biochemical and Biophysical Research Communications.* 470:35–40.

Velikonja, A., P. Kramar, D. Miklavcic, and A. Macek Lebar. 2016. Specific electrical capacitance and voltage breakdown as a function of temperature for different planar lipid bilayers. *Bioelectrochemistry.* 112:132–137.

Vernier, P.T. 2011. Mitochondrial membrane permeabilization with nanosecond electric pulses. *Conference Proceedings: IEEE Engineering in Medicine and Biology Society.* 2011: 743–745.

Vernier, P.T., Y. Sun, L. Marcu, S. Salemi, C.M. Craft, and M.A. Gundersen. 2003. Calcium bursts induced by nanosecond electric pulses. *Biochemical and Biophysical Research Communications* 310:286–295.

White, J.A., P.F. Blackmore, K.H. Schoenbach, and S.J. Beebe. 2004. Stimulation of capacitative calcium entry in HL-60 cells by nanosecond pulsed electric fields. *Journal of Biological Chemistry* 279:22964–22972.

Winstanley, D., T. Blalock, N. Houghton, and E.V. Ross. 2012. Treatment of sebaceous hyperplasia with a novel 1,720-nm laser. *Journal of Drugs in Dermatology: JDD.* 11:1323–1326.

11 Calcium Electroporation – A Novel Treatment to Overcome Cancer-Mediated Immune Suppression

Ryan Burke and P. Thomas Vernier

CONTENTS

According to the World Health Organization, cancer is the second leading cause of death worldwide, accounting for roughly 1 in every 6 deaths (Cancer, 2019). The current standard of care involves surgery, chemotherapy, radiotherapy, or a combination of these. Certain cancers such as lung, pancreatic, brain, stomach, and esophageal remain highly resistant to conventional forms of treatment, which is compounded by a lack of early detection/diagnosis. These cancers represent a major challenge, since no current treatment option has demonstrated long-term benefit for patients with advanced disease. Even where remission is achieved, there is a high recurrence rate that is often associated with metastases.

The poor prognosis of certain cancers has been linked to a lack of an effective immune response (Fridman, Zitvogel, Fridman-Sautès, & Kroemer, 2017), and in this context, the tumor microenvironment has received significant attention. The term "immune-editing" has been used to describe the selection process by which a malignant cell develops into a tumor by altering its microenvironment to prevent immune recognition (Teng, Galon, Fridman, & Smyth, 2015). Specific biochemical events have been identified which are related to the immune-suppressive environment.

Under normal physiological conditions, the extracellular environment is more alkaline, and the cytosol is more acidic (Spugnini & Fais, 2017). This is reversed in established tumors, resulting from a combination of hypoxic conditions and altered metabolism (Huber et al., 2017). The acidic tumor microenvironment has been shown to enhance regulatory immune function, while impairing effector immune function. The result is unchecked proliferation and metastases. The reversed proton gradient observed in tumors has additional consequences. It has been shown to fundamentally alter membrane lipid composition and rigidity which has been linked to multidrug resistance in many cancers (Peetla, Vijayaraghavalu, & Labhasetwar, 2013). In addition to the structural changes in the plasma membrane, the acidic pH conditions have been shown to alter the function of many common chemotherapeutic agents by protonating them, rendering them less cytotoxic (Rauch, 2009).

Several studies have demonstrated a significant relationship between prognosis and immune cell infiltration in many tumor types (Huber et al., 2017; Takeuchi & Nishikawa, 2016). Developing strategies to overcome tumor-mediated immune suppression thus has the potential, not only to increase treatment efficacy, but also to control disease recurrence.

Calcium electroporation (CaEP) is one such therapy. It has been shown in several preclinical and clinical studies to be effective at treating a variety of cancer types (Frandsen, Gibot, Madi, Gehl, & Rols, 2015). An extension to electrochemotherapy (ECT), which combines the membrane-permeabilizing effects of locally applied electric fields with the systemic or local administration of membrane-impermeant chemotherapeutic agents, CaEP utilizes supraphysiological concentrations of calcium, alone or in combination with anticancer drugs, to induce immunogenic cell death.

CaEP and ECT have important advantages over conventional treatments, including low-cost and rapid outpatient treatment with a significantly reduced side-effect profile (Frandsen & Gehl, 2018). CaEP and ECT both have stronger lethal effects on malignant cells than on surrounding normal cells (Frandsen et al., 2017). One important difference between CaEP and ECT is the immune-recruitment profile. Both treatments have demonstrated an enhanced immune response immediately following treatment, but CaEP has shown immune infiltration that persists over a longer time (Falk, Forde, et al., 2017). This chapter will discuss CaEP as a novel cancer treatment with the potential to overcome tumor-mediated immune suppression. We will also discuss ways that this new therapy can be improved, with the primary goal being single-treatment eradication of tumors and induction of a systemic immune response capable of increasing patient survival and reducing risk of recurrence.

PART I – CAEP

CaEP is an extension of traditional ECT that substitutes supraphysiological concentrations of calcium for the chemotherapeutic agent. Results from preclinical trials have demonstrated great success, comparable to that observed with traditional ECT (Falk, Forde, et al., 2017). Although the exact mechanism of action remains to be elucidated, CaEP has shown to be effective on a variety of cancer cell lines (Frandsen et al., 2015).

Like other therapies, its quantitative *in vitro* efficacy is dependent on the cell type (Frandsen et al., 2017), but more important is the fact that normal tissues both *in vitro* and *in vivo* are significantly less vulnerable to this treatment (Frandsen & Gehl, 2018).

Whereas the precise mechanisms through which CaEP kills cells is unknown, studies have shown that CaEP is followed by an acute depletion of intracellular adenosine triphosphate (ATP) followed by necrotic cell death (Frandsen et al., 2015, 2012). A dose-dependent *in vitro* analysis using transformed Chinese hamster lung fibroblast (DC-3F), human leukemia cell (K-562), and Lewis lung carcinoma (murine) cells demonstrated that, in all cases, a significant reduction in survival occurred around $0.5\,mM$ $CaCl_2$ and plateaued around $1\,mM$ $CaCl_2$ (Frandsen et al., 2012). In the same study, using stably transfected human small cell lung cancer cells (H69) in mice, it was found that in 89% of CaEP-treated mice, tumor eradication was observed. When EP was used alone, an initial decrease in tumor volume was observed, but this was reversed around day 7 post treatment.

The results above were shown to be directly related to a reduction in intracellular ATP concentration that was not recovered even $8\,h$ following CaEP. Significant intracellular ATP reduction was confirmed in a separate study using a 3D spheroid model (Frandsen et al., 2015). Here the effects were compared among malignant colorectal adenocarcinoma (HT29), bladder transitional cell carcinoma (SW780), breast adenocarcinoma (MDA-MB231), and nonmalignant primary normal human dermal fibroblasts (HDF-n) cell lines. Interestingly, despite all cells experiencing an acute ATP depletion following treatment, only the malignant cell spheroids underwent a reduction in tumor volume, suggesting that ATP depletion is a secondary effect not associated with the tumor-specific lethal effects of CaEP.

To further explore how ATP levels are associated with the antitumor effects of CaEP, an *in vivo* study looked at the effects of metformin on CaEP lethality (Frandsen & Gehl, 2017). Metformin was used because of its role on mitochondrial bioenergetics resulting in impaired ATP production (Andrzejewski, Gravel, Pollak, & St-pierre, 2014; Hur, Lee, & Lee, 2015; Rena, Hardie, & Pearson, 2017). Mice in the metformin treatment condition were given an intraperitoneal injection of 250 mg/kg of metformin prior to treatment and received $2\,mg/mL$ in their drinking water for 2 weeks following treatment. If ATP depletion is responsible for cell death, it is reasonable to expect that CaEP combined with metformin is more lethal than CaEP alone. Results from this study did not support the ATP depletion hypothesis, and no significant difference in tumor volume was found between mice treated with CaEP and those treated with CaEP and metformin.

The above study also included an *in vitro* component, where the authors investigated the calcium concentration required to kill 50% of the cells (EC_{50}). Interestingly, cancer cell types displayed significant variability in their EC_{50}. For example, this value was more than $2 \times$ and $3 \times$ higher in HT29 colon cancer cells ($1.25\,mM$) than it was for H69 lung cancer ($0.56\,mM$) and MDA-MB231 breast cancer ($0.37\,mM$).

These results suggest that the capacity to cope with high concentrations of calcium differs among cancer cell types. Understanding *why* will be crucially important for the further development and optimization of CaEP treatment protocols. We will explore this in greater detail in the next section, beginning with a discussion of why calcium is toxic to cells and how cells have evolved to cope with it and to utilize the unique properties of calcium for various physiological processes.

PART II – HOW CAEP KILLS CANCER

IIA – CALCIUM TOXICITY

Traditional chemotherapeutic agents, most of which directly damage DNA or RNA or interfere with enzymes involved in their replication (My Cancer Genome, 2019), act against cancer cells and tumors through a completely different set of mechanisms than calcium, an abundant ion found in every cell and tissue and accounting for roughly 2% of our body weight (Bronner & Pansu, 1999). A clue to calcium's mechanism of action can be found by looking at the ion concentration gradients across cell membranes (Table 11.1).

Concentration gradients for other ions range from approximately 2- to 30-fold; the gradient for calcium across the plasma membrane is roughly 10,000-fold. Why is the third most abundant metal on Earth, being readily available to the primitive evolving cell, excluded from the cell interior? Why have cells evolved mechanisms to ensure that calcium is kept out of the cell, at the expense of considerable amounts of energy? The primary reason for this can be attributed to the properties of the cell's energy currency, phosphate. Compared with magnesium, calcium binds less tightly to water and, upon interaction with a phosphate molecule, liberated from ATP hydrolysis, will precipitate to form an insoluble salt (Clapham, 2007). Therefore, to cope with this danger, cells have evolved ways to control the concentration of free Ca^{2+} through channels dedicated to its extrusion or compartmentalization.

Mechanisms for maintaining nM cytosolic calcium include extrusion via the plasma membrane Ca^{2+}-ATPase (PMCA) and the Ca^{2+}/Na^+ exchanger (NCX) (Carafoli & Krebs, 2016). The former is a pump that uses the energy from hydrolysis of one ATP molecule for each calcium ion extruded (Bruce, 2018), while the latter is an antiporter which uses the energy in the electrochemical gradient of sodium to remove calcium, where three sodium ions are imported per export of a single calcium ion (Giladi, Tal, & Khananshvili, 2016). Alternatively, cytosolic Ca^{2+} can be transported into mitochondria through the mitochondrial Ca^{2+} uniporter (MCU)

TABLE 11.1
Typical Concentrations of Ions Located Inside and Outside Mammalian Cells

Ion	Intracellular (mM)	Extracellular (mM)
K^+	139	4
Na^+	12	145
Cl^-	4	116
HCO_3^-	12	29
Mg^{2+}	0.8	1.5
Ca^{2+}	<0.0002	1.8

Source: Adapted from Lodish et al. (2000).

or stored in the endoplasmic reticulum (ER) via the sarcoplasmic/ER Ca²⁺-ATPase (SERCA) (Kamer & Mootha, 2015; Raguimova et al., 2018).

Although maintaining a low intracellular concentration of calcium may have been a requirement for survival in the evolution of the early cell, this segregation has additional consequences. From the concentrations in Table 11.1, the Nernst potential for calcium is 122 mV. Recall, the Nernst potential (Equation 11.1) describes the voltage that would exist across the membrane for a given ionic gradient (Sawyer, Hennebry, Revill, & Brown, 2017). Stated differently, it describes the electric pressure resulting from the concentration gradient.

$$V_m = \frac{RT}{zF} \ln \frac{[X]o}{[X]i}. \tag{11.1}$$

Equation 11.1 – Nernst equation describing voltage across a membrane for a given chemical gradient. Here, R refers to the gas constant (8.315 VC/Kmol), T is temperature in (K), z is the ion valence, F refers to Faraday's constant (9.648×10^4 C/mol), and X refers to the concentration of a given ion (o) outside and (i) inside the cell.

When we combine the existing concentration gradient for calcium with its unusual coordination chemistry, the Ca²⁺ ion becomes an excellent candidate for a messenger molecule. Specifically, properties such as charge, ionic radius, polarizability, hydration energy, and hydration radius make Ca²⁺ readily accommodated by complex geometries observed in proteins that evolved with the cell (Carafoli & Krebs, 2016).

Indeed, calcium certainly is among the most important messengers in cell physiology, regulating various pathways including gene transcription, cell proliferation and migration, and even cell death (Raffaello, Mammucari, Gherardi, & Rizzuto, 2016; Tang et al., 2015; Wei et al., 2009). In a larger context, calcium forms a significant portion of our skeleton, is intimately involved in the formation of memories, and is essential for the muscular contraction required for movement of our bodies and the beating of our hearts. It shouldn't be too surprising, then, that alterations in calcium signaling are associated with a variety of pathological conditions.

Alterations in calcium signaling have been directly linked to multiple hallmarks of cancer, including evasion of apoptotic death, sustained angiogenesis, and enhanced potential for replication and invasion (Prevarskaya, Skryma, & Shuba, 2010). Several reviews have highlighted the alterations in calcium signaling observed in a variety of cancers (Cui, Merritt, Fu, & Pan, 2017; Li & Xiong, 2011; So, Saunus, Roberts-Thomson, & Monteith, 2018).

In light of this knowledge, calcium channels have been targeted in the development of new cancer treatments. Inhibition of SERCA has been investigated in various cancer models. Thapsigargin, an inhibitor of SERCA, is known to deplete intracellular calcium stores by preventing reuptake of cytosolic calcium (Sehgal et al., 2017), leading to induction of apoptosis. Results from several studies have shown that SERCA inhibition using thapsigargin either sensitizes or directly kills LNCaP and PC3 prostate cancer (Huang, Wang, & Wang, 2018; Sehgal et al., 2017), EC109 and TE12 esophageal cancer (Ma et al., 2016), MCF-7 breast cancer (Sehgal et al., 2017), and A549 lung adenocarcinoma (Wang et al., 2014) cells.

Its widespread efficacy in producing a rapid induction of apoptosis is also its major limitation. Thapsigargin's potent and irreversible SERCA inhibition produces the same effect in all mammalian cells, thus limiting its use in a clinical environment (Andersen, López, Manczak, Martinez, & Simonsen, 2015). To overcome this limitation, an analogue was developed, mipsagargin, which incorporates a prostate-specific membrane antigen present in a variety of tumor cell lines. Results from this study show a significant regression in a panel of cancer cell lines including prostate, breast, renal cell carcinoma, and bladder cancer, with limited toxicity in normal, healthy tissues (Denmeade et al., 2012).

The plasma membrane calcium-ATPase (PMCA1–4) pumps are a family of ATP-driven pumps which play an important role in calcium homeostasis. PMCA1 is ubiquitously expressed, while PMCA2–4 are expressed in a cell- and tissue-dependent manner (Padányi et al., 2016). Modified expression of these channels has been observed in malignant cells, although not in a straightforward manner. Upregulation of PMCA1 and PMCA2 and downregulation of PMCA4 has been reported in breast cancer cell lines (Curry, Roberts-Thomson, & Monteith, 2011; Monteith, Davis, & Roberts-Thomson, 2012; Padányi et al., 2016). PMCA4 is also downregulated in colon carcinoma (Cui et al., 2017; Curry et al., 2011; Padányi et al., 2016), while PMCA1 has been found to be upregulated in oral squamous cell carcinoma (Cui et al., 2017).

Because of the variable expression of the PMCA channels, inhibition of specific isoforms produces different responses. For example, knockdown of the PMCA2 channel significantly reduces proliferation and potentiates cell death in breast cancer (Curry, Roberts-Thomson, & Monteith, 2016); knockdown of PMCA1 or PMCA4 produces no effect (So et al., 2018). In BRAF mutant melanoma cells, treatment with vemurafenib was found to upregulate a variant of the PMCA4 channel and slow migration and metastasis (Hegedűs et al., 2017).

The importance of calcium channels in cancer is well established, but their role in CaEP has not yet been clarified. One would expect, for example, that PMCA and sodium–calcium exchanger (NCX) channels would be critical in the restoration of baseline calcium levels following treatment. Attempting to shed light on this matter, one study compared the effects of CaEP on a human rhabdomyosarcoma cell line (RD) to a normal murine muscle cell line (C2C12), uncovering a somewhat paradoxical difference between normal and cancerous cells (Szewczyk et al., 2018). This study found that, in normal cells, an upregulation of the NCX with no change in PMCA following treatment, whereas a downregulation in both PMCA and NCX channels was observed following CaEP of the cancerous cells.

The rapid depletion of intracellular ATP following treatment has been shown to be consistent among normal and cancerous cell types (Frandsen & Gehl, 2018). It is not surprising then that normal cells upregulate the NCX following CaEP. Switching to favor the electrogenic NCX, which is ATP independent, serves as a survival tactic to cope with cytotoxic calcium levels. Explaining what survival strategy could be offered by downregulating both PMCA and NCX in CaEP-treated sarcoma cells is more difficult.

One explanation could arise from an often-overlooked feature of the PMCA. Depending on the tissue, the PMCA acts to varying degrees as an antiporter, importing from 1 to 3 H^+ for each Ca^{2+} exported (DeSantiago et al., 2007; Stafford, Wilson, Oceandy, Neyses, & Cartwright, 2017; Thomas, 2009). Under normal conditions, this may be inconsequential. In response to supraphysiological concentrations of Ca^{2+}, however, importing that many protons may result in a harmful shift in both intra- and extracellular pH (Boczek et al., 2014; Chen & Chesler, 2015). As we will explore in the following section, tumors have a reversed proton gradient compared to healthy cells, which confers many advantages for survival and proliferation. Disruption of this condition would have significant consequences for the tumor, which might be exploited to enhance CaEP effects.

IIB – REVERSAL OF pH = REVERSAL OF TUMOR-MEDIATED IMMUNE SUPPRESSION

ECT introduces toxic drugs into tumor cells in a similar manner to CaEP, but the latter may have additional effects related to reversal of the proton gradient observed in tumor cells. Interestingly, an observation made nearly 100 years ago suggests an explanation for the differential effects observed between CaEP and ECT on normal and malignant tissues. In 1924, Otto Warburg reported that cancer cells have an impaired ability to generate ATP through oxidative phosphorylation (Warburg, Posener, & Negelein, 1924). One consequence of this is that the lactate produced from ATP production by glycolysis will reduce the intracellular pH of tumor cells. From these observations Warburg developed the hypothesis that this metabolic shift is a causative factor in carcinogenesis (Warburg, 1956), an idea which was heavily contested.

In the 1970s, when new technology allowed for the measurement of intracellular and extracellular pH, Warburg's hypothesis further lost traction when it was found that cancer cell pH shifts are the opposite of his prediction (Kaminskas, 1978; Wilhelm, Schulz, & Hofmann, 1971). Despite the production of lactate, the intracellular pH of cancer cells is alkaline, and the extracellular pH is acidic. This proton reversal is an adaptation present in all malignant tumors (Harguindey et al., 2013), where the average pH_i values for tumors are 7.12–7.70 compared to 6.99–7.05 in normal cells. Likewise, the pH_e in malignant cells is 6.20–6.90 compared to 7.30–7.40 for normal cells (Harguindey et al., 2013).

This reversal of pH gradient acts as an important regulator of the metabolic switch from oxidative phosphorylation to glycolysis in malignant cells. For example, even in the presence of oxygen, where oxidative phosphorylation would be the dominant energetic pathway, a small increase in intracellular pH stimulates the binding of hexokinase to voltage-dependent anion channels (VDAC) on the mitochondrial membrane, inhibiting oxidative phosphorylation (Calderon-Montano et al., 2011).

This would stimulate glycolysis via phosphofructokinase (PFK), which is activated under conditions of reduced ATP (inhibition of oxidative phosphorylation) and is also highly sensitive to pH, such that an increase of 0.1–0.3 units can modulate its activity from a dormant to a fully active state (Mathupala, Ko, & Pedersen, 2009). Likewise, an alkaline pH favors the conversion of pyruvate to lactate through lactate dehydrogenase which is required to regenerate NAD^+ used in glycolysis (Webb, Chimenti, Jacobson, & Barber, 2011).

The maintenance of this intracellular alkalinity is supported by the upregulation of specific proteins, including Na$^+$/H$^+$ exchangers (NHEs), carbonic anhydrases (CAIX, CAXII), transporters, lactate/H$^+$ symporters (monocarboxylate transporters 1 and 2 (MCT1/2), and intracellular H$^+$ pumps (Parks, Chiche, & Pouysségur, 2013). Indeed, these structures have been shown to play a pivotal role in many cancers (Amith & Fliegel, 2013, 2017; Parks et al., 2013; Sanhueza et al., 2016; Spugnini et al., 2014). Moreover, they have also been shown to be upregulated as part of a metabolic switch to favor glycolysis by various carcinogens and oncogenes (Harguindey et al., 2013; Webb et al., 2011). Whereas this metabolic switch is less efficient in terms of ATP production, it has been shown to confer a significant proliferative advantage, increasing the capacity of rapidly proliferating cells to import vital nutrients required for cell replication (Vander Heiden, Cantley, & Thompson, 2009).

Modulating pH as an Adaptive Response for Tumor Survival

If proton reversal is a trait common to all malignant tumors, and if a variety of carcinogens and oncogenes upregulate protein channels involved in the maintenance of an intracellular alkalinity, it would be expected that this would confer some evolutionary adaptive advantage. Recent research has demonstrated that alterations of the pH in the tumor microenvironment as well as the intracellular environment contribute to tumor survival through modulation of immune cell function and through resistance to multiple types of chemotherapeutic agents. The following section will explore the consequences of alterations in pH homeostasis for tumor cell survival.

Drug Resistance and Tumor Microenvironment pH

One of the key features of many cancers is their resistance to current treatments, like chemotherapy. It has been suggested that a combination of the acidic microenvironment and modifications of lysosome quantity and morphology can explain the multidrug resistance observed in many types of cancer (Taylor et al., 2015). For example, certain chemotherapeutics such as anthracyclines, anthraquinones, and vinca alkaloids are weak bases and have been shown to have reduced uptake in the pH environment associated with solid tumors (Wojtkowiak, Verduzco, Schramm, & Gillies, 2011). Not only does the acidic environment reduce uptake efficiency of weak-base chemotherapeutics, but those that readily diffuse across the plasma membrane are taken up and concentrated in lysosomes where they are protonated and no longer able to diffuse across the lysosome membrane (Zhitomirsky & Assaraf, 2016).

This lysosome sequestration has been demonstrated with a variety of weak-base chemotherapeutics, and repeated exposure of these drugs leads to drug resistance by altering the numbers and volume of lysosomes within cancer cells (Gotnik et al., 2015; Zhitomirsky & Assaraf, 2014, 2016). Interestingly, this drug resistance isn't limited to weak bases. Cisplatin, a weak acid was tested on various cancer cell lines, including MCF7 (human breast cancer), Me30966 and Me501 (human metastatic melanomas), and SW480 (human colon carcinoma) (Federici et al., 2014). All cell lines showed a significant reduction of drug uptake at the acidic pH values observed in a solid tumor microenvironment. This reduced uptake characteristic of multidrug-resistant cancer cells has been associated with several factors all resulting

from changes in intra- and extracellular pH. In fact, drug-resistant cells have different membrane lipid composition, membrane fluidity, structural order, and lipid pack-
ing density than the parent drug-sensitive cells (Peetla et al., 2013; Rauch, 2009).

Developing methods to counter the acidic tumor microenvironment through proton pump inhibitors (PPIs) has been proposed as a method for overcoming multidrug
resistance in tumors (De Milito, Marino, & Fais, 2012; Lu, Tian, & Guo, 2017).
Preclinical investigations have revealed that pretreatment of a variety of tumor cell
lines with PPIs sensitizes chemoresistant tumors to subsequent therapy and enhances
tumor destruction without additional side effects (Azzarito, Venturi, Cesolini, &
Fais, 2015; Federici et al., 2016; Ferrari et al., 2013). These results have been further
substantiated in a clinical trial [NCT01069081], where intermittent high doses of PPI
combined with a docetaxel + cisplatin regimen significantly enhanced the objective
response rate (68% versus 47%) compared to those not receiving PPI intervention
(Wang et al., 2015).

Modulate Immune Function

Overcoming immune evasion is an equally challenging obstacle for developing
new and effective cancer treatments. Like multidrug resistance, the reversed proton
gradient observed in malignant tissues can be linked directly to their capacity to
escape immune recognition. It has been shown that antitumor effector cells, such as
natural killer (NK) and T-cells, experience a loss of function followed by apoptosis
when exposed to the acidic conditions associated with the tumor microenvironment
(Huber et al., 2017). In CD8+ T-lymphocytes, this loss of function was associated
with reduced cytolytic activity and cytokine secretion, downregulation of CD25
(IL-2Rα) and TCRs (T-cell receptors), and reduced STAT5 and ERK activity following TCR activation (Calcinotto et al., 2012). When buffered at a physiological pH, all
measures of activity were restored.

Both T-cells and NK cells have been shown to be significantly impaired by the
acidic tumor microenvironment. One difference, however, appears to be that NK
cells are irreversibly damaged from this exposure. When first cultured at low pH,
NK cells were shown to be unable to recover function when the culture medium
was buffered to a physiological pH (Calcinotto et al., 2012). When colorectal cancer
cells were taken from human patients and compared to healthy donors, this loss of
function was shown to be due to a combination of reduced concentration of NK cells,
along with a significant reduction in cytokine production and inability to degranulate
(Husain, Huang, Seth, & Sukhatme, 2013; Rocca et al., 2012).

Whereas immune-effector cells are impaired, the opposite is true for immune-
suppressive cells such as myeloid-derived suppressor cells (MDSCs) and regulatory
T-cells (TREGs), which become activated in the same pH range (Huber et al., 2017).
In fact, although the mechanism is not fully understood, MDSCs are not only activated by but also attracted to the low pH in the tumor microenvironment (TME)
(Huber et al., 2017; Safari et al., 2018). Once MDSCs accumulate in the TME, they
potently suppress an immune response through a number of activities that include
downregulating T-cell differentiation, directly preventing maturation of NK cells
and dendritic cells (DCs), and promoting the differentiation and proliferation of
TREGs (Brand et al., 2016; Safari et al., 2018).

In addition to direct activation and proliferation stimulated by MDSCs, TREGs are also stimulated by the low pH in the TME. This ability of TREGs to thrive in the TME is explained by their mixed metabolic phenotype, utilizing a combination of glycolysis, oxidative phosphorylation, and fatty acid oxidation for their energy requirements, in contrast with effector T-cells which predominantly utilize glycolysis (Herbel et al., 2016; Pearce, Poffenberger, Chang, & Jones, 2013). Like MDSCs, TREGs are potent immune-suppressive cells that can directly lyse effector T-cells, produce immune-suppressing cytokines and metabolites, deplete costimulatory signals on DCs, promote formation of M2-macrophages, and promote tumor angiogenesis (Facciabene, Motz, & Coukos, 2012; Frydrychowicz, Boruczkowski, Kolecka-Bednarczyk, & Dworacki, 2017; Wang, Franco, & Ho, 2017; Whiteside, 2014).

Thus, the acidic TME facilitates evasion of immune recognition through multiple direct and indirect mechanisms. Overcoming the immune-suppressing effects of the TME using PPI has been investigated. Using B16 melanoma models, both *in vitro* and *in vivo*, pretreatment with PPI therapy was associated with a rapid rise in intratumoral pH and subsequent increase in effector function (Bellone et al., 2013). The effect of PPIs appears to be quite specifically dependent on the acidic microenvironment of tumors. Because they are lipophilic and weakly basic, PPIs readily diffuse across the plasma membrane and accumulate in acidic vesicles where they are converted into their active, inhibitory sulfonamide form (Bellone et al., 2013). Neutralizing the acidic TME produces different effects on different tumor cell lines. A significant reduction in growth rate and metastases, along with enhanced CD8+ T-cell infiltration, is observed in prostate tumors (Ibrahim-Hashim et al., 2011, 2012) and in Yumm1.1 melanoma tumors (Pilon-Thomas et al., 2016; Robey et al., 2009), but no effect on tumor growth is seen in B16 melanoma (Pilon-Thomas et al., 2016; Robey et al., 2009) or in immune-deficient mouse models (Robey et al., 2009).

Combining pH modulation with immune-checkpoint inhibitors promotes a synergistic antitumor effect. For example, a combination of bicarbonate therapy with either programed cell death receptor 1 (PD-1)/programmed cell death ligand 1 (PD-L1) or cytotoxic T-lymphocyte antigen (CTLA-4) inhibitors significantly improves the anti-tumor response in several tumor models (Pilon-Thomas et al., 2016). Despite promising experimental evidence, two retrospective analyses found no significant relationship between PPI use with anti-PD-1/anti-PD-L1 and patient survival (Mukherjee, Ibrahimi, et al., 2018; Mukherjee, Khalid, et al., 2018). Authors from these studies recognize the limitations of their results and state that dosage, compliance, and type of proton pump inhibitor were not taken into account. Further research will be required to elucidate the potential for combined PPI–immune-checkpoint-inhibitor therapy.

To summarize up to this point, we have explored some of the preclinical literature investigating CaEP. We also discussed why calcium is toxic to cells and the mechanisms that have evolved to remove it from the cytosol when concentrations become too high. We saw that tumor cells and normal cells respond quite differently to the supraphysiological calcium doses following CaEP. Very importantly, calcium extrusion channels are downregulated in tumor cells, which may be at least part of the reason that CaEP can produce a long-lasting immune response. Specifically, this could be explained by the relationship between tumor-mediated immune suppression and pH. Here we propose a mechanism to explain how CaEP elicits an immune response.

Proposed Mechanism of CaEP Immune Stimulation

- The delivery of electric pulses permeabilizes the cell membrane, resulting in a rapid influx of Ca^{2+} into the cell.
- The cell recognizes threat and devotes significant energy to removing calcium present in it.
- PMCA and NCX transport are the two main calcium extrusion methods. The NCX will not function initially, because membrane permeabilization eliminates ion concentration gradients, including the Na^+ gradient used by the NCX antiporter to extrude Ca^{2+}.
- The PMCA pump uses ATP to import protons and extrude Ca^{2+}, depleting ATP, acidifying the cytosol, and alkalinizing the TME.
- As cells undergo calcium-overload-induced necrotic cell death, which is known to stimulate immune recruitment, the associated changes in pH reverse tumor-mediated immune suppression.

In the final section, we will explore the current state of the-art. Several clinical trials using CaEP are ongoing or have been recently completed. We will also explore options for combination therapies that may further enhance the immune-stimulating response.

PART III – CURRENT STATE OF THE ART AND FUTURE DIRECTIONS FOR CAEP RESEARCH

Following the initial CaEP publication in 2012, it was apparent that this treatment had significant potential. Rapidly accumulating evidence has demonstrated its efficacy on multiple cancer cell lines while sparing normal tissues. This has led to its application in multiple clinical trials, two of which have recently been completed.

Data from a phase II trial compared the effects of CaEP with ECT on cutaneous metastases from breast cancer and malignant melanoma (Falk, Matthiessen, Wooler, & Gehl, 2018). Each metastasis was treated once, with either CaEP or ECT and then followed up 6 months later. For CaEP, there was an objective response in 72% (13/18) and a complete response in 66% (12/18) of metastases. ECT treatment resulted in an objective response in 84% (16/19) and a complete response in 68% (13/19), with no significant difference between CaEP and ECT based on this study.

Side effects were reported more often with ECT treatment than CaEP. For the ECT-treated metastases, ulceration was reported in 68% (13/19), itching in 26% (5/19), exuding in 10% (2/19), and hyperpigmentation in 26% (5/19) of metastases. For the CaEP-treated metastases, ulceration was reported in 38% (7/18), itching in 5% (1/18), exuding in 11% (2/18) of metastases with no incidence of hyperpigmentation. The results from this study demonstrated that CaEP is as effective as ECT in treating cutaneous metastases with a reduced side-effect profile.

Six of the seven patients included above were treated for breast cancer that progressed with metastases outside the treatment side during the 1-year follow-up, regardless of CaEP or ECT treatment, suggesting the treatment effect was local and did not induce a systemic immune response. The seventh was a patient with malignant melanoma who received ECT treatment 4 months prior to inclusion in the study. The case report describing this patient is described in Falk, Lambaa, et al., (2017). This patient had over 100 melanoma metastases covering a large surface of their left lower limb. Four months post ECT, they were treated with the combination CaEP + ECT after new metastases were discovered outside the treatment area. One year following combination treatment, complete remission was observed for both treated and untreated metastases suggesting that a systemic immune response was elicited.

A second clinical study completed using CaEP was released this year, studying the effectiveness at treating recurrent head and neck cancers in six patients (Plaschke et al., 2019). None of the patients reported any severe adverse effects resulting from treatment. MRI was used in this study to measure changes in the tumor volume, and PET was used to measure metabolic activity and inflammation. This study showed objective responses in 50% of patients treated with CaEP (3/6), with one of them in complete remission 1 year following treatment.

After the 2-month follow-up, four of the six patients were referred to systemic chemotherapy, one was referred to treatment with radiotherapy, leaving one to continue with follow-up due to a good response. The next step will be to combine CaEP with other treatments to improve long-term survival. The greatest obstacle we are faced with is finding a way to overcome the tumor-mediated immune suppression to prevent disease recurrence.

Several studies have looked at combining immune-checkpoint inhibitors with ECT in clinical trials. As monotherapies, checkpoint inhibitors have had limited success due to a severe side-effect profile associated with an exacerbated immune response (Chae et al., 2018; Seidel, Otsuka, & Kabashima, 2018). The benefit of combining immune-checkpoint inhibitors with electroporation-based treatments is the ability to introduce smaller doses systemically or directly into tumor.

A retrospective analysis assessed a combination of ECT with ipilimumab on patients with melanoma. Ipilimumab is a checkpoint inhibitor that blocks the negative regulator of the CTLA-4, thus augmenting T-cell activation and proliferation (Mozzillo et al., 2015). The results from this study were quite promising. An objective response rate of 67% was observed, which was lower than the 87% of objective response the same authors found in a similar study using only ECT (Caracò et al., 2013). Although a lower objective response was observed, the overall survival was much higher, with 86% survival at 12 months versus only 40% in the ECT-only study.

A second clinical investigation compared the effects of ECT with either ipilimumab or a PD-1 inhibitor (permbrolizumab or nivolumab) on patients with melanoma (Heppt et al., 2016). Whereas CTLA-4 binds with a protein (B7) on antigen presenting cells, PD-1/PDL-1 is expressed on the surface of T-cell, lymphocytes, and certain subsets of dendritic cells; thus inhibition is expected to produce a broader effect (Gardiner et al., 2015). Results from this multicenter analysis confirmed the hypothesis. The systemic objective response was 19% for the ECT + ipilimumab versus 40% for the ECT + PD-1 treatment. No severe adverse effects were reported for the PD-1-treated group, whereas 25% of the ipilimumab group reported severe systemic reactions, which included endocrine-related abnormalities, colitis, peritonitis/ascites, and acute kidney injury.

PD-1 treatment using Nivolumab was combined with ECT in a recently published case report (Karaca, Yayla, Erdem, & Gürler, 2018). The patient who suffered from metastatic melanoma had received several different treatments including interferon-α, temozolamide, vemurafenib and ipilimumab, none of which were successful in preventing further metastases. After the disease spread to the brain, the patient was operated and received whole-brain radiotherapy. Nivolumab was then administered and patient was exposed to a standard ECT protocol. Remarkably, complete response was achieved with no evidence of disease after 4 years of follow-up.

Clinical trials combining CaEP with immune-checkpoint inhibitors are required. There is evidence that combining checkpoint inhibitors could also be more effective, demonstrating significant therapeutic advantages in various cancer types (Chae et al., 2018). More recently, the adenosine receptor (A2AR) has been described as the next-generation checkpoint inhibitor (Leone, Lo, & Powell, 2015).

Adenosine, a by-product of ATP catabolism, is the ligand for the A2AR and is known for having immune-stimulating and immune-suppressing activity. When ATP is released from damaged cells, it acts as a potent inductor of various immune cells including dendritic cells, NK cells, and T-cells through its binding action to P2X receptors (Sachet, Liang, & Oehler, 2017). On the other hand, when ATP is degraded to adenosine through ectoenzymes CD39 and CD73, it is known to suppress T-cell activation via the A2AR (Ohta, 2016).

The importance of the A2AR has been evaluated *in vitro* on A375 melanoma, A549 lung, and MRMT1 breast cancer cell lines (Gessi et al., 2017). Using a selective A2AR agonist CGS-21680, a concentration-dependent induction of cell proliferation was observed in all cell types. Conversely, the effect was potently blocked using the selective A2AR antagonist TP455. To date, two clinical studies have published results using A2ARs antagonists.

The first is an ongoing phase I/Ib study looking at the A2AR CPI-444 either alone or in combination with the anti-PDL-1 drug atezolizumab (atezo) in patients with diverse incurable cancers including non-small-cell lung cancer, melanoma, renal cell carcinoma, triple-negative breast cancer, urinary bladder cancer, prostate cancer, colorectal cancer, and head and neck cancer (Emens et al., 2017). Ninety-six patients have been treated with CPI-444 (n = 52) or with a combination of CPI-444 + atezo (n = 44). With a median follow-up of 16 weeks, an objective response of 38% has been observed for the monotherapy and 39% with the combination therapy. No severe adverse effects have been observed in patients receiving CPI-444 alone, with two patients experiencing grade 3/4 reactions.

The second clinical trial was a phase I/II investigation of the A2AR antagonist NIR-178 in patients with advanced non-small-cell lung cancer (Chiappori et al., 2018). Twenty-four patients were treated with doses ranging from 80–640 mg. After 44 weeks, 71% (17/24) patients demonstrated an objective response including one complete response (480 mg) and one partial response (80 mg). It was found to be well tolerated, with grade 3 adverse events in 8% (pneumonitis) and 4% (nausea) of the patients and no grade 4 events.

Finally, no studies have looked at the effects of combining pH modulators such as PPIs in the context of ECT or CaEP. Considering the supporting role that pH can have on tumor-mediated immune suppression and chemotherapy resistance, this would make an excellent target, alone or in combination with immune-checkpoint inhibition, to enhance CaEP by stimulating a systemic immune reaction.

PART IV – SUMMARY

Cancer is responsible for nearly 20% of deaths worldwide. The current standard of care has failed to address the major obstacle, immune evasion, preventing long-term treatment success in many cancer types. CaEP represents a novel treatment option to overcome this hurdle. With less than a decade of research into this treatment, it has already demonstrated considerable potential in multiple preclinical and clinical models. One of the most intriguing features of CaEP is it appears to act on certain vulnerabilities in tumor cells not present in healthy tissues. We have proposed a mechanism by which CaEP exerts its effects which are both cytotoxic and immune stimulating. The final section of this chapter is dedicated to highlighting some of the recent clinical findings and describing potential avenues to be explored in future research. Combination therapies incorporating CaEP with immune-checkpoint inhibitors and pH modulators may be the key to stimulating a long-lasting, systemic immune response with the potential to not only treat existing tumors, but also prevent future recurrence.

REFERENCES

Amith, S. R., & Fliegel, L. (2013). Regulation of the N+/H+ Exchanger (NHE1) in breast cancer metastasis. *Cancer Research*, *1*(8), 1259–1265. doi: 10.1158/0008-5472. CAN-12-4031.

Amith, S. R., & Fliegel, L. (2017). Na+/H+ exchanger-mediated hydrogen ion extrusion as a carcinogenic signal in triple-negative breast cancer etiopathogenesis and prospects for its inhibition in therapeutics. *Seminars in Cancer Biology*, 1–27. doi: 10.1016/j. semcancer.2017.01.004.

Andersen, T. B., López, C. Q., Manczak, T., Martinez, K., & Simonsen, H. T. (2015). Thapsigargin-from Thapsia L. to Mipsagargin. *Molecules*, *20*(4), 6113–6127. doi: 10.3390/molecules20046113.

Andrzejewski, S., Gravel, S., Pollak, M., & St-pierre, J. (2014). Metformin directly acts on mitochondria to alter cellular bioenergetics. *Cancer and Metabolism*, *2*(12), 1–14.

Azzarito, T., Venturi, G., Cesolini, A., & Fais, S. (2015). Lansoprazole induces sensitivity to suboptimal doses of paclitaxel in human melanoma. *Cancer Letters*, *356*(2), 697–703. doi: 10.1016/j.canlet.2014.10.017.

Bellone, M., Calcinotto, A., Filipazzi, P., De Milito, A., Fais, S., & Rivoltini, L. (2013). The acidity of the tumor microenvironment is a mechanism of immune escape that can be overcome by proton pump inhibitors. *OncoImmunology*, *2*(1), 9–11. doi: 10.4161/onci.22058.

Boczek, T., Lisek, M., Ferenc, B., Kowalski, A., Stepinski, D., Wiktorska, M., & Zylinska, L. (2014). Plasma membrane Ca2+-ATPase isoforms composition regulates cellular pH homeostasis in differentiating PC12 cells in a manner dependent on cytosolic Ca2+ elevations. *PLoS one*, *9*(7), 1–15. doi: 10.1371/journal.pone.0102352.

Brand, A., Singer, K., Koehl, G. E., Kolitzus, M., Schoenhammer, G., Thiel, A., ... Kreutz, M. (2016). LDHA-associated lactic acid production blunts tumor immunosurveillance by T and NK cells. *Cell Metabolism, 24*, 657–671. doi: 10.1016/j.cmet.2016.08.011.

Bronner, F., & Pansu, D. (1999). Recent advances in nutritional science nutritional aspects of calcium. *The Journal of Nutrition, 129*(1), 9–12. doi: 10.1590/S0102-67202007000100010.

Bruce, J. I. E. (2018). Metabolic regulation of the PMCA : Role in cell death and survival. *Cell Calcium, 69*, 28–36.

Calcinotto, A., Filipazzi, P., Grioni, M., Iero, M., De Milito, A., Ricupito, A., ... Rivoltini, L. (2012). Modulation of microenvironment acidity reverses anergy in human and murine tumor-infiltrating T lymphocytes. *Cancer Research, 72*(11), 2746–2756. doi: 10.1158/0008-5472.CAN-11-1272.

Calderon-Montano, J. M., Burgos-Moron, E., Perez-Guerrero, C., Salvador, J., Robles, A., & Lopez-Lazaro, M. (2011). Role of the intracellular pH in the metabolic switch between oxidative phosphorylation and aerobic glycolysis - relevance to cancer. *Webmed Central - Cancer, 2*(3), 1–10.

Cancer. (2019). Retrieved from www.who.int/news-room/fact-sheets/detail/cancer.

Caracò, C., Mozzillo, N., Marone, U., Simeone, E., Benedetto, L., Di Monta, G., ... Ascierto, P. A. (2013). Long-lasting response to electrochemotherapy in melanoma patients with cutaneous metastasis. *BMC Cancer, 13*(September 2012), 2–5. Retrieved from www.embase.com/search/results?subaction=viewrecord&from=export&id=L52896722%5Cn; www.biomedcentral.com/1471-2407/13/564%5Cn; doi: 10.1186/1471-2407-13-564; https://bmccancer.biomedcentral.com/articles/10.1186/1471-2407-13-564#citeas.

Carafoli, E., & Krebs, J. (2016). Why calcium? How calcium became the best communicator. *Journal of Biological Chemistry, 29*(40), 20849–20857. doi: 10.1074/jbc.R116.735894.

Chae, Y. K., Arya, A., Iams, W., Cruz, M. R., Chandra, S., Choi, J., & Giles, F. (2018). Current landscape and future of dual anti-CTLA4 and PD-1/PD-L1 blockade immunotherapy in cancer; lessons learned from clinical trials with melanoma and non-small cell lung cancer (NSCLC). *Journal for ImmunoTherapy of Cancer, 6*(1). doi: 10.1186/s40425-018-0349-3.

Chen, H.-Y., & Chesler, M. (2015). Autocrine boost of NMDAR current in hippocampal CA1 pyramidal neurons by a PMCA-dependent, perisynaptic, extracellular pH shift. *Journal of Neuroscience, 35*(3), 873–877. doi: 10.1523/JNEUROSCI.2293-14.2015.

Chiappori, A., Williams, C. C., Creelan, B. C., Tanvetyanon, T., Gray, J. E., Haura, E. B., ... Antonia, S. J. (2018). Phase I/II study of the A2AR antagonist NIR178 (PBF-509), an oral immunotherapy, in patients (pts) with advanced NSCLC. *Journal of Clinical Oncology, 36*(15). doi: 10.1200/JCO.2018.36.15_suppl.9089.

Clapham, D. (2007). Calcium signaling. *Cell, 131*, 1047–1058. doi: 10.1016/j.cell.2007.11.028.

Cui, C., Merritt, R., Fu, L., & Pan, Z. (2017). Targeting calcium signaling in cancer therapy. *Acta Pharmaceutica Sinica B, 7*(1), 3–17. doi: 10.1016/j.apsb.2016.11.001.

Curry, M. C., Roberts-Thomson, S. J., & Monteith, G. R. (2011). Plasma membrane calcium ATPases and cancer. *BioFactors, 37*(3), 132–138. doi: 10.1002/biof.146.

Curry, M. C., Roberts-Thomson, S. J., & Monteith, G. R. (2016). Biochemical and Biophysical Research Communications PMCA2 silencing potentiates MDA-MB-231 breast cancer cell death initiated with the Bcl-2 inhibitor ABT-263. *Biochemical and Biophysical Research Communications, 478*(4), 1792–1797. doi: 10.1016/j.bbrc.2016.09.030.

De Milito, A., Marino, M. L., & Fais, S. (2012). A rationale for the use of proton pump inhibitors as antineoplastic agents. *Current Pharmaceutical Design, 18*(10), 1395–1406. doi: 10.2174/138161212799504911.

Denmeade, S. R., Mhaka, A. M., Rosen, D. M., Brennen, W. N., Dalrymple, S., Dach, I., ... Isaacs, J. T. (2012). Engineering a prostate-specific membrane antigen–activated tumor endothelial cell prodrug for cancer therapy. *Science Translational Medicine, 4*(140), 1–22. doi: 10.1126/scitranslmed.3003886.Engineering.

DeSantiago, J., Batlle, D., Khilnani, M., Dedhia, S., Kulczyk, J., Duque, R., … Rasgado-Flores, H. (2007). Ca2+/H+ exchange via the plasma membrane Ca2+ ATPase in skeletal muscle. *Frontiers in Bioscience*, *12*, 4641–4660.

Emens, L. A., Powderly, J., Fong, L., Brody, J., Forde, P. F., Hellmann, M., … Laport, G. (2017). CPI-444, an oral A2A receptor (A2AR) antagonist, demonstrates clinical activity in patients with advanced solid tumors. In *American Association for Cancer Research Annual Meeting* (pp. 1–21).

Facciabene, A., Motz, G. T., & Coukos, G. (2012). T Regulatory Cells: Key Players in Tumor Immune Escape and Angiogenesis. *Cancer Research*, *72*(9), 2162–2171. doi: 10.1158/0008-5472.CAN-11-3687.T.

Falk, H., Forde, P. F., Bay, M. L., Mangalanathan, U. M., Hojman, P., Soden, D. M., & Gehl, J. (2017). Calcium electroporation induces tumor eradication, long-lasting immunity and cytokine responses in the CT26 colon cancer mouse model. *OncoImmunology*, *6*(5), 1–8. doi: 10.1080/2162402X.2017.1301332.

Falk, H., Lambaa, S., Johannesen, H. H., Wooler, G., Venzo, A., & Gehl, J. (2017). Electrochemotherapy and calcium electroporation inducing a systemic immune response with local and distant remission of tumors in a patient with malignant melanoma – a case report. *Acta Oncologica*, *56*(8), 1126–1131. doi: 10.1080/0284186X.2017.1290274.

Falk, H., Matthiessen, L. W., Wooler, G., & Gehl, J. (2018). Calcium electroporation for treatment of cutaneous metastases ; a randomized double- blinded phase II study, comparing the effect of calcium electroporation with electrochemotherapy. *Acta Oncologica*, *57*(3), 311–319. doi: 10.1080/0284186X.2017.1355109.

Federici, C., Lugini, L., Marino, M. L., Carta, F., Iessi, E., Azzarito, T., … Fais, S. (2016). Lansoprazole and carbonic anhydrase IX inhibitors sinergize against human melanoma cells. *Journal of Enzyme Inhibition and Medicinal Chemistry*, *31*(May), 119–125. doi: 10.1080/14756366.2016.1177525.

Federici, C., Petrucci, F., Caimi, S., Cesolini, A., Logozzi, M., Borghi, M., … Fais, S. (2014). Exosome release and low pH belong to a framework of resistance of human melanoma cells to cisplatin. *PLoS One*, *9*(2). doi: 10.1371/journal.pone.0088193.

Ferrari, S., Perut, F., Fagioli, F., Brach Del Prever, A., Meazza, C., Parafioriti, A., … Fais, S. (2013). Proton pump inhibitor chemosensitization in human osteosarcoma: From the bench to the patients' bed. *Journal of Translational Medicine*, *11*(1), 1–7. doi: 10.1186/1479-5876-11-268.

Frandsen, S. K., & Gehl, J. (2017). Effect of calcium electroporation in combination with metformin in vivo and correlation between viability and intracellular ATP level after calcium electroporation in vitro. *PLoS One*, *12*(7), 1–12. doi: 10.1371/journal.pone.0181839.

Frandsen, S. K., & Gehl, J. (2018). A review on differences in effects on normal and malignant cells and tissues to electroporation-based therapies : A focus on calcium electroporation. *Technology in Cancer Research and Treatment*, *17*, 1–6. doi: 10.1177/1533033818788077.

Frandsen, S. K., Gibot, L., Madi, M., Gehl, J., & Rols, M.-P. (2015). Calcium electroporation: Evidence for differential effects in normal and malignant cell lines, evaluated in a 3D spheroid model. *PLoS One*, *10*(12), 1–11. doi: 10.1371/journal.pone.0144028.

Frandsen, S. K., Gissel, H., Hojman, P., Tramm, T., Eriksen, J., & Gehl, J. (2012). Direct therapeutic applications of calcium electroporation to effectively induce tumor necrosis. *Cancer Research*, *72*(6), 1336–1341. doi: 10.1158/0008-5472.CAN-11-3782.

Frandsen, S. K., Kruger, M. B., Mangalanathan, U. M., Tramm, T., Mahmood, F., Novak, I., & Gehl, J. (2017). Normal and malignant cells exhibit differential responses to calcium electroporation. *Cancer Research*, *77*(16), 4389–4402. doi: 10.1158/0008-5472.CAN-16-1611.

Fridman, W.-H., Zitvogel, L., Fridman-Sautès, C., & Kroemer, G. (2017). The immune contexture in cancer. *Nature Reviews Clinical Oncology, 14*, 717–734. doi: 10.1038/nrclinonc.2017.101.

Frydrychowicz, M., Boruczkowski, M., Kolecka-Bednarczyk, A., & Dworacki, G. (2017). The dual role of treg in cancer. *Scandinavian Journal of Immunology, 86*(6), 436–443. doi: 10.1111/sji.12615.

Gardiner, R. E., Jahangeer, S., Forde, P. F., Ariffin, A. B., Bird, B., Soden, D. M., & Hinchion, J. (2015). Low immunogenicity in non-small cell lung cancer; do new developments and novel treatments have a role? *Cancer and Metastasis Reviews, 34*(1), 129–144. doi: 10.1007/s10555-015-9550-8.

Gessi, S., Bencivenni, S., Battistello, E., Vincenzi, F., Colotta, V., Catarzi, D., … Varani, K. (2017). Inhibition of A2A adenosine receptor signaling in cancer cells proliferation by the novel antagonist TP455. *Frontiers in Pharmacology, 8*(December), 1–13. doi: 10.3389/fphar.2017.00888.

Giladi, M., Tal, I., & Khananshvili, D. (2016). Structural features of ion transport and allosteric regulation in Sodium-Calcium Exchanger (NCX) proteins. *Frontiers in Physiology, 7*(FEB), 1–13. doi: 10.3389/fphys.2016.00030.

Gotnik, K. J., Browterman, H. J., Labots, M., de Haas, R. R., Dekker, H., Honeywell, R. J., … Verheul, H. M. (2015). Lysosomal sequestration of sunitinib: a novel mechanism of drug resistance. *Clinical Cancer Research, 17*(23), 7337–7346. doi: 10.1158/1078-0432. CCR-11-1667.Lysosomal.

Harguindey, S., Arranz, J. L., David, J., Orozco, P., Rauch, C., Fais, S., … Reshkin, S. J. (2013). Cariporide and other new and powerful NHE1 inhibitors as potentially selective anticancer drugs – an integral molecular/biochemical/metabolic/clinical approach after one hundred years of cancer research. *Journal of Translational Medicine, 11*(282), 1–17.

Hegedűs, L., Garay, T., Molnár, E., Varga, K., Bilecz, Á., Török, S., … Enyedi, Á. (2017). The plasma membrane Ca2+ pump PMCA4b inhibits the migratory and metastatic activity of BRAF mutant melanoma cells. *International Journal of Cancer, 140*(12), 2758–2770. doi: 10.1002/ijc.30503.

Heppt, M. V., Eigentler, T. K., Kähler, K. C., Herbst, R. A., Göppner, D., Gambichler, T., … Berking, C. (2016). Immune checkpoint blockade with concurrent electrochemotherapy in advanced melanoma: A retrospective multicenter analysis. *Cancer Immunology, Immunotherapy, 65*(8), 951–959. doi: 10.1007/s00262-016-1856-z.

Herbel, C., Patsoukis, N., Bardhan, K., Seth, P., Weaver, J. D., & Boussiotis, V. A. (2016). Clinical significance of T cell metabolic reprogramming in cancer. *Clinical and Translational Medicine, 5*(29), 1–23. doi: 10.1186/s40169-016-0110-9.

Huang, F., Wang, P., & Wang, X. (2018). Thapsigargin induces apoptosis of prostate cancer through cofilin-1 and paxillin. *Oncology Letters, 16*(2), 1975–1980. doi: 10.3892/ol.2018.8833.

Huber, V., Camisaschi, C., Berzi, A., Ferro, S., Lugini, L., Triulzi, T., … Rivoltini, L. (2017). Cancer acidity: An ultimate frontier of tumor immune escape and a novel target of immunomodulation. *Seminars in Cancer Biology*, 1–38. doi: 10.1016/j.semcancer.2017.03.001.

Hur, K. Y., Lee, M., & Lee, M. (2015). New mechanisms of metformin action: Focusing on mitochondria and the gut. *Journal of Diabetes Investigation, 6*, 600–609. doi: 10.1111/jdi.12328.

Husain, Z., Huang, Y., Seth, P., & Sukhatme, V. P. (2013). Tumor-derived lactate modifies antitumor immune response: Effect on myeloid-derived suppressor cells and NK cells. *Journal of Immunology, 191*, 1486–1495. doi: 10.4049/jimmunol.1202702.

Ibrahim-Hashim, A., Cornell, H. H., Coelho Ribeiro, M. D. L., Abrahams, D., Cunningham, J., Lloyd, M., … Gillies, R. J. (2011). Reduction of metastasis using a non-volatile buffer. *Clin Exp Metastasis, 28*, 841–849. doi: 10.1007/s10585-011-9415-7.

Ibrahim-Hashim, A., Cornnell, H. H., Abrahams, D., Lloyd, M., Gillies, R. J., & Gatenby, R. A. (2012). Systemic buffers inhibit carcinogenesis in TRAMP mice. *Journal of Urology, 188*(2), 624–631. doi: 10.1016/j.juro.2012.03.113.Systemic.

Kamer, K. J., & Mootha, V. K. (2015). The molecular era of the mitochondrial calcium uniporter. *Nature Reviews Molecular Cell Biology, 16*(9), 545–553. doi: 10.1038/nrm4039.

Kaminskas, E. (1978). The pH-dependence of sugar transport and of glycolysis in cultured Ehrlich ascites-tumour cells. *Biochemistry Journal, 172*, 453–459.

Karaca, B., Yayla, G., Erdem, M., & Gürler, T. (2018). Electrochemotherapy with anti-PD-1 treatment induced durable complete response in heavily pretreated Metastatic melanoma patient. *Anti-Cancer Drugs, 29*(2), 190–196. doi: 10.1097/CAD.0000000000000580.

Leone, R. D., Lo, Y. C., & Powell, J. D. (2015). A2aR antagonists: Next generation checkpoint blockade for cancer immunotherapy. *Computational and Structural Biotechnology Journal, 13*, 265–272. doi: 10.1016/j.csbj.2015.03.008.

Li, M., & Xiong, Z. G. (2011). Ion channels as targets for cancer therapy. *International Journal of Physiology Pathophysiology and Pharmacology, 3*(2), 156–166. doi: 10.1115/1.4026364.

Lodish, H., Berk, A., Zipursky, S., Matsudaira, P., Baltimore, D., & Darnell, J. (2000). Intracellular ion environment and membrane electric potential. In *Molecular Cell Biology* (4th ed.) New York: W H Freeman and company. Retrieved from www.ncbi.nlm.nih.gov/books/NBK21627/.

Lu, Z. N., Tian, B., & Guo, X. L. (2017). Repositioning of proton pump inhibitors in cancer therapy. *Cancer Chemotherapy and Pharmacology, 80*(5), 925–937. doi: 10.1007/s00280-017-3426-2.

Ma, Z., Fan, C., Yang, Y., Di, S., Hu, W., Li, T., … Yan, X. (2016). Thapsigargin sensitizes human esophageal cancer to TRAIL-induced apoptosis via AMPK activation. *Scientific Reports, 6*(September), 1–17. doi: 10.1038/srep35196.

Mathupala, S. P., Ko, Y. H., & Pedersen, P. L. (2009). Hexokinase-2 bound to mitochondria: Cancer's stygian link to the "Warburg effect" and a pivotal target for effective therapy. *Seminars in Cancer Biology, 19*(1), 17–24. doi: 10.1016/j.semcancer.2008.11.006. Hexokinase-2.

Monteith, G. R., Davis, F. M., & Roberts-Thomson, S. J. (2012). Calcium channels and pumps in cancer: Changes and consequences. *Journal of Biological Chemistry, 287*(38), 31666–31673. doi: 10.1074/jbc.R112.343061.

Mozzillo, N., Simeone, E., Benedetto, L., Curvietto, M., Giannarelli, D., Gentilcore, G., … Ascierto, P. A. (2015). Assessing a novel immuno-oncology-based combination therapy: Ipilimumab plus electrochemotherapy. *OncoImmunology, 4*(6), 1–8. doi: 10.1080/2162402X.2015.1008842.

Mukherjee, S., Ibrahimi, S., Khalid, B., Roman, D., Zhao, D., & Aljumaily, R. (2018). Do proton pump inhibitors modulate the efficacy of anti-PD-1/PD-L1 therapy ? A retrospective study. *Journal of Oncology Pharmacy Practice*, 3–5. doi: 10.1177/1078155218771152.

Mukherjee, S., Khalid, B., Ibrahimi, S., Morton, J. M., Roman, D., Zhao, Y. D., & Aljumaily, R. (2018). Efficacy of PD-1/PD-L1 therapy: Do proton pump inhibitors affect the outcome? *Journal of Clinical Oncology, 36*(5).doi: 10.1200/JCO.2018.36.5_suppl.208.

My Cancer Genome. (2019). Retrieved from www.mycancergenome.org/content/molecular-medicine/pathways/cytotoxic-chemotherapy-mechanisms-of-action.

Ohta, A. (2016). A metabolic immune checkpoint: Adenosine in tumor microenvironment. *Frontiers in Immunology, 7*(MAR), 1–11. doi: 10.3389/fimmu.2016.00109.

Padányi, R., Pászty, K., Heged, L., Varga, K., Papp, B., Penniston, J. T., & Enyedi, Á. (2016). Multifaceted plasma membrane Ca2+ pumps : From structure to intracellular Ca2+ handling and cancer. *Biochimica et Biophysica Acta, 1863*, 1351–1363.

Parks, S. K., Chiche, J., & Pouysségur, J. (2013). Disrupting proton dynamics and energy metabolism for cancer therapy. *Nature Reviews Cancer, 13*(September), 611–623. doi: 10.1038/nrc3579.

Pearce, E. L., Poffenberger, M. C., Chang, C., & Jones, R. G. (2013). Fueling immunity: Insights into metabolism and lymphocyte function. *Science, 342*(6155), 1–30. doi: 10.1126/science.1242454.Fueling.

Peetla, C., Vijayaraghavalu, S., & Labhasetwar, V. (2013). Biophysics of cell membrane lipids in cancer drug resistance: Implications for drug transport and drug delivery with nanoparticles. *Advanced Drug Delivery Reviews, 65*(0), 1–29. doi: 10.1016/j.addr.2013.09.004.Biophysics.

Pilon-Thomas, S., Kodumudi, K., El-Kenawi, A., Russell, S., Weber, A., Luddy, K., … Gillies, R. J. (2016). Neutralization of tumor acidity improves antitumor responses to immunotherapy. *Cancer Research, 76*(6), 1381–1390. doi: 10.1158/0008-5472.CAN-15-1743. Neutralization.

Plaschke, C. C., Gehl, J., Johannesen, H. H., Fischer, B. M., Kjaer, A., Lomholt, A. F., & Wessel, I. (2019). Calcium electroporation for recurrent head and neck cancer : A clinical phase I study. *Laryngoscope Investigative Otolaryngology*, (January). doi: 10.1002/lio2.233.

Prevarskaya, N., Skryma, R., & Shuba, Y. (2010). Ion channels and the hallmarks of cancer. *Trends in Molecular Medicine, 16*(3), 107–121. doi: 10.1016/j.molmed.2010.01.005.

Raffaello, A., Mammucari, C., Gherardi, G., & Rizzuto, R. (2016). Calcium at the center of cell signaling: Interplay between endoplasmic reticulum, mitochondria and lysosomes. *Trends in Biochemical Sciences, 41*(12), 1035–1049. doi: 10.1186/s40945-017-0033-9. Using.

Raguimova, O. N., Smolin, N., Bovo, E., Bhayani, S., Autry, J. M., Zima, A. V., & Robia, S. L. (2018). Redistribution of SERCA calcium pump conformers during intracellular calcium signaling. *Journal of Biological Chemistry, 293*(28), 10843–10856. doi: 10.1074/jbc.RA118.002472.

Rauch, C. (2009). Toward a mechanical control of drug delivery. On the relationship between Lipinski' s 2nd rule and cytosolic pH changes in doxorubicin resistance levels in cancer cells : A comparison to published data. *European Biophysics Journal : EBJ, 38*, 829–846. doi: 10.1007/s00249-009-0429-x.

Rena, G., Hardie, D. G., & Pearson, E. R. (2017). The mechanisms of action of metformin. *Diabetologia, 60*, 1577–1585. doi: 10.1007/s00125-017-4342-z.

Robey, I. F., Baggett, B. K., Kirkpatrick, N. D., Roe, D. J., Dosescu, J., Sloane, B. F., … Gillies, R. J. (2009). Bicarbonate increases tumor pH and inhibits spontaneous metastases. *Cancer Research, 69*(6), 2260–2268. doi: 10.1158/0008-5472.CAN-07-5575. Bicarbonate.

Rocca, Y. S., Roberti, M. P., Arriaga, J. M., Amat, M., Bruno, L., Pampena, M. B., … Levy, E. M. (2012). Altered phenotype in peripheral blood and tumor-associated NK cells from colorectal cancer patients. *Innate Immunity, 19*(1), 76–85. doi: 10.1177/1753425912453187.

Sachet, M., Liang, Y. Y., & Oehler, R. (2017). The immune response to secondary necrotic cells. *Apoptosis, 22*(10), 1189–1204. doi: 10.1007/s10495-017-1413-z.

Safari, E., Ghorghanlu, S., Ahmadi-khiavi, H., Mehranfar, S., Rezaei, R., & Motallebnezhad, M. (2018). Myeloid - derived suppressor cells and tumor : Current knowledge and future perspectives. *Journal of Cell Physiology*, 1–16. doi: 10.1002/jcp.27923.

Sanhueza, C., Araos, J., Naranjo, L., Toledo, F., Beltran, A., Ramirez, M., … Sobrevia, L. (2016). Sodium/proton exchanger Isoform 1 regulates intracellular pH and cell proliferation in human ovarian cancer. *BBA - Molecular Basis of Disease*, 1–42. doi: 10.1016/j.bbadis.2016.10.013.

Sawyer, J. E. R., Hennebry, J. E., Revill, A., & Brown, A. M. (2017). The critical role of logarithmic transformation in Nernstian equilibrium potential calculations. *Advances in Physiology Education*, *41*, 231–238. doi: 10.1152/advan.00166.2016.

Sehgal, P., Szalai, P., Olesen, C., Praetorius, H. A., Nissen, P., Christensen, S. B., ... Møller, J. V. (2017). Inhibition of the sarco/endoplasmic reticulum (ER) Ca2+-ATPase by thapsigargin analogs induces cell death via ER Ca2+ depletion and the unfolded protein response. *Journal of Biological Chemistry*, *292*(48), 19656–19673. doi: 10.1074/jbc.M117.796920.

Seidel, J. A., Otsuka, A., & Kabashima, K. (2018). Anti-PD-1 and anti-CTLA-4 therapies in cancer: Mechanisms of action, efficacy, and limitations. *Frontiers in Oncology*, *8*(March), 1–14. doi: 10.3389/fonc.2018.00086.

So, C. L., Saunus, J. M., Roberts-Thomson, S. J., & Monteith, G. R. (2018). Calcium signalling and breast cancer. *Seminars in Cell and Developmental Biology*, (November). doi: 10.1016/j.semcdb.2018.11.001.

Spugnini, E. P., & Fais, S. (2017). Proton pump inhibition and cancer therapeutics: A specific tumor targeting or it is a phenomenon secondary to a systemic buffering? Enrico. *Seminars in Cancer Biology*, 1–8. doi: 10.1016/j.semcancer.2017.01.003.

Spugnini, E. P., Sonveaux, P., Stock, C., Perez-Sayans, M., De Milito, A., Avnet, S., ... Fais, S. (2014). Proton channels and exchangers in cancer. *Biochimica et Biophysica Acta*, 1–12. doi: 10.1016/ j.bbamem.2014.10.015.

Stafford, N., Wilson, C., Oceandy, D., Neyses, L., & Cartwright, E. J. (2017). The plasma membrane calcium ATPases and their role as major new players in human disease. *Physiological Reviews*, *97*(3), 1089–1125. doi: 10.1152/physrev.00028.2016.

Szewczyk, A., Gehl, J., Daczewska, M., Saczko, J., Frandsen, S. K., & Kulbacka, J. (2018). Calcium electroporation for treatment of sarcoma in preclinical studies. *Oncotarget*, *9*(14), 11604–11618. doi: 10.18632/oncotarget.24352.

Takeuchi, Y., & Nishikawa, H. (2016). Roles of regulatory T cells in cancer immunity. *International Immunology*, *28*(8), 401–409. doi: 10.1093/intimm/dxw025.

Tang, S., Wang, X., Shen, Q., Yang, X., Yu, C., Cai, C., ... Zou, F. (2015). Mitochondrial Ca2+ uniporter is critical for store-operated Ca2+ entry-dependent breast cancer cell migration. *Biochemical and Biophysical Research Communications*, *458*, 186–193. doi: 10.1016/j.bbrc.2015.01.092.

Taylor, S., Pierluigi, E., Assaraf, Y. G., Azzarito, T., Rauch, C., & Fais, S. (2015). Microenvironment acidity as a major determinant of tumor chemoresistance: Proton pump inhibitors (PPIs) as a novel therapeutic approach. *Drug Resistance Updates*. doi: 10.1016/j.drup.2015.08.004.

Teng, M. W. L., Galon, J., Fridman, W.-H., & Smyth, M. J. (2015). From mice to humans : Developments in cancer immunoediting. *The Journal of Clinical Investigation*, *125*(9), 3338–3346. doi: 10.1172/JCI80004.become.

Thomas, R. C. (2009). The plasma membrane calcium ATPase (PMCA) of neurones is electroneutral and exchanges 2H+ for each Ca2+ or Ba2+ ion extruded. *Journal of Physiology*, *587*(2), 315–327. doi: 10.1113/jphysiol.2008.162453.

Vander Heiden, M. G., Cantley, L. C., & Thompson, C. B. (2009). Understanding the Warburg effect: The metabolic requirements of cell proliferation. *Science*, *324*(5930), 1029–1033. doi: 10.1126/science.1160809.

Wang, B. Y., Zhang, J., Wang, J. L., Sun, S., Wang, Z. H., Wang, L. P., ... Hu, X. C. (2015). Intermittent high dose proton pump inhibitor enhances the antitumor effects of chemotherapy in metastatic breast cancer. *Journal of Experimental and Clinical Cancer Research*, *34*(1), 1–12. doi: 10.1186/s13046-015-0194-x.

Wang, F., Liu, D., Xu, H., Li, Y., Wang, W., Liu, B., & Zhang, L. (2014). Thapsigargin induces apoptosis by impairing cytoskeleton dynamics in human lung. *The Scientific World Journal*, *2014*, 1–7. doi: 10.1155/2014/619050.

Wang, H., Franco, F., & Ho, P.-C. (2017). Metabolic regulation of tregs in cancer: Opportunities for immunotherapy. *Trends in Cancer, 3*(8), 583–592. doi: 10.1016/j.trecan.2017.06.005.

Warburg, O. (1956). On the origin of cancer cells. *Science, 123*(3191), 309–314.

Warburg, O., Posener, K., & Negelein, E. (1924). Ueber den Stoffwechsel der tumoren. *Biochemische Zeitschrift, 152*, 319–344.

Webb, B. A., Chimenti, M., Jacobson, M. P., & Barber, D. L. (2011). Dysregulated pH: A perfect storm for cancer progression. *Nature Reviews Cancer, 11*(September), 671–677.

Wei, C., Wang, X., Chen, M., Ouyang, K., Song, L.-S., & Cheng, H. (2009). Calcium flickers steer cell migration. *Nature, 457*(7231), 901–905. doi: 10.1038/nature07577.Calcium.

Whiteside, T. (2014). Induced regulatory T cells in inhibitory microenvironments created by cancer. *Expert Opinion on Biological Therapy, 14*(10), 1411–1425. doi: 10.1517/14712598.2014.927432.Induced.

Wilhelm, G., Schulz, J., & Hofmann, E. (1971). pH-dependence of aerobic glycolysis in Ehrlich ascites tumour cells. *FEBS Letters, 17*(1), 158–162.

Wojtkowiak, J. W., Verduzco, D., Schramm, K. J., & Gillies, R. J. (2011). Drug resistance and cellular adaptation to tumor acidic pH microenvironment. *Molecular Pharmacology, 8*(6), 2032–2038. doi: 10.1021/mp200292c.Drug.

Zhitomirsky, B., & Assaraf, Y. G. (2014). Lysosomal sequestration of hydrophobic weak base chemotherapeutics triggers lysosomal biogenesis and lysosome-dependent cancer multidrug resistance. *Oncotarget, 6*(2), 1143–1156.

Zhitomirsky, B., & Assaraf, Y. G. (2016). Lysosomes as mediators of drug resistance in cancer. *Drug Resistance Updates, 24*, 23–33. doi: 10.1016/j.drup.2015.11.004.

12 A History of Pulsed Electromagnetic Fields

Marko S. Markov, James T. Ryaby,
Nianli Zhang, and Erik I. Waldorff

CONTENTS

INTRODUCTION

The use of electric and magnetic fields for treatment of injuries and pathologies has a long history. There are records that physicians from ancient Greece, China, Japan, and Europe successfully applied natural magnetic materials in their daily practice with the first record being found in the book of William Gilbert *The Magnet and the Big Magnet Earth* published in 1600. Contemporary magnetotherapy begun in Japan immediately after the World War II by introducing both magnetic and electromagnetic fields (EMFs) generated by various waveshapes of the supplied currents. In the 1950s, this modality quickly moved to Europe, first in Romania and the former Soviet Union. During the period 1960–1985, nearly all European countries designed and manufactured their own magnetotherapeutic systems. In parallel with engineering, basic science and clinical research became developed. As result, professional societies, such as BEMS (Bioelectromagnetic Society), have been organized, and the number of publications has increased. The first book on magnetotherapy, written by N. Todorov, was published in Bulgaria in 1982 summarizing the experience of utilizing magnetic fields for treatment of 2,700 patients having 33 different pathologies (Todorov 1982).

In the USA during the 1970s, the team of Andrew Bassett introduced a new approach for treatment of delayed fractures, which employed a very specific biphasic low-frequency signal (Bassett et al. 1974, 1977). This signal was approved by the USA Food and Drug Administration (FDA) for treatment of only nonunion/delayed fractures in the USA. A decade later, the FDA approved the use of pulsed radiofrequency electromagnetic field (PRF) for treatment of pain and edema in superficial soft tissues.

Today there is an abundance of experimental and clinical data which suggest that various exogenous magnetic field at surprisingly low level can have a profound effect on a large variety of biological systems. It is now commonly accepted that weak EMFs are capable of initiating various healing processes from delayed fractures and pain relief to multiple sclerosis and Parkinson's disease (Rosch and Markov 2004, Barnes and Greenebaum 2007, Markov 2015).

This historic overview will discuss only the modalities that utilize a specific type of time-varying EMFs known as PEMFs. Therefore, studies that report the successful application of static magnetic fields as well as electrical stimulation will remain outside the scope of this chapter. It should be noted, that, thus far, the medical community approach to magnetotherapy is as to an adjuvant therapy, especially for the treatment of a variety of musculoskeletal injuries.

EMF SIGNALS

Signal development for therapeutic use follows the development of electrical engineering that first provided sine wave with frequency of 60 Hz in North America and 50 Hz in the rest of the world. In the 1930s, the 27.12 MHz continuous sine wave was proposed for deep tissue heating in fighting various forms of cancer (Ginsburg 1934).

The next step was to move from symmetrical sine waves to other waveshapes. First was the engineering of asymmetrical waveforms by means of rectification. These types of signals basically flip-flop the negative part of the sine wave into a positive part, which creates a pulsating sine wave. Textbooks usually show the rectified signal as a set of ideal semisine waves. It should be taken into account that each signal-generating system has specific characteristics, such as impedance, which prevent the generation of the ideal waveshape. Therefore, the ideal form is distorted by a short DC-type component between two consecutive semisine waves. This form of the signals reportedly has been tested for treatment of low back pain (Harden et al. 2007) and reflex sympathetic disorder (Ericsson et al. 2004). The most impressive were results obtained in animal experiments where PEMFs initiate antiangiogenic effects (Williams et al. 2001, Markov 2004a). Authors reported that, in a range of amplitudes for 120 pulses per second, the 15 mT amplitude is the most effective in prevention of the formation of new blood vessels in growing tumors. As a result, the tumor cannot expand the blood vessel network, and this causes starvation and death of the tumor.

In early days of PEMF application, devices were engineered when a sine wave signal is modulated by another signal which might be of low or high frequency. For example, devices with two high-frequency signals in which the interference of both signals resulted in magnetic field interference were used for therapeutic purposes (Todorov 1982).

Most of the PEMF devices had been designed by the intuition of engineers. The design of a device based on the ion cyclotron resonance (ICR) theory proposed by Liboff (1985) should be pointed out as an exemption of such practice. The signal, an alternating sinusoidal magnetic field with a frequency of 76.6 Hz (a combination of Ca^{2+} and Mg resonance frequencies), has oscillating characteristics but oscillates only as a positive signal (Liboff et al. 1987).

The next step is represented by a set of devices which utilize unipolar or bipolar rectangular signals. However, due to the electrical characteristics of the specific generator, these signals can never be rectangular. The maximum value of the signal cannot be reached instantaneously, and the signal amplitude will need a certain time to decrease from maximum to zero. This causes a short delay both in raising the signal up and in its decay to zero. The rise time or slew rate of such a signal could be of extreme importance because the large value of dB/dt could induce significant changes in the magnetic field and induce electric current into the target tissue. Some authors consider that the rate of raising time dB/dt is the factor responsible for the observed biological effects. Since these signals are always somewhat trapezoidal, one can ask why engineers do not design trapezoidal waveshapes instead of rectangular ones? It is easy to say, but from engineering and manufacturing points of view, it is very hard to achieve. So, since the proposal of Kotnik and Miklavcic (2006) for the implementation of such signal shapes, no device generating such types of signals has been constructed.

An interesting fact is that the first clinical signal approved by the FDA for treatment of nonunion or delayed fractures (Bassett 1974, 1977) exploited the pulse-burst approach. Having a repetition rate of 15 bursts per second, this asymmetrical signal (with a long positive and very short negative component) has been successfully used for healing nonunion bone fractures. It was assumed that the cell would ignore the short opposite polarity pulse and respond only to the envelope of the burst which had a duration of 5 ms, enough to induce sufficient amplitude in the kilohertz frequency range.

Interestingly enough, more contemporary devices utilize signals that consist of single narrow pulses separated by long "signal-off" intervals. This approach allows modification of not only the amplitude of the signal, but also the duty cycle (time on/time off) as well. This allows large variations in the selection of the appropriate for specific pathology signal parameters.

The radiofrequency signal, originally proposed by Ginsburg in 1934 (Ginsburg 1934) for deep tissue heating and later approved by the FDA for treatment of pain and edema in superficial soft tissues (Diapulse) utilizes the 27.12 MHz in pulsed mode. Thus, having 65 short bursts and 1,600 pauses between pulse bursts, the signal does not generate heat for 30 min of use. The important issue here is that even if the signal initiates some heating, during the "silence" time, the heat might dissipate.

PROVEN CLINICAL BENEFIT

It has been reported that PEMF is beneficial in bone unification; reduction of pain, edema, and inflammation; increasing blood circulation; and stimulation of immune and endocrine systems. The use of PEMF for treatment of soft tissues results in accelerated healing of arterial or venous skin ulcers, diabetic ulcers, pressure ulcers as well as surgical and burn wounds. In addition, since cells involved in wound repair are electrically charged, some endogenous EMF signals facilitate migration of specific cells to the wound area (Lee et al. 1993), thereby restoring normal electrostatic and metabolic conditions. At any injury site of the musculoskeletal system, an injury current occurs (Canaday and Lee 1991). Since the main goal of any therapy is to restore normal function of the organism, electromagnetic modalities appear to be the most suitable to compensate for the injury currents. PEMFs have also been beneficial in the treatment of chronic pain associated with connective tissue (cartilage, tendon, ligaments, and bone) injury and joint-associated soft tissue injury (Rosch and Markov 2004, Barnes and Greenebaum 2007, Parker and Markov 2015).

One indirect possibility of achieving the clinical effects is the effects of PEMFs on signal transduction pathways. It is now well accepted that the cellular membrane is a primary target for magnetic field action (Adey 2004). It has been reported that selected EMFs are capable of affecting the signal transduction pathways via alteration of ion binding and transport. In a series of studies of myosin phosphorylation dependent on calcium/calmodulin (CaM), Markov's group demonstrated that specific magnetic fields, PEMFs, and 27.12 MHz PRFs could modulate Ca^{2+} binding to CaM to a twofold enhancement in Ca^{2+}-binding kinetics in a cell-free enzyme preparation (Markov 1993, 1994, Markov et al. 1992a,b,).

A meta-analysis performed on randomized clinical trials using PEMFs on soft tissues and joints showed that both PEMFs and PRFs were effective in accelerating healing of skin wounds (Canedo-Dorantes et al. 2002, Mayrovitz 2015) and soft tissue injury (Foley-Nolan et al. 1990, Vodovnik and Karba 1992) as well as providing symptomatic relief in patients with osteoarthritis and other joint conditions (Fitzsimmons et al. 1994, Zizic et al. 1995, Ryaby 1998).

IN SEARCH OF MECHANISMS OF ACTION

The biophysical mechanism(s) of interaction of weak electric and magnetic fields with biological systems has been intensively studied first by basic science. Both experimental and theoretical data have been collected worldwide in the search for potential mechanisms of action. Starting with electrochemical information transfer [30], a number of mechanisms have been proposed, such as ICR, ion parametric resonance (IPR), free radical concept, heat shock proteins, etc. The linear physicochemical approach (Pilla 1974) employs an electrochemical model of the cell membrane to assess the EMF parameters for which bioeffects might be expected. It was assumed that nonthermal EMF may directly affect ion binding and/or transport and possibly alter the cascade of biological processes related to tissue growth and repair.

In search of mechanisms that would explain the clinical benefit of the EMF, various theories have been proposed. Pilla believed that "Today there is abundance of in vitro and in vivo data obtained in the laboratory research as well as clinical evidence that time-varying magnetic fields of various configurations can generate beneficial effects for various conditions, such as chronic and acute pain, chronic wounds and recalcitrant bone fractures. This has been achieved with low intensity, non-thermal, non-invasive time-varying EMF, having various configurations within a broad frequency range" (Pilla 2007). However, Pilla missed one keyword "some". By not specifying that only some or selected PEMFs could initiate specific plausible therapeutic effects, we simply say that all magnetic fields could achieve the goals, which is incorrect.

The most important question in magnetotherapy is: Which are the signals that could be effective and at what conditions? Are any signal parameters better than others? It should be pointed out that many EMF signals used in research and as therapeutic modalities are chosen in some arbitrary manner. It depends on the specific interest of the provider of medical service, not on serious biological and clinical evidence.

In most cases, the selection of the specific device is based on the very general considerations and unfortunately on the presumption that "more means better". For the EMF therapy, this principle cannot lead to beneficial results. During the origin and evolution of life, living creatures adapted to specific physical conditions and developed specific mechanisms to accept or reject an applied stimulus.

In the late 1970s, three research laboratories which were unknown to one another had published papers which introduced the concept of "biological windows". The "biological windows" could be identified by the amplitude, frequency, and their combinations. It has been shown that at least three amplitude windows exist: at 50–100 mT (5–10 Gauss), 15–20 mT (150–200 Gauss), and 45–50 mT (450–500 Gauss) (Markov 2005). Using cell-free myosin phosphorylation to study a variety of signals has shown that the biological response depends strongly on the parameters of the applied signal, confirming the validity of the last two "windows" (Markov 2004b,c). The "biological windows" concept and related research indicate that, in search of the mechanisms of action and therefore in choosing the appropriate therapeutic devices, the principle "more is better" is not valid.

The electrochemical information transfer hypothesis, developed by Pilla, postulated that one possible mechanism of action of the EMF on the cell membrane could be via modulation of the rate of ion binding to receptor sites. Several distinct types of electrochemical interactions can occur at cell surfaces, but the most important nonspecific electrostatic interactions involve water dipoles and hydrated (or partially hydrated) ions at the lipid bilayer/aqueous interface of a cell membrane as well as voltage-dependent ion/ligand binding (Pilla 1997).

The clinical application of PEMF should be based upon the "biological windows" mechanism that does not consider thermal energy a necessity. In order to resolve the thermal noise problems in the ICR model, Lednev (1991) formulated an IPR model which was further developed during the 1990 (Blanchard and Blackman 1994, Blackman et al. 1995). In this quantum approach, an ion in the binding site of a macromolecule is considered to be a charged harmonic oscillator.

Any discussion of the possibility for EMFs to cause biological/clinical effects involves a discussion of the problem of thermal noise ("kT"). Physicists and physical chemists have rejected the possibility that static and low-frequency magnetic fields may cause biological effects because of the "thermal noise" hypothesis.

The biophysical dogma pushed into discussion in the middle of the 1980s and lingering to this day is that, unless the amplitude and frequencies of an applied electric field are sufficient to produce tissue heating, no biological effect should be expected. This was a serious obstacle in the search for biological mechanisms and therapeutic applications of weak EMF signals. Even a recent update of the ICNIRP (International Commission of Nonionizing Radiation Protection) guidelines claims that the only possible biological effect of EMFs is thermal effect (presented at the BioEM 2018 meeting in Portoroz, Slovenia, June 24–29, 2018).

However, it has been proven that the vast majority of PEMF signals employed clinically do not directly cause a physiologically significant temperature rise in the target cell/tissue area. Furthermore, many experimental and clinical studies show that the preferential responses to the specific waveform cannot be explained on the basis of power or energy transfer alone. It was proposed more than 40 years ago that PEMF signals could be configured to be bioeffective by matching the amplitude spectrum of the *in-situ* electric field to the kinetics of ion binding (Pilla 1974). Models have been developed which show that nonthermal PEMF could be configured to modulate electrochemical processes at the electrified interfaces of macromolecules by assessing its detectability–background voltage fluctuations in a specific ion-binding pathway (Pilla 2007). This has allowed the a priori configuration of nonthermal pulse-modulated radio frequency signals which have been demonstrated to modulate CaM-dependent nitric oxide (NO) signaling in many biological systems (Pilla 2013).

We would like to stress that, for any model of biophysical mechanism of weak EMF bioeffects, the most important issue is the signal detection at the molecular/cellular/tissue target in the presence of thermal noise, i.e., signal to thermal noise ratio (SNR). However, this has nothing to do with the "thermal mechanism" of action as it is related to the thermodynamic problem of detection of plausible EMF signals.

Numerous animal and *in vitro* studies, as well as clinical experience, suggest that the initial status of the EMF-sensitive target pathway determines whether a physiologically meaningful bioeffect could be achieved. For example, when a broken bone is treated with PEMF, the surrounding soft tissues receive the same dose as the fracture site, but the physiologically important response occurs only in the injured bone tissue, while the soft tissues surrounding the bone are not affected (Pilla 2007).

This is a crucially important observation, indicating that EMFs are more effective for the injured tissues that are out of equilibrium. Therefore, experiments with healthy volunteers are not always indicative of the potential response of patients who are victims of injury or disease. It is obvious that healthy organisms have much larger compensational ability than the diseased organism, and for that reason, the response (if it even occurs) is much smaller. What is more important is that the system in equilibrium does not need stimulation.

This view was supported by Nindl et al. (2002) in an investigation of Jurkat cells in which the state of the cells was found to be important in regards to the response to EMF: normal T-lymphocytes neglect the applied PEMF, while response has been observed in cells stimulated by other factors. In other words, it might be approximated by the pendulum effect – the larger the deviation from equilibrium, the stronger is the response (Nindl et al. 2002, Markov et al. 2006). For example, it was observed in an *in vitro* study that the initial conditions of lymphocytes are important in terms of the biological effects of EMFs on those cells.

As it was shown earlier, there are various devices and signals in use for PEMF therapy, with semisine waves, rectangular or triangular in shape, which are more effective compared to continuous sine waves. In addition, for rectified sinusoidal signals, more research is needed to clarify the importance of the short DC component between the consecutive semisine waves, although Markov (2004a) found that the durations of this DC component are associated with different biological responses in several outcomes.

The novel computer technologies allow the computerized control of the signal and maintenance of the parameters of the signal during the whole treatment session. They also allow the development of user-friendly software packages with prerecorded programs, as well as with the ability to modify programs depending on patients' needs. With appropriate sensors, the feedback information could be recorded and used during the course of therapy. Last but not least is the possibility to store the data for the treatment of individuals in a large database and to further analyze the cohort of data for a particular study or disease.

PHYSICAL AND BIOPHYSICAL DOSIMETRY OF THE SIGNALS AND THEIR CLINICAL IMPORTANCE

As mentioned, there is an abundance of devices in the market, and most of them have not even gone through the rigorous screening of the FDA and other governmental regulatory offices. In most cases, the Internet offers devices that exploit different physical parameters of signals and promise clinical benefits for nearly the entire spectrum of medical problems. It should be stressed that each of these devices needs to pass through engineering testing followed by physical and biophysical dosimetry prior to being used for treatment of patients.

Let us distinguish physical and biophysical dosimetry in evaluating the potential benefit of using a specific device. Usually manufacturers provide very limited information about the devices they offer. We would like to recommend to any clinician who plans to use magnetotherapeutic devices to first perform a test of the physical parameters of the signal-generating system. It should be by measurement, not by calculation or computation.

It has been suggested by Markov and Valberg (Markov 1994, Valberg 1995) that a set of parameters should be considered, such as

- type of the field
- component (electric or magnetic)
- intensity or induction
- gradient dB/dt
- vector dB/dx
- frequency
- pulse shape
- localization
- duration of exposure
- depth of penetration.

While the first seven parameters are more or less connected with the device, the last three are related to the medical application. However, it is important to say that physical dosimetry, which basically takes care of these three parameters and describes the device, needs to be complemented with the knowledge about the values of these parameters at the site of application. In other words, biophysical dosimetry needs to control the values of the described parameters at the target site or at least in close proximity to the target. **It should be remembered that the acting signal is not**

what is emitted from the device but what is received by the target tissue. We would like to alert the readers that SAR (specific absorption rate) cannot be used for the characterization of any device, because computing this value requires knowledge of the absorption process. Therefore, SAR **must always be associated to the target, not the device.**

CONCLUSIONS

PEMF stimulation in clinical practice has been shown as a plausible modality in resolving various medical problems. The mechanisms of action are definitely non-thermal, related to restoration of biological equilibrium in the tissues subjected to therapy. The choice of PEMF device and protocol of treatment should be based upon the biophysical dosimetry.

REFERENCES

Adey, W. R. (2004). Potential therapeutic application of non-thermal electromagnetic fields: Ensemble organization of cells in tissue as a factor in biological tissue sensing. In Roscht, P. J., Markov, M. S. (Eds.) *Bioelectromagnetic Medicine*. Taylor & Francis Group: Marcel Dekker, NY, 1–15.

Barnes, F. S. and B. Greenebaum (2007). *Handbook of Biological Effects of Electromagnetic Fields*. CRC Press: Boca Raton, FL.

Bassett, C. A. L., R. J. Pawluk, and A. A. Pilla (1974). Acceleration of fracture repair by electromagnetic fields. *Annals of the New York Academy of Sciences* **238**: 242–262.

Bassett, C. A. L., A. A. Pilla, and R. J. Pawluk (1977). A non-surgical salvage of surgically-resistant pseudoarthroses and non-unions by pulsing electromagnetic fields. *Clinical Orthopaedics* **124**: 117–131.

Blackman, C. F., J. P. Blanchard, S. G. Benane, and D. E. House (1995). The ion parametric resonance model predicts magnetic field parameters that affect nerve cells. *Federation of American Societies for Experimental Biology Journal* **9**: 547–551.

Blanchard, J. P. and C. F. Blackman (1994). Clarification and application of an ion parametric resonance model for magnetic field interactions with biological systems. *Bioelectromagnetics* **15**: 217–238.

Canaday, D. J. and R. C. Lee (1991). Scientific basis for clinical applications of electric fields in soft-tissue repair. In Brighton, C. T., Pollack, S. R. (Eds.) *Electromagnetics in Biology and Medicine*. Taylor & Francis, San Francisco Press Inc.: San Francisco, CA, 275–291.

Canedo-Dorantes, L., R. Garcia-Cantu, R. Barrera, I. MendezRamirez, V. H. Navarro, and G. Serrano (2002). Healing of chronic arterial and venous leg ulcers with systemic electromagnetic fields. *Archives of Medical Research* **33**: 281–289.

Ericsson, A. D., C. F. Hazlewood, M. S. Markov, and F. Crawford (2004). Specific biochemical changes in circulating lymphocytes following acute ablation of symptoms in Reflex Sympathetic Dystrophy (RSD). In Kostarakis, P. (Ed.) *Proceedings of Third International Workshop on Biological Effects of EMF*. KOS, Greece, 683–688. ISBN 960-233-151-8.

Fitzsimmons, R. J., J. T. Ryaby, F. P. Magge, and D. J. Baylink (1994). Combined magnetic fields increase net calcium flux in bone cells. *Calcified Tissue International* **55**: 376–380.

Foley-Nolan, D., C. Barry, R. J. Cougyhlan, P. O'Connor, and D. Roden (1990). Pulsed high frequency (27 MHz) electromagnetic therapy for persistent neck pain: A double blind placebo-controlled study of 20 patients. *Orthopedics* **13**: 445–451.

Ginsburg, A. J. (1934). Ultrashort radio waves as a therapeutic agent. *Medical Record* **19**: 1–8.

Harden, R. N., T. A. Rembel, T. T. Houle, J. E. Long, and M. S. Markov (2007). Prospective, randomized, single-blind, sham treatment-controlled study on the safety and efficacy of an electromagnetic field device for the treatment of chronic low back pain. *Pain Practice* **7**: 248–255.

Kotnik, T. and D. Miklavcic (2006). Theoretical analysis of voltage inducement on organic molecules. In Kostarakis, P. (Ed.) *Proceedings of Forth International Workshop Biological Effects of Electromagnetic Fields*, 217–266. ISBN: 960-233-172-0.

Lednev, V. V. (1991). Possible mechanism for the influence of weak magnetic fields on biological systems. *Bioelectromagnetics* **12**: 71–75.

Lee, R. C., D. J. Canaday, and H. Doong (1993). A review of the biophysical basis for the clinical application of electric fields in soft-tissue repair. *The Journal of Burn Care and Rehabilitation* **14**: 319–355.

Liboff, A. R. (1985). Cyclotron resonance in membrane transport. In Chiabrera, A., Nicolini, C., Schwan, H. P. (Eds.) *Interactions between in Interactions Between Electromagnetic Fields and Cells*. Plenum Press: New York, 281–396.

Liboff, A. R., F. J. Fozek, M. L. Sherman, B. R. Mcleod, and S. D. Smith (1987). Ca^{2+}-45 cyclotron resonance in human lymphocytes. *Journal of Bioelectricity* **6**: 13–22.

Markov, M. S. (1993). Ambient range sinusoidal and DC magnetic fields affect myosin phosphorylation in a cell-free preparation. In Blank, M. (Ed.) *Electricity and Magnetism in Biology and Medicine*. San Francisco Press: San Francisco, CA, 323–327.

Markov, M. S. (1994). Biological effects of extremely low frequency magnetic fields. In Ueno, S. (Ed.) *Biomagnetic Stimulation*. Plenum Press: New York, 91–103.

Markov, M. S. (2004a). Magnetic and electromagnetic field therapy: Basic principles of application for pain relief. In Rosch, P. J., Markov, M. S. (Eds.) *Bioelectromagnetic Medicine*. Marcel Dekker: New York, 251–264.

Markov, M. S. (2004b). Myosin light chain phosphorylation modification depending on magnetic fields: Theoretical. *Electromagnetic Biology and Medicine* **23**: 55–74.

Markov, M. S. (2004c). Myosin phosphorylation: A plausible tool for studying biological windows. Ross Adey Memorial Lecture. In Kostarakis, P. (Ed.) *Proceedings of Third International Workshop on Biological Effects of EMF*, KOS, Greece, 4–8 October, 1–9. ISBN 960-233-151-8.

Markov, M. S. (2005). Biological windows: A tribute to W. Ross Adey. *The Environmentalist* **25** (2–3): 67–74.

Markov, M. S. (2015). *Electromagnetic Fields in Biology and Medicine*. CRC Press: Boca Raton, FL.

Markov, M. S., J. T. Ryaby, J. J. Kaufman, and A. A. Pilla (1992a). Extremely weak AC and DC magnetic field significantly affect myosin phosphorylation. In Allen, M. J., Cleary, S. F., Sowers, A. E., Shillady, D. D. (Eds.) *Charge and Field Effects in Biosystems-3*. Birkhauser: Boston, MA, 225–230.

Markov, M. S., J. T. Ryaby, S. Wang, and A. A. Pilla (1992b). Modulation of myosin phosphorylation rates by weak (near ambient) DC magnetic fields. *In Proceedings of 18th Annual Northeast Bioengineering Conference*, Kingston, RI, March 1992. IEEE: New York, 63–64.

Markov, M. S., G. Nindl, C. F. Hazlewood, and J. Cuppen (2006). Interactions between electromagnetic fields and immune system: Possible mechanisms for pain control. In Ayrapetyan, S., Markov, M. (Eds.) *Bioelectromagnetics: Current Concepts, NATO Advanced Research Workshops Series*. Springer: Berlin, Germany, 213–226.

Mayrovitz, H. N. (2015). Electromagnetic fields for soft tissue wound healing. In: Markov, M. S. (Ed) *Electromagnetic Fields in Biology and Medicine*. CRC Press, Boca Raton FL, 231–252.

Nindl, G., M. T. Johnson, E. E. Hughes, and M. S. Markov (2002). Therapeutic electromagnetic field effects on normal and activated Jurkat cells. *International Workshop of Biological Effects of Electromagnetic Fields*, Rhodes, Greece, 1 October 2002, 167–173. ISBN: #960-86733-3-X.

Parker, R. and M. S. Markov (2015). Electromagnetic fields in the treatment of tendon injury in human and veterinarian medicine. In: Markov, M. S. (Ed.) *Electromagnetic Fields in Biology and Medicine*. CRC Press, Boca Raton, FL, 233–252.

Pilla, A. A. (1974). Mechanisms of electrochemical phenomenon in tissue growth and repair. *Bioelectrochem Bioenergetics* 1: 227–243.

Pilla, A. A. (1997). A dynamical system/Larmor precession model for weak magnetic field bioeffects: ion binding and orientation of bound water molecules. *Bioelectrochemistry and Bioenergetics* 43: 239–249.

Pilla, A. A. (2007). Mechanisms and therapeutic application of time varying and static magnetic fields. In Barnes, F., Greenebaum, B. (Eds.) *Biological and Medical Aspects of Electromagnetic Fields*. CRC Press: Boca Raton, FL, 351–411.

Pilla, A. A. (2013). Nonthermal electromagnetic fields: From first messenger to therapeutic applications. *Electromagnetic Biology and Medicine* 32: 123–136.

Rosch, P. J. and M. S. Markov (2004). *Bioelectromagnetic Medicine*. Marcel Dekker: New York.

Ryaby, J. T. (1998). Clinical effects of electromagnetic and electric fields on fracture healing. *Clin Orthopaedics* 355(Suppl): 205–215.

Todorov, N. (1982). *Magnetotherapy*. Meditzina i Physcultura Publishing House: Sofia, 106 pp.

Valberg, P. (1995). How to plan EMF experiments. *Bioelectromagnetics* 16: 396–401.

Vodovnik, L. and R. Karba (1992). Treatment of chronic wounds by means of electric and electromagnetic fields. *Medical and Biological Engineering and Computing* 30: 257–266.

Williams, C. D., M. S. Markov, W. E. Hardman, and I. L. Cameron (2001). Therapeutic electromagnetic field effects on angiogenesis and tumor growth. *Anticancer Research* 21: 3887–3892.

Zizic, T., P. Hoffman, D. Holt, J. Hungerford, J. O'Dell, and M. Jacobs (1995). The treatment of osteoarthritis of the knee with pulsed electrical stimulation. *The Journal of Rhetunatology* 22: 1757–1761.

13 A Brief History of Orthofix Medical Inc.'s Biostimulation

Erik I. Waldorff, Nianli Zhang, and James T. Ryaby

CONTENTS

ORTHOFIX MEDICAL INC.'S PEMF HISTORY

What is now Orthofix Medical Inc. began in Verona, Italy, from the work of orthopedic researcher Giovanni De Bastiani of the University of Verona. Toward the end of the 1970s, De Bastiani proposed the concept of "dynamization", based on the natural ability of a bone to repair itself. He developed a modular system of external axial frame devices that could be fitted to a bone, allowing micromovement at the fracture site to stimulate bone healing. Together with a group of surgeons and an industrial engineer, De Bastiani founded Orthofix in 1980 in order to continue the development of these devices and to bring them to market.

In parallel, American Medical Electronics (AME) was founded ca. 1982 in Minnesota by Joe Mooibroek and cofounders Stu Johnson (Operations) and John Erickson (Engineering). All were from the Minneapolis–St. Paul area medical device industry, specifically pacemakers. Shortly after incorporation, the company relocated to Addison, TX.

The initial "big idea" for AME's product was to make a more convenient/portable version of the EBI (electrobiology) external fracture healing device, by getting approval for a shorter treatment time and by making it more efficient by utilizing a ferromagnetic (metglas) core material so it could operate on a rechargeable battery pack. The resulting device was the PhysioStim device.

Following a prospective clinical study, the PhysioStim device was approved by the FDA in February 1986 for the treatment of an established nonunion acquired secondary to trauma, excluding vertebrae and all flat bones, where the width of the nonunion defect is less than half the width of the bone to be treated (PMA 1986, Garland, Moses et al. 1991).

In all, five different PhysioStim device models were developed to treat nonunions at different anatomical sites: tibia, ulna/radius, humeral head, hip, and the clavicle.

In parallel to the development of the PhysioStim device, AME conducted a randomized, double-blind, prospective IDE study of their second device, SpinalStim. Following the clinical trial, the FDA approved SpinalStim on February 7, 1990, as a noninvasive electromagnetic bone growth stimulator indicated as an adjunct to spinal fusion to increase the probability of fusion success and as a nonoperative treatment for the salvage of failed spinal fusion, where a minimum of 9 months has elapsed since the last surgery (Mooney 1990, PMA 1990, Mooney 1993, Simmons, Mooney et al. 2004).

Subsequent to the FDA approval of SpinalStim, AME sought to develop a more efficient version of the PhysioStim and SpinalStim signals in order for the devices to operate acceptably on primary (replaceable/disposable) batteries. This would enable AME to transition to a single-patient-use business model, eliminating refurbishment operations and enabling economies of scale in manufacturing. In addition, the battery pack would be smaller enabling better patient comfort. The more efficient signals were achieved using a biphasic pulsed electromagnetic field (PEMF) signal which derived its efficiency from the fact that since the induced magnetic field is proportional to the change in current and not the magnitude of the current, by driving the current through the coil in both directions (rather than increasing and decreasing the current in one direction only), the peak current could be reduced by half which reduced power losses in the coil by a factor of 4. In addition, the magnetic field energy was recovered through the same coil winding that generated it: i.e., there was no separate secondary for field recovery, reducing/eliminating coupling inefficiencies with the recovery winding. This arrangement also made manufacturing of the coil simpler and less expensive.

The FDA approval for the biphasic versions of PhysioStim and SpinalStim was achieved through PMA supplements based on spectral analysis and comparison with the previously approved signal version.

In 1992, in parallel to AME's increased FDA-approved product portfolio, Orthofix acquired Novamedix Ltd. and A-V Impulse System technology and was listed for the first time on NASDAQ (as OFIX). In 1995, Orthofix finally acquired AME and its U.S. manufacturing and distribution network, including the SpinalStim and PhysioStim bone growth stimulation technology.

Following the acquisition of AME, Orthofix conducted a randomized, controlled, prospective, multicenter investigational device exemption (IDE) clinical study for their third PEMF device, CervicalStim. Following the CervicalStim premarket approval (PMA) submission, the FDA approved CervicalStim on December 23, 2004, as a noninvasive, adjunct treatment option for cervical spine fusion surgery in patients at high risk for nonfusion (PMA 2004, Foley, Mroz et al. 2008).

More recently Orthofix, under the guidance of Chief Scientific Officer James T. Ryaby, PhD, has invested in several preclinical studies looking at translating PEMF science into new clinical indications (Waldorff, Zhang et al. 2017). This has resulted in the initiation of several new PEMF IDE clinical studies, namely as an adjunct to enhance union in conservatively treated type II fractures of the odontoid process (Sep 2014, Trial #: NCT02281994), for the treatment of osteoarthritis of the knee (July 2016, Trial #: NCT02436590), and most recently as an adjunctive treatment to surgical repair of full thickness rotator cuff tears (January 2018, Trial #: NCT03339492).

Today, Orthofix is a global medical device company focused on musculoskeletal healing products and value-added services. The Company's mission is to improve patients' lives by providing superior reconstruction and regenerative musculoskeletal solutions to physicians worldwide. Headquartered in Lewisville, Texas, the Company has two strategic business units: Orthofix Spine and Orthofix Extremities. Orthofix products are widely distributed via the Company's sales representatives and distributors. In addition, Orthofix is collaborating on research and development activities with leading clinical organizations such as MTF Biologics, the Orthopedic Research and Education Foundation, and the Texas Scottish Rite Hospital for Children.

REFERENCES

Foley, K. T., T. E. Mroz, P. M. Arnold, H. C. Chandler, Jr., R. A. Dixon, G. J. Girasole, K. L. Renkens, Jr., K. D. Riew, R. C. Sasso, R. C. Smith, H. Tung, D. A. Wecht and D. M. Whiting (2008). "Randomized, prospective, and controlled clinical trial of pulsed electromagnetic field stimulation for cervical fusion." *Spine J* 8(3): 436–442.

Garland, D. E., B. Moses and W. Salyer (1991). "Long-term follow-up of fracture nonunions treated with PEMFs." *Contemp Orthop* 22(3): 295–302.

Mooney, V. (1990). "A randomized double-blind prospective study of the efficacy of pulsed electromagnetic fields for interbody lumbar fusions." *Spine (Phila Pa 1976)* 15(7): 708–712.

Mooney, V. (1993). "Pulsed electromagnetic fields: an adjunct to interbody spinal fusion surgery in the high risk patient." *Surg Technol Int* 2: 405–410.

PMA (1986). P850007: Physio-Stim.

PMA (1990). P850007/S6: Spinal-Stim.

PMA (2004). P030034: Cervical-Stim.

Simmons, J. W., Jr., V. Mooney and I. Thacker (2004). "Pseudarthrosis after lumbar spine fusion: nonoperative salvage with pulsed electromagnetic fields." *Am J Orthop (Belle Mead NJ)* 33(1): 27–30.

Waldorff, E. I., N. Zhang and J. T. Ryaby (2017). "Pulsed electromagnetic field applications: A corporate perspective." *J Orthop Translat* 9: 60–68.

14 History of the Frank Reidy Research Center for Bioelectrics

Stephen J. Beebe

The Frank Reidy Research Center for Bioelectrics (FRRCBE) got off to an unexpected start in the mid-1990s. One of the fundamental concepts of a center was present from the very beginning – merging diverse disciplines into common collaborations as a nucleus for interdisciplinary ventures. The fundamental concept of collaborations that developed was based on pulsed power and the use of ultrashort electric pulses (usEPs) in the nanosecond (10^{-9}s) range and ultimately in the picosecond (10^{-12}s) range. A high-powered physics discipline was about to find new beginnings in very different landscapes of microbiology, molecular biology, and finally medicine. How could a technology first developed in World War II for use in radar and used in particle accelerators, fusion research, electromagnetic pulses, and high-power pulsed lasers be used to treat cancer? So, physics and engineering fused with biology and medicine and a new field called Bioelectrics evolved over time into a center devoted to understanding how short bursts of high-powered electric fields could impact cells and change their behavior, especially to eliminate cancer cells. More recently, it was found that these USPs could induce immune responses, providing a potential immunotherapy. If it was a surprise that pulsed power could kill cancer, it was a shock (no pun intended) that pulsed power could induce immune responses.

Unexpectedly, at least one other aspect of a center had already begun with a scientific collaboration between Dr. Karl Schoenbach and me in the mid to late 1990s. Karl's degree was in physics, and his work with pulsed power involved energy storage and opening switch technology (Schoenbach et al., 1984). I was an Associate Professor in Pediatrics and Physiological Sciences at Eastern Virginia Medical School (EVMS) with a Biochemistry thesis (Beebe et al., 1984). I had come to EVMS after receiving a PhD in Medical Sciences in a Pharmacology Department with a Biochemistry thesis, a postdoctoral fellowship in the Department of Molecular Physiology and Biophysics at the Howard Hughes Institute at Vanderbilt studying hormone signal transduction (Beebe et al., 1985), and a Fulbright Scholar in Oslo, Norway, using molecular cloning to enhance biochemical studies(Beebe et al., 1990). Coming from this background of relatively pure basic biological sciences, I didn't know that pulsed power even existed, let alone what it was. I met

Dr. Karl Schoenbach, a professor in the Department of Electrical and Computer Engineering at Old Dominion University (ODU), sometime in 1993–1994. Actually, Karl first asked Dr. Barbara Hargrave, an Associate Professor in Biology at ODU, if she would be interested to work with him, but her schedule did not allow that collaboration. I had met Barbara at EVMS while she was working on reproduction project, and she thought I might be interested, so she suggested Karl get in touch with me. So, Karl asked me if I would be interested to work with him using pulsed power in microbiology. I had always been interested in doing things that were different and new, so I was curious about this new technology. Karl explained that in pulsed power, energy is accumulated and released in very short bursts giving immediate high power with low energy. These were high-intensity electric fields, some as high as 300 kV/cm. He added several other intriguing concepts. He reiterated that these nanosecond pulses were high power and low energy – that is they were nonthermal. Even though the electric fields were so intense, the pulse durations were so short or fast that they did not generate significant levels of heat. The idea of high power and low energy sounded unique and at first seemingly contradictory, because I had equated power and energy. While the concept of high power and low energy was fascinating, even more alluring was the theory that when these pulses were extremely short and the rise and fall of the pulses were very short or fast, these pulses could affect intracellular structures, which meant they could affect cell functions. Since my postdoctoral days, my primary interests have been investigating signal transduction mechanisms – that is, how extracellular signals like hormones activate intracellular signals like enzymes and transcription factors. This concept of electric fields getting inside cells and affecting intracellular structures seemed like a unique way to investigate intracellular functions.

While this was new to me, pulsed power had been around for decades. Pulsed power was the focus of Karl's training at Texas Tech before coming to ODU. He saw an opportunity to extend this technology into biological fields such as bacterial decontamination, ridding nuisance species from water lines and ship hauls, and other possible environmental applications. As we continued to work together, my understanding and interests in cell functions and cancer began to fuse with Karl's understanding and interests in how pulsed power can get inside of cells. Thus, USPs seemed like a unique stimulus from outside the cell that would affect the inside of a cell. This was similar to a hormone, but an electric one.

While the above ideas for applications of pulsed power in biology and medicine are actualities in the present, they were only possible, imagined, or unforeseen applications in the early days of pulsed power collaborations. These ideas could lead to patents, but ODU did not have patenting experience and didn't have a budget to secure patents, which can be costly. Dr. James Koch, President of ODU at the time, had been petitioned by several faculty for patenting possibilities but only had one lawyer without extensive patenting knowledge. Dr. Koch had recently met Frank Reidy and asked him to meet some of these faculty members and consider if any of their requests were worthy of patent consideration by the university. One of those faculty members was Dr. Karl Schoenbach. While this did not constitute the beginning of a center, it included yet another concept common to centers, which was a *consideration* for translating research into practice. A *commitment* to do this would come several years later.

Mr. Frank Reidy has led a very eventful and interesting life. While Dr. Koch had only met Frank, it was not surprising that he could recognize an entrepreneur when he saw one. Little did he likely know that Frank was also a philanthropist. When Frank and his family settled in Virginia, he did not have a single thought of starting the FRRCBE.

Frank and Karl agreed to meet every Friday to discuss Karl's work and patenting ideas. Frank admitted that although he was an engineer, it took some time before he could get a picture of what pulsed power was and what Karl had in mind. And I suspect that, through conversations with Frank, Karl's ideas became more crystalized, as is often the case when ideas are exchanged between different mind-sets. It was sometime during these visits in Karl's laboratories that I met Frank. Frank was clearly interested in what was going on. He would watch students working, ask pointed questions, and make comments from his perspective. He actually seemed to enjoy these interactions.

In addition to adding a solid biology component to the developing research ideas, Frank saw something beyond that in my presence. I was on the faculty of EVMS and had been on the faculty of the Jones Institute for Reproductive Medicine before moving to Physiological Sciences and Pediatrics. The Jones Institute had significant patenting successes. Frank anticipated that EVMS could be helpful when it came to intellectual property. In fact, EVMS was very helpful to secure the initial patent. Karl and Frank spent a lot of time planning strategies to make the base patent a solid and secure document upon which no one could impinge. In addition, I spent hours talking with patent lawyers conveying what I knew about these ultrashort pulses did to cells, especially apoptosis, death processes, cancer, and what we referred to as "intracellular electromanipulation". While electric fields with long durations were known to electroporate the plasma membrane, we didn't want to assume that these shorter pulses would only electroporate intracellular membranes, so we settled on intracellular manipulations. The patent costs were high, but during the many years that followed, there were several attempts to bypass this patent for medical or cosmetic applications that were unsuccessful. This base patent became the cornerstone of the present Pulse Biosciences, Inc., who are presently taking the technology to dermatological and cancer clinical trials.

As our research progressed through the first year, there were other aspects common to the formation of a center that were obvious but not yet considered as such. Working in institutions of higher learning, including a state-supported university and a medical school, both with graduate programs, there were educational opportunities for students and other professionals to establish collaborative relationships and to initiate and sustain a learning environment. As biologists began to think like engineers and physicists, who began to think like biologists, it was easier to think more outside the boxes from which we generally communicated and strategized. These were valuable benefits in having an interdisciplinary group. We clearly spoke different technical languages, and our thinking was based on totally different foundations of knowledge. The concept of a center is fulfilled in part by putting these multidisciplinary individuals in the same building; sharing the same workspace, drinking fountain, and coffee stations; and recruiting young students into the same building with common conference rooms to spawn the next generation for new ideas to the ongoing studies.

In the earliest days of my collaboration with Karl, my lab was on the EVMS campus several miles south of ODU. Karl's lab was on the easternmost side of the ODU campus. Pulse power generators were built and housed in labs at EVMS near important equipment such as a fluorometer and flow cytometer. This required that we have new space for these studies. Space in the university is at a premium, so asking for new space becomes a major request and usually means taking it from someone else. After a meeting of the "space minds", Dr. Dieter Bartchat, an associate professor of Physiological Sciences, was interested enough in our work to let us use space in his lab. This was also complicated because this new space and the flow cytometer, which we heavily used, were in a different building than my primary labs. That meant we carried cells from my lab to the new space for pulse treatment and then down the hall to the flow cytometer. This worked out relatively well, except we didn't have needed communications with the engineers other than weekly meetings, which would come later with building for what would become the FRRCBE.

There was already a well-established field for using electric fields in biology with relatively long pulses using reversible electroporation for the delivery of impermeable cancer drugs and DNA and other molecules that were not permeable to the plasma membrane. Reversible electroporation is also used in medical applications for the delivery of chemotherapeutic drugs into tumor cells, for gene therapy, and for transdermal drug delivery. Electroporation used relatively low electric fields, generally <1 kV/cm, with pulse durations in the millisecond (ms) to microsecond (μs) range. Because successful delivery of molecules such as DNA required cell survival, electroporation pulses were designed to transiently electroporate or permeabilize cells such that they would recover to express the delivered molecules. The new Bioelectrics now included pulses with electric fields as high as 300 kV/cm and pulse duration as low as and lower than 10 nanoseconds (ns). Because, nanosecond pulses generated such high power and were initially designed to decontaminate bacteria and to eliminate cancer, they were designed to prevent survival. Nevertheless, nanosecond pulses can exploit both sides of nature. That is, at lower electric fields, nanosecond pulses can stimulate cell mechanisms such as activating dendritic cells (Guo et al., 2018), and at higher electric fields, nanosecond pulses can terminate cell mechanisms (Beebe et al., 2002, 2003).

There have been a number of terms used for describing electric pulses with nanosecond durations and intense electric fields. These include intracellular electromanipulation (IEM), nanosecond pulsed electric fields (nsPEFs), ultrashort nanosecond high-field electric pulses, nanoelectropulse, electroperturbation, submicrosecond intense pulsed electric fields (sm/i-PEF), nanosecond electric pulses (nsEPs), usEPs, ultrashort pulsed electric fields (USPEFs), nanoelectroablation, pulsed power ablation (PPA), and, most recently, nanopulse stimulation (NPS) or nanopulse electrostimulation (NPES). The use of NPS was initiated by Pulse Biosciences, Inc. to avoid the specific term "ablation" because nanosecond pulses can do more than ablate tissues. Thus, NPS can be considered as a hormetic stimulus, eliciting a biphasic response such that at low intensity it can stimulate while at higher intensities it can inhibit or be cytotoxic. While the low-intensity-induced stimulation has been noticed for some time, it has only been a focus of recent studies. Considering the stimuli used in the early studies at EVMS and ODU, pulse durations were 10, 60, or 300 ns with electric field as high as 300 kV/cm. Pulse generators with these durations were used in a multiuniversity research initiative (MURI) funded by the Air Force Office of Scientific Research (ASOSR) with Professor Karl Schoenbach as principle investigator.

While pulsed power seems like a single entity, there are many ways to build nanosecond pulse devices. While many individuals had hands-on experience in building these devices, Dr. Jürgen Kolb, who was a member of the Center until 2014, and Shu Xiao, now Dr. Shu Xiao, a former student of Karl's and an Associate Professor of Electrical and Computer Engineering and a member of the Center, were the main engineering designers. Their names appear in many papers, but not just because they constructed pulsers. While trained in Physics and Engineering, their quick learning and broad intellect brought distinct and unique contributions to many studies. Shu has become the main engineer in the FRRCBE. He is in high demand and is always willing to help, but he has his own projects besides building devices for others. He has also designed pulsers for Pulse Biosciences, Inc. who are using them in clinical trials. Shu's main interest is the effects of picosecond pulses on cells. Picosecond pulses are also of great interest because they can be delivered noninvasively by antennae. Shu's studies with picosecond pulses will be presented later.

The fundamental theory that was hypothesized for biological cell responses with nanosecond pulses comes from the charging time constant of the plasma membrane. As pointed out in the initial paper, "the charging time constant is a measure of the time during which the cell interior is exposed to the applied pulsed electric field" (Schoenbach et al., 2001). This is equivalent to the statement that "the outer membrane becomes increasingly transparent for oscillating electric fields when the angular frequency of the oscillation exceeds a value given by the inverse of the charging time". Schwan (1985) had shown this high-frequency effect, but Karl appears to be the first to experimentally consider it for determining intracellular effects in biological cells. That shorter pulses (Semenov et al., 2013a) with shorter rise–fall times (Beebe et al., 2012) affect intracellular structures and functions has been supported by a number of studies. That nanosecond pulse could affect intracellular structures without affecting the plasma membrane required some amendment. Effects of shorter nanosecond pulses on plasma membranes were different than classical electroporation pulses. Shorter nanosecond pulses induced lipid nanopores with channel-like conductances (Pakhomov et al., 2009). These nanopores were smaller than propidium iodide (PI), which was commonly used to determine membrane permeability. As nanopulses became larger with more intense conditions, this lipid nanopores lost their unique conductances and opened into classical electroporation pore.

Our initial studies used electric fields for decontamination of bacteria and biofouling of nuisance species and later for biological and medical purposes. There were a number of other engineers who worked with Karl and other biologists who worked with me. These individuals made exceptional and innovative contributions to the development of this new field of Bioelectrics. The initial studies with pulsed power began by determining if the nanosecond pulses could kill bacteria. With the help of an undergraduate student in my lab Mr. Don Byers, who went on to become a physician, and Karl's students and colleagues, we quickly found that the nanosecond pulses could kill bacteria. These manuscripts were published early in IEEE journals: one in *Transactions on Plasma Science* (Schoenbach et al., 1997) and another in *Transactions on Dielectrics and Electrical Insulation* (Schoenbach et al., 2000). There were also several conference papers in Power Modulation Conferences and Symposia in 1995–1997. These studies focused on our earliest work together on bacterial decontamination and biofouling of nuisance species with Frank Peterkin, Fred Dobbs, and Raymond Alden among other students and postdoctoral fellows. It was found that for lysing of bacteria, or stunning of aquatic species, the electrical energy required from the pulsed electric field technique (PEFT) was decreased when the pulse duration is reduced (Schoenbach et al., 1997). Although the electric field must be increased as the pulse duration is decreased, the total energy decreases with decreasing pulse duration. The optimum range for biofouling prevention occurred with pulse durations in the microsecond range, and that for bacterial decontamination occurred in the tens of nanoseconds range. This energy minimum or maximum in efficacy could be explained by taking into account the time required for electrical charging of the cell membrane. Another important point was that thermal effects on organisms allowed the separation of nonlinear electrical effects from thermal effects. Applications for pulsed power could then be used to control the population of organisms, such as bacteria, or to stun them, e.g., brine shrimp, over a certain field-dependent time interval. It was also realized that pulsed power could be used for not only the environment but also medical applications. It was predicted that the nonlinear effects could open up applications for pulsed power by modifying cell structures in a controlled way without lysing cells. This turned out to be correct.

Another paper focused on bacterial decontamination and reviewed fundamental aspects of effects of pulsed electric fields on biological cells (Schoenbach et al., 2000). This included considerations of electric field distribution, amplitude, current density, pulse shape, pulse duration, repetition rate, and energy efficiency. Following from the simple model presented here, the cytoplasm of a cell would be exposed to the same applied electric field for pulse durations that are shorter compared to the membrane charging time, if the pulse rise–fall times were short compared to charging time constant – something pointed out previously. However, for the conditions described, the electrical power reaching the cytoplasm for applied electric fields of 100 kV/cm would be 100 MW/cm^3. Using a megawatt-hour as a measure of energy, one MW would be equivalent to about 300 gallons of gasoline, enough to drive from New York to San Francisco! However, this is the power produced for that 75 ns of the membrane time constant. The time frame for applying these pulses is several orders of magnitude less than that for common pulsed electric field application. The demonstration of intracellular effects with this technology in mammalian cells was in progress when this paper was published.

These events and others preceding them initiated specific funding for this project in 1998 from the Air Force Office of Scientific Research (AFOSR) with a project called "Pulsed Electric Field Effects on Biological Cells". These funds and several other grants that followed were provided by Dr. Bob Barker, PhD at AFOSR. Karl met Bob when Karl was chairing an International Conference on Plasma Science in 1991.

Bob's interests in Karl's work began a long-term relationship between the two. Dr. Barker's enthusiastic interest in and support to Karl's new uses of pulsed power and the team that he was building was in many ways the beginning of Bioelectrics. While a physical structure was some years away, Bob provided fundamental elements of a center by encouraging the discoveries that were evolving, fostering collaborations and supporting education within the research.

Bob visited Karl's lab at ODU in early 1997. Karl showed him a movie he had made of "zapping" brine shrimp or *Artemia salina*. This was featured in the biofouling control part of the project (Schoenbach et al., 1997). It always made significant impressions on viewers. The film showed brine shrimp swimming or treading as they suspended themselves in sea water in a cuvette. When a stunning PEF was applied, the brine shrimp stopped moving and drifted to the bottom of the cuvette. After several minutes, depending on the intensity of the pulse, the brine shrimp began swimming again moving up from the bottom of the cuvette and continued normal swimming movement.

By this time, we had begun exposing human cells to ultrashort PEFs. Karl then told Bob of the electrical model that predicted the ultrashort pulses could get into the interior of these cells and told him about the work we were doing that killed cancer cells. Like the rest of us, Bob was especially excited about using pulsed power as a cancer therapy. But we had a long way to go before this was to become reality to us. Following this and other meetings, Bob became an even greater supporter of this research. He encouraged this technology further by supporting a series of meetings nicknamed "ElectroMed". The ElectroMed symposia did as much or more than any other contribution to promote the field of Bioelectrics by increasing its visibility worldwide and strengthening the knowledge and possible application of pulsed power.

Karl Schoenbach organized and ODU sponsored the First International Symposium on Nonthermal Medical/Biological Treatments Using Electromagnetic Fields and Ionized Gases or ElectroMed '99 in Norfolk Virginia. I organized and EVMS sponsored the second meeting called ElectroMed 2001 in Portsmouth, Virginia. Michael Murphy of the U.S. Air Force Research Laboratories sponsored ElectroMed 2003 held in San Antonia, Texas. The last meeting, ElectroMed 2005, was held in Portland, Oregon, and sponsored by MicroEnergy Technologies. The ElectroMed meetings focused on research into nonthermal biomedical effects of electromagnetic fields and ionized gases and generated great interest in electrical pulses with duration down to one billionth of a second with voltages exceeding ten thousand volts. These meetings also showcased the idea of utilizing electrical interactions with biological cells without heating such that high-frequency components in the usEPs provided a pathway to the interior of cells. Another technological area that rapidly developed after the first ElectroMed meeting was the generation of cold ionized gases. These cold plasmas are charged particles and radicals that have applications in bacterial and chemical decontamination.

Dr. Bob Barker was the real founder of ElectroMed and the driving force, supporter, and benefactor of using usEPs on their own and to generate ionized gases as another developing technology. Bob championed this technology, working closely with Karl, Steve Buescher, and me over most of the rest of Dr. Barker's life, which ended in his courageous fight against cancer in 2013. He was intrigued that pulsed power technology, which he knew was used in radar, particle accelerators, and high-powered physics, could be used to treat cancer. We would have been even more thrilled if he had lived long enough for us to show that pulsed power could also induce an immune response. It was Bob's curiosity, intuition, insight, and foresight as well as his trust in Karl Schoenbach that sustained the direction for the use of ultrashort pulses and ionized gasses that continues today in the FRRCBE.

After the ElectroMed conference series, attendees of these meetings later attended meetings of the Bioelectromagnetics Society and/or symposia of the Bioelectrochemistry and Bioenergetics Society. In September 2015, the first World Congress on Electroporation and Pulsed Electric Fields in Biology, Medicine and Food & Environmental Technologies was held in Portoroz, Slovenia. The second Congress was held in Norfolk, Virginia in September 2017 and the third Congress took place in Toulouse, France in September 2019. The 4th World Congress will be held in Copenhagen, Denmark in September 2021. These meetings focus on basic research and developing applications based on electroporation and the use of pulsed electric fields of high intensity.

While the earliest papers focused on bacteria and small organisms, work had already begun on effects of nanosecond pulses on mammalian cells. The first paper is generally considered the seminal paper for this new use of pulsed power. It ushered in new and unique applications of electric fields expanding concepts of Bioelectrics, especially intracellular effects. In a sense, nanosecond pulses were considered an extension of conventional electroporation, which used much longer pulses in the micro- to millisecond range (Schoenbach et al., 2001). This paper and two that followed presented experimental data that supported the hypothesis that ultrashort pulses with short rise and fall times could affect intracellular structures. Several manuscripts that followed provided significant information for the base patent and several others that followed here at ODU and several elsewhere provided the bases for the spin-off company called Pulse Biosciences, Inc. (PLSE on the NASDAQ). This company is now taking pulsed power technology to clinical applications in humans for benign skin diseases as well as skin cancers.

The first paper on human cells that demonstrated intracellular effects of usEPs and led the way for the development of this technology had a notable story. It involved a bit of serendipity or at least a bit of a surprise in how the findings came about. While studies with bacteria were ongoing, I was working with Dr. E. Stephen Buescher, a pediatric physician and an active researchers in the Center for Pediatric Research (CPR). The CPR was a research program in the Department of Pediatrics in the Children's Hospital of the King's Daughters, which was associated with EVMS. Steve was not only a practicing physician, but also an active researcher. Steve had studied medicine and did a pediatric residency at Johns Hopkins before serving as a Clinical Associate and Medical Staff Fellowship in an infectious disease program at the NIH. He also practiced in a program of infectious diseases at the University of Texas, Houston, before coming to Children's Hospital of the King's Daughters (CHKD) and EVMS, where he practiced clinical pediatric infectious diseases. When I talked to Steve about this project and when Karl briefed Steve on some of the specifics of pulsed power technology, especially how it affected intracellular structures and functions, Steve was onboard with a unique enthusiasm that only he could display. In addition to his clinical duties, Steve was an expert in human blood phago-cytic cells with an active and imaginative intellect and excellent skills in micros-copy, among others. Steve and I were working on a project with a PhD student and resident physicians working with Steve using human neutrophils and characterizing how cAMP and PKA modulated apoptosis in neutrophils (Parvathenani et al., 1998). So, these neutrophil preparations were an obvious starting point for investigating cell responses to nanosecond pulses.

Jingdong Deng, one of Karl's students, built a pulser on Steve's microscope, and Steve installed stainless steel electrodes to study effects of nanosecond pulses on human neutrophil preparations with 60 ns pulses. Thus, effects could be observed in real time. Steve's technician, Ms. Pamela Hair, was observing nanosecond pulse effects of these neutrophil preparations isolated from healthy human volunteers. The neutrophils were loaded with the dye calcein, which crosses the cell membrane but is then trapped inside live cells. In addition, because of its membrane impermeability, it is excluded from intracellular membranous structures. Pam noticed that, when cells were pulsed, there was a subset of cells in this preparation that lit up or sparkled. She called them sparkler cells. Upon closer examination, these sparkler cells were not neutrophils, but eosinophils that were incompletely separated from neutrophils. So, the major effects observed in this study came from cells that contaminated our neutrophil preparations. In subsequent studies, enriched eosinophils were pre-pared and studied in response to three and five 60 ns pulses with electric fields of 36 and 53 kV/cm (3.6 and 5.3 MV/m). Eosinophils contained relatively large gran-ules that contained cationic proteins that bound the highly anionic calcein. When nanosecond pulses breached the intracellular granule membranes, calcein exhibited enhanced fluorescence. When exposed to these 60 ns pulses, 60%–84% of cells were classified as sparkler cells. These eosinophil preparations contained 13%–35% neu-trophils, which did not show sparkler characteristics. Thus, it was the presence of the anionic proteins in eosinophils that allowed enhanced fluorescence as calcein crossed the intracellular granular membranes. However, this did not mean that neu-trophils were not affected by these pulses but meant that their granules were smaller or less susceptible to sparkler characteristics. The sparkler cells were shrunken com-pared to control or nonsparkler cells (Schoenbach et al., 2001), which is common to cells undergoing apoptosis. However, this was not sufficient evidence that these cells were undergoing apoptosis. However, this was suspected because nanosecond pulses were inducing apoptosis in other cells in ongoing studies.

The findings presented in this paper led to new considerations for how electric fields could be used to treat biological cells (Schoenbach et al., 2002). New applica-tions for pulsed power were contemplated that led to this new technology that could control membrane transport processes and other functions in biological cells. These ideas were more specifically reported in two subsequent papers that built on the pre-vious reports (Schoenbach et al., 2001) showing how intracellular effects could have cytotoxic effects on biological cells. However, these cytotoxic effects had a new twist to them (Beebe et al., 2002, 2003).

The first three manuscripts using mammalian cells (Schoenbach et al., 2001; Beebe et al., 2002; 2003) provided the fundamental material needed to secure the base pat-ent of the pulsed power technology developed by ODU and EVMS using nanosecond electric pulses to kill cancer. They also provided the foundation for what was to become the FRRCBE. While these studies provided some important experimental evidence for one of the major hypotheses related to pulsed power technology on biological cells, we also made some predictions, many of which came to be substantiated by experimental evidence. The first one was related to induction to apoptosis. This was especially important at the time because it was the first known mechanism of genetically regulated or programmed cell death (Kerr et al., 1972; Kroemer et al., 2005). Also, cancers were known to intercept and block apoptosis mechanisms as a means to survive (Hanahan and Weinberg, 2000, 2011). Many of these mechanisms that evade apoptosis are centered on the mitochondria. We also suggested that perturbations of mitochondria could be responsible for apoptosis induction, as indicated by release of cytochrome c into the cytoplasm as evidence for intrinsic apoptosis (Beebe et al., 2003; Ren et al., 2012). We also suggested the possibility for locally killing tumor cells, papillomas and nevi by apoptosis. Cancer treatment turned out to be the most investigated application for nanosecond pulses and the first one that is presently being investigated in clinical trials for basal cell carcinoma. More on this is discussed below. We also proposed the possibility for sculpting tissues. As I will discuss later, there were efforts by a company called Cellutions as a division of the Innovation Company that invested in this technology to reduce cellulite and adipose tissue. We also suggested that, in conjunction with classical electroporation, USPEF could be used to modify nuclear, mitochondrial, or other intracellular membranes to enhance the delivery of genes or drugs for therapeutic purposes. This idea was shown to have credence (Guo et al., 2014). Finally, results from this work and others discussed below led to the base patent that was fundamental to take this technology to successful nationwide clinical trials for benign skin conditions and more recently national clinical trials for cancer treatment (Nuccitelli, 2019).

Demonstrating that USPEF could kill bacteria and modify intracellular vesicles in human eosinophils raised the possibility that this same technology, called nanosecond pulsed electric fields (nsPEFs), could induce apoptosis in unwanted cells (Beebe et al., 2002). In the mid to late 1990s, the scientific community had caught up with a finding from Kerr and colleagues that cells could die by programmed mechanisms (Kerr et al., 1972). It was obvious that cells died naturally as a counterbalance to proliferation, but cell death was considered to occur by "autolysis", and physiological cell death was called "necrobiosis". Alternatively, cells could die by noxious stimuli such that cellular homoeostatic mechanisms were irreversibly interrupted, which appear as coagulative necrosis to histologists viewing cellular corpses. However, these processes had not been characterized. Kerr and coworkers realized that cell death was not instantaneous and that the kinetics of the cell death process were genetically programmed and had significant importance whereby they called it "apoptosis". Apoptosis was from Greek to describe the "dropping-off" or "falling-off" of petals from flowers or leaves from trees. While we know now that there are many different cell death mechanisms, apoptosis versus necrosis, which just means dead, was all that was known about cell death. Because we had already been characterizing apoptotic cell death in cAMP analog-stimulated neutrophils and had developed a caspase assay, which was not yet commercially available, we used these and other apoptosis markers to determine if nsPEFs induced apoptosis in

Jurkat and HL-60 cells, which were already used in other studies. In these studies, human tumor cell suspensions and mouse tumor tissues were exposed to nsPEFs with durations of 10, 60, and 300 ns and electric fields ≤300 kV/cm. The underlying molecular mechanisms of how nsPEFs actually caused cancer cells to die were still unknown. It was shown these pulses initiate cellular responses that are distinctly different than responses to electroporation pulses.

Additional steps were made that clearly showed that nsPEFs induced apoptosis in Jurkat cells and did it by intrinsic mechanisms – that is from the inside. One of those was induction of cytochrome c release from mitochondria into the cytoplasm (Beebe et al., 2003). Like cell permeabilization and caspase activation in the previous paper (Beebe et al., 2002) and repeated in this paper, nsPEF induced cytochrome c release and annexin-V externalization, another apoptosis marker. All of these cell responses were independent of energy and directly proportional to charging effects. When conditions were 10, 60, and 300 ns durations at 150, 60, and 26 kV/cm electric fields (1–2 J/cc), respectively, caspase activation and cytochrome c release were greater as the pulse duration increased, indicating that these nsPEF-induced responses were energy independent. Using a scaling law that later defined effects of nsPEFs as charging effects expressed in Vs/cm {electric field (E as V/cm) × pulse duration (τ in seconds) × pulse number (n) or $E\tau n^{1/2}$} (Schoenbach et al., 2009), caspase activation and cytochrome c release were directly proportional to the Vs/cm.

That mitochondria were affected was consistent with nsPEF affecting intracellular structures and functions. Cytochrome c release from mitochondria indicated a point of no return in apoptotic cell death progression (Green and Amarante-Mendes, 1998). This was the point at which cells resigned themselves to undergo regulated cell death, in this case by apoptosis. This no-return commitment was orchestrated by mitochondria. Since apoptosis was a natural cell death mechanism in most cells, one of the successful mechanisms of cancer was to interfere with apoptotic cell death activity, especially in mitochondria, as means for survival. That apoptosis could be induced in cancer cells and that mitochondria could be affected, an organelle that controlled life and death, indicated an important ability of nsPEFs to influence the cell death machinery through a point-of-no-return and demonstrated a critical target for cancer therapy.

The finding that nsPEFs induced apoptosis was quite interesting. My lab had been working with human Jurkat and HL-60 cells as rather common cell lines in the lab for several years. Both were commonly used, so each had a rather rich history of characterization. Of course, we were looking for apoptosis, because that was the only cell death mechanism known other than frank necrosis, which is now called "accidental cell death" (ACD) and is generally opposed to apoptosis or programmed/regulated cell death (Kroemer et al., 2005). It was fortuitous that we chose Jurkat and HL-60 cells for these studies. Subsequently, others have also shown that Jurkat and HL-60 cells undergo apoptosis in response to nsPEFs (Morotomi-Yano et al., 2014). Most other cells that we and others have assayed for caspase activity are negative. Our present *in vivo* tumor models of rat N1-S1 hepatocellular carcinoma (HCC) and mouse 4T1-luc breast cancer models do not undergo apoptosis in response to nsPEFs (Beebe et al., 2018). Others have shown that nsPEF-induced cell death is dependent on cell type (Morotomi-Yano et al., 2014), culture (Pakhomova et al., 2013;

Morotomi-Yano et al., 2014), and electric field even in Jurkat cells (Ren et al., 2012). While we showed apoptosis in melanoma cells (Ford et al., 2010), this was not confirmed (Rossi et al., 2019). We had also shown apoptosis in the tumor microenvironment in animal tumor models (Chen et al., 2010, 2012; Chen et al., 2014). However, it was not determined that these were tumor cells undergoing apoptosis as opposed to other host cells. While it was clear that nsPEFs could induce apoptosis *in vivo*, it was not clear that these apoptotic cells were tumor cells. Thus for cell types in culture, it was by chance that the two cell types we chose to determine nsPEF-induced apoptosis actually responded to nsPEF by apoptosis induction. It appears that many cell types do not undergo apoptosis in response to nanosecond pulses.

At the turn of the century, understanding the effects of nanosecond pulses on cells began to escalate. Karl's engineering lab at ODU and my cell/molecular signal transduction lab at EVMS were joined by laboratories of Steve Buescher, Mike Stacey, and Peter Blackmore at EVMS. Each group utilized their own unique expertise working on individual projects all connected to providing mechanistic understanding of how nanosecond pulses interacted with cellular structures and functions and how this technology might find a value in the marketplace. This progress was made easier since we had received several grants with Karl as principle investigator from the AFOSR. The ElectroMed conferences in 1999 and 2001 had created some significant excitement around the world about new applications for this technology. In 2002, we received a $5 million MURI entitled "Subcellular Responses to Narrowband and Wideband Radiofrequency Radiation from AFOSR" with Bob Barker at the helm. Karl was the overall principle investigator. This was in cooperation with 11 co-principle investigators from ODU, EVMS, MIT/Harvard, the University of Wisconsin, the University of Texas Health Science Center, and Washington University. Dr. Barker's contribution of federal support from the MURI provided partnerships with other universities, while applications for patents provided more of a commitment for transforming research to medical applications. In 2002, I also received a $250,000 grant from the American Cancer Society for this project as a possible cancer treatment. In addition to Bob's support, there were funds from the Whitaker Foundation and the Center for Innovative Technology to support ElectroMed. Both ODU and EVMS were especially engaged in the project, each school providing significant funding support. By 2003, about $9M had come to support projects using pulsed power with nanosecond pulses and ionized gases. EVMS, which was actively involved in community outreach as a private institution, was active in getting spots for Karl and me for interviews on local news stations, talk shows, and newspaper articles. It was an exciting time to know that the university was reaching out to lay people who were becoming interested in the work that was going on in Norfolk, Virginia. By this time, the major new stories were related to possibilities for treating cancer. While this was clearly becoming a real possibility, it had only been shown in a small preclinical study in mice. But the best was yet to come.

By 2000–2001, ODU and EVMS began considering a joint venture to begin a Center for Bioelectrics. The president of ODU, Dr. Roseann Runte, provided Dr. Schoenbach the right to form a Center for Bioelectrics. This was a departure from the general academic approach since the Center was separate from the Department of Electrical and Computer Engineering, where Karl held an appointment.

ODU and EVMS established a memorandum of understanding (MOU) and began finding a "home" for this technology. EVMS applied and received two grants from the Department of Housing and Urban Development (HUD), and the space on the fifth floor of Norfolk's Public Health Building was allocated to house the Center. In essence, everything that establishes a center was present, including a building. Our research was certainly interdisciplinary involving faculty, students, and staff from various academic departments. There were engineers, physicists, biologists, a physician, and Frank Reidy working and learning together. We were conducting innovative basic research with a unique stimulus to understand fundamental mechanisms of pulsed power effects on biological cells and how it could be used for environmental and medical applications. The final touch to establish a Center for Bioelectrics was a $5 million donation from Mr. Frank Reidy, and the Center was thus christened the Frank Reidy Research Center for Bioelectrics.

The intrigue of the power of nsPEFs to modify or target internal cellular organelles such as mitochondria suggested its capacity to manipulate other organelles. This interest spread to the nucleus very early considering what nsPEFs could do to mammalian and human cells. Since the nucleus was the largest organelle in the cell and had a double cellular membrane, with a more permeable outer membrane, it seemed possible that electric fields affecting intracellular structures could modify the nucleus and/or DNA. A strand of double helix DNA surrounded by a perforated membrane with a bipolar pulse passing through became the logo for this center.

Dr. Mike Stacey, another colleague in the Center for Pediatric Research at EVMS, became interested in this project. Mike had come to EVMS as a postdoctoral fellow to work with another faculty member and served as an assistant professor in the Center for Pediatric Research. Mike had a solid background in genetics, so the nucleus was a reasonable scientific home for him. Mike designed studies to examine effects of nsPEF on DNA structure and cell survival on 11 cell lines of different origins and growth characteristics. He included both cells grown in suspension and adherent growing cells and included four human cell lines with different genetic instability syndromes that were characterized by defective DNA repair and checkpoint control mechanisms, hypersensitivity to various genotoxic agents, syndromes showing increased incidence of cancers (Stacey et al., 2003).

There were a number of interesting findings in this study. In general, cells grown in suspension were found to be much more vulnerable to nsPEFs treatment than adherent cells. Adherent cells have different membrane structures and a more extensive cytoskeletal structure that may be able to distribute capacitances that prevent electric field build-up. This vulnerability appeared to be related to DNA damage, which was found to be immediate in Jurkat cells as determined by the comet assay. This was done in a way to eliminate the occurrence of downstream DNA damage response. While nsPEFs exert impressively high power, it was questioned whether that power could damage DNA. DNA damage could be observed as chromosomal aberrations. However, the damage was not as great as that in these cells exposed to ionizing radiation. This suggested that the mechanisms of cell response to damage may be different than ionizing radiation. Cells treated with nsPEF had a lower mitotic index, and these were cells that showed decreased cell survival, suggesting that surviving cells accumulated in the cell cycle at a point before mitosis, presumably to repair damaged DNA before resuming the cell cycle. Surprisingly, it took surviving cells 1 week to re-enter mitosis after exposure to nsPEFs. Cells not showing significant decreased cell survival, primarily adherent cells, did not show significant decrease in mitotic indices, suggesting that the cells had not been damaged sufficiently by nsPEF to activate cell cycle checkpoints. Exposing cells that survived a previous exposure demonstrated similar cell death profiles as the initial treatment, indicating that effects induced by nsPEF may not be completely reversible.

Given that nanosecond pulses affected intracellular structures and functions, calcium became a molecule of interest since it is a multipurpose molecule, stored in and transported out of the endoplasmic reticulum (ER), transported across the plasma membrane, and modulated by mitochondria, among other activities. Dr. Buescher and Dr. Peter Blackmore, associate professors in Physiological Sciences were both interested in effects of nanosecond pulses on cellular calcium. Peter was a Howard Hughes associate when I was a postdoctoral fellow at Vanderbilt. We arrived at EVMS about the same time, me from Norway after leaving Vanderbilt (Beebee et al., 1985) and he directly from Vanderbilt. Peter had investigated calcium regulation in hepatocytes for years. We had published several papers together on hormone regulation at Vanderbilt, and we were collaborating on calcium mobilization in human sperm (Blackmore et al., 1990). My discussion with Peter about nanosecond pulses interested him, and we began looking at calcium mobilization in cells in response to nanosecond pulses.

Steve's approach primarily measured effects of submicrosecond intense pulsed electric fields (sm/i-PEFs) on human neutrophils using 60 and 300 ns pulses with electric fields 12–60 kV/cm (Buescher et al., 2004). These studies were carried out under the microscope using Fluo-3 as a calcium indicator. The study demonstrated some basic calcium-mobilizing concepts showing mobilization of calcium released from intracellular stores as well as across the plasma membrane. It also showed that an initial calcium response to sm/i-PEFs desensitized responses to a second application. Under the conditions tested, sm/i-PEF had no effects on neutrophil phagocytosis. However, sm/i-PEF suppressed a neutrophil's spontaneous H_2O_2 production and chemotaxis.

Funded by Dr. Barker's MURI, Jody White, an EVMS PhD student with Dr. Blackmore and me, carried out calcium mobilization studies in HL-60 cells in response to nanosecond pulsed electric fields (nsPEFs) using real-time fluorescent microscopy with Fluo-3 and fluorometry with Fura-2 (White et al., 2004). nsPEFs induced a two- to sevenfold increase in intracellular calcium (60 ns, 4–15 kV/cm) without an increase in PI, suggesting calcium release from intracellular stores. She found that nsPEFs and uridine triphosphate (UTP), a purinergic agonist, mobilized calcium irrespective of whether extracellular calcium was present or not. UTP targets the (inositol trisphosphate) IP_3 channels, which releases calcium from the ER. It was later shown that nsPEFs induce phosphatidylinositol 4,5-bisphosphate (PIP$_2$) hydrolysis or depletion from the plasma membrane and an accumulation of IP_3 in the cytoplasm (Tolstykh et al., 2013). Since PI did not cross the plasma membrane, calcium influx was suggested to be due to capacitative calcium entry (CCE). CCE occurs when increases in intracellular calcium release stimulate opening of store-operated channels in the plasma membrane allowing influx of calcium into the cell for replenishment of the internal stores. However, as discovered later by Dr. Andrei Pakhomov's group, investigators in the FRRCBE, nsEPs induced voltage-sensitive, inward-rectifying membrane pores with a maximum size of about 1 nm, which were impermeable to PI (Pakhomov et al., 2009). Using a different approach, Dr. P. Thomas Vernier' group reached the same conclusions (Vernier et al., 2006), before he joined the FRRCBE in 2010. Experimental observations and molecular dynamics (MD) simulations indicated with the nanoelectropulse-induced electric potential across the lipid bilayer that there is a tight association between phosphatidylserine (PS) externalization and membrane pore formation. That is, the pulse appears to drive externalization of PS through pores in the plasma membrane during pulse generation.

In a later study carried out in the FRRCBE by Dr. Iurii Semenov and colleagues in Pakhomov's group, nsPEF-induced calcium mobilization appeared differently in Chinese hamster ovary (CHO) cells, which lack voltage-gated calcium channels. Results showed that nsPEFs induced calcium influx and/or calcium release from the ER, likely by formation of nanopores in the plasma membrane and/or ER, respectively. Once a critical threshold was reached, a positive feedback mechanism called calcium-induced calcium release amplified the response with enhanced mobilization of intracellular calcium (Semenov et al., 2013b).

By 2004, with the increase in activities, new findings about nanosecond pulse affects biological cells and how they may be used, ODU began to strengthen their approaches to patenting. Dr. Mohammad Karim, a well-recognized scientist and researcher in Applied Optics, took the position of Vice President of Research at ODU. He hired Zohir Handy to direct the ODU Office of Technology Licensing who began to educate the ODU faculty about intellectual property and licensing. Zohir's background in Engineering and Business Administration was an asset for developing and advancing patenting for Bioelectrics. With the strong support of President Roseann Runte and Dr. Karim, the Center for Bioelectrics gained stature in the university and the community. Zohir established a strong working relationship with Dr. Bob Williams, who headed the Office of Technology Transfer at EVMS, and the two universities shared an expanding patenting portfolio.

In 2004, the Innovation Factory licensed intellectual property for ODU/EVMS patents and formed a company called Cellutions to use pulsed power and nanosecond pulses to develop proprietary medical devices and procedures to treat cellulite, a condition that affects 90% of postadolescent females. The strategy was to create a proprietary technology and/or procedure that could be performed in an outpatient setting as a minimally invasive "lunchtime" treatment. The company raised nearly $14 million as a proof of concept for the feasibility in humans. Adipose tissue in Zucker rats was treated with nanosecond pulses. Although no studies were published for proprietary reasons, it was shown that nanosecond pulses could induce apoptosis in adipose tissue. In order to determine effects on the skin, trials were carried out where adipose tissue was treated with nanosecond pulses in the mid-gut region of women who were going to have abdominoplasty (tummy tuck). At various times after nanosecond pulse treatment, abdominoplasty was performed, and the treated tissues were analyzed. These studies indicated that within 4–6 weeks after treatment, there was not scarring. Skin discoloration generally returned to normal in 6–8 weeks. Clinical trials were run and showed that there was a reduction in cellulite in human adipose tissue, resulting in an "orange-peel" effect typical of cellulite, which was satisfactory to 60%–70% of patients. Local anesthetic was applied before pulsing with no pain to or complaints from the patients. Unfortunately, additional rounds of funding were not successful, at least in part, because of the financial recession around 2008.

As work rapidly progressed during the MURI grant, we gained a much greater understanding of what and how the nanosecond pulses affected cells. My group continued to determine mechanisms of action. It was clear that there were a number of targets for these nanosecond pulses including the plasma membrane, ER, nucleus, mitochondria, and signaling pathways leading to apoptosis. While several studies indicated that nsPEFs would act like hormones on the outside of the cells and modify functions inside the cells, such as calcium mobilization, perhaps the greatest interest was to capitalize on nsPEF potential to induce cell death, especially cancer cell death. The first study to determine if nsPEFs could eliminate or slow the growth of ectopic tumors was in mice with B10.2 fibrosarcoma tumors (Beebe et al., 2002). In one study, tumor tissue was removed, minced, and assayed for caspase-3 activity, which was present, indicating cells within the tumor microenvironment underwent apoptosis. We also treated the tumors directly with nsPEFs. In retrospect, we were concerned about deleterious effects on mice; this was the first time such high power was applied to living mammalian organisms. We now know the tumor treatment conditions were considerably deficient. The pulse durations were 300 ns and electric fields at 75 kV/cm. Two needle electrodes were used, and only two treatments over 8 days included five or seven pulses each. Now we know that optimal conditions for most tumors include 100 or 300 ns durations, electric fields at 50 kV/cm, and as many as 1,000 pulses for a single treatment (Chen et al., 2014; Lassiter et al., 2018; Guo et al., 2018) or 300 pulses on three alternate days (Chen et al., 2012). In any case, in this initial study, tumor sizes were reduced by 60% in terms of size and weight. The treatment had no ill effects on the mice as they recovered quickly from anesthesia.

Now that there was a physical Center for Bioelectrics, a number of new scientists joined from ODU's Department of Biology. In addition, Dr. Runte funded several student scholarships for Frank and Karl to fill new positions. The MURI also funded several students, and all received master's or PhD degrees based on their work with nanosecond pulses. By then, Karl had new positions to fill. Karl telephoned Dr. Richard Nuccitelli to see if he had any postdoctoral fellows that would like to apply for the position. Rich had a company called RPN Research that was funded by a Small Business Innovation Research (SBIR) grant investigating electric fields in wounds. Rich was interested in endogenous ionic currents in cells and tissues and ionic regulation of cell activation. Rich told Karl that he himself would like to apply and took the position soon after. Rich maintained his company but changed its name to BioElectroMed Corp. Rich led a team of 10–12 scientists to conduct the first real preclinical trial for nsPEF treatment of tumors in mice (Nuccitelli et al., 2006). The study demonstrated an effective localized tumor ablation treatment that would shrink melanoma tumors in mice with two treatments using 300 ns pulse durations and electric fields that were >20 kV/cm. A later study showed that 300 pulses with 300 ns durations and 40 kV/cm could completely eliminate tumors in 47 days (Nuccitelli et al., 2009). Melanoma treatments were optimized with 2,000 pulses with 100 ns durations and 30 kV/cm (Nuccitelli et al., 2010).

In 2005, Dr. Andrei Pakhomov arrived at the FRRCBE and added patch-clamp and electrophysiology techniques that enhanced our understanding of pulse power effects on eukaryotic cell plasma membranes. His first study combined patch-clamp approaches with nsPEFs. In 2009, Dr. Olga Pakhomova joined the Center and enhanced the molecular biology component for realizing how nsPEFs induce their effects, including cell death. She has an MS in Microbiology from Moscow State University and a PhD in Biophysics/Radiation Biology from the Russian Academy of Medical Sciences. Olga's knowledge in molecular biology and Andrei's expertise in electrophysiology have provided an excellent complement or balance of knowledge and skills that has greatly enhanced our awareness and familiarity of nsPEF effects on cell membrane ion transport, calcium mobilization, regulated cell death, and cell-to-cell communication.

Drs. Andrei Pakhomov and Olga Pakhomova brought a new dimension to the studies of nanosecond pulse technology using electrophysiological methodologies including patched clamp studies and molecular cell biology, respectively. Andrei's earlier studies had been on effects of microwaves and millimeter waves on nerves and heart tissues (Pakhomov et al., 1998, 2000). The early focus was on intracellular effects of nanosecond pulses being different from effects of conventional electroporation. However, Andrei's favorite landscape was the plasma membrane and membranes close by. He prolifically and completely investigated the physical phenomena of nanosecond pulsed effects on cell membranes and their protein channels and, in many cases, compared them to conventional electroporation technology, thus extending what had been done on "conventional" electroporation decades ago.

One of his earliest contributions was the use of patch-clamp methods for the first time with nanosecond pulses to show that, when cells were pulsed with a single 60 ns 12 kV/cm pulse, there was a decrease in plasma membrane resistance and loss of the plasma membrane potential without uptake of PI. This suggested that the plasma membrane had been permeabilized with relatively long-lasting pores smaller than PI (Pakhomov et al., 2007). Andrei's group also found these nanopores were actually channel-like as voltage-sensitive and inward-rectifying membrane pores, mostly impermeable to PI. This suggested that pore sizes were no bigger than about 1 nm. The ion-channel-like properties were specific to these nanopores but disappeared if they broke into larger, conventional electroporation pores permeable to PI (Pakhomov et al., 2009). They also used a novel approach to identify nanopores by loading cells with a thallium Tl(+)-sensitive fluorophore FluxOR. In so doing, they were able to show uptake of thallium without uptake of larger dyes, demonstrating in yet another way that nanopores were <1–1.5 nm. Furthermore, thallium uptake was not sensitive to ion channel blockers or chelation of calcium, so these nanopores were not protein channels (Bowman et al., 2010).

Using whole patch-clamp techniques, Andrei and his collaborators also began to look at the plasma membrane much more closely providing more comprehensive analysis of nanosecond pulses on membranes not only on lipid nanopores, but also on protein channels. Andrei's electrophysiological skills should not be underestimated because nanosecond pulse effects on lipid nanopores and protein channels were overlapping. Nevertheless, with his physiological prowess and methodologies, it was possible to resolve their different properties. They found that nanosecond pulses inhibited voltage-gated (VG) Na$^+$ and Ca^{2+} channels with concurrent leak channels that were most likely nanopores or larger permeabilization pores. Considering differences in their responses with electric field magnitude and kinetics after exposure, leak currents and VG ion channels were independent of one another (Nesin et al., 2012). An investigation into the mechanism(s) of the ion channel inhibition considered and ruled out two possibilities. They showed that inhibition was not due to a decrease in the membrane Na+ gradient caused by influx through nanosecond-pulse-induced nanopores. Furthermore, the inhibition was not due to Ca^{2+}-dependent downregulation of the VG channels. They suggested a Ca^{2+}-independent downregulation of the VG Na$^+$ channel caused by effects on lipids in the membrane or a direct effect on channels themselves (Nesin and Pakhomov, 2012).

In another work, Andrei's group did go intracellularly to investigate calcium mobilization, not only influx across the plasma membrane, but also release from the ER. In a study that supported the hypothesis that shorter pulses have greater intracellular effects, Dr. Iurii Semenov in Andrei's group and his coworkers showed that 10 ns pulses induced greater calcium release from the ER than 60 ns pulses, which released more calcium from the ER than 300 ns pulses (Semenov et al., 2013a). They also demonstrated that the nanosecond-pulse-induced calcium mobilization was different in CHO cells, which lacked VG calcium channels. Results showed that nsPEFs induced calcium influx and/or calcium release from the ER, likely by formation of nanopores in the plasma membrane and/or ER, respectively. Once a critical threshold was reached, a positive feedback mechanism called calcium-induced calcium release amplified the response by enhancing the mobilization of intracellular calcium (Semenov et al., 2013b). This demonstrated direct electric field effects and biological responses to them.

In his continued collaboration with Dr. Olga Pakhomova, Andrei exploited a novel approach to attach cells to glass coverslips coated with indium tin oxide (ITO), which allowed fast buffer exchanges after treatments. Olga found that cells pulsed with 300 ns pulses, which gradually began to take up PI and YO-PRO-1, exhibited a considerable uptake of PI in the presence of calcium (Pakhomova et al., 2014). They later showed that the expression of VG Ca^{2+} channels (VGCCs) in HEK293 cells, compared to those cells without VGCCs, makes them much more susceptible to permeabilization by nanosecond pulses. Interestingly, this occurred at membrane potentials below those for opening VGCCs. At this point, it appears they began to think that nanosecond pulses might have more severe effects on VGCCs (Hristov et al., 2018).

Olga and her coworkers also demonstrated that cells could be sensitized to nanosecond pulses. They found that lethal membrane permeabilization could be enhanced severalfold by splitting a high-rate long exposure into two fractions separated by 1–5 min. In this way, sensitization could possibly reduce electric fields and/or the pulse number needed for tumor ablation therapy (Pakhomova et al., 2011). Dr. Claudia Muratori, a new FRRCBE faculty member, showed that the ablation volume of a squamous carcinoma cell 3D culture was increased by splitting the treatment into two fractions (Muratori et al., 2016) and the tumor volume in squamous carcinoma tumors was reduced in mice (Muratori et al., 2017).

In collaboration with Dr. Bennett Ibey's group in San Antonio, Pakhomov group's most recent mission took them to the bipolarity of nanosecond pulses. They found that nanosecond bipolar pulses were distinctly different than bipolar conventional millisecond pulses. Compared to monopolar pulses, bipolar pulses significantly reduced effects of Ca^{2+} mobilization, PI uptake, annexin-V binding, and cell survival; this was in addition to delivering twice the energy. The cancellation effect decreased with an increase in delay between the two bipolar pulses (Ibey et al., 2014; Pakhomov et al., 2014, 2018). This was unexpected since cancellation was not predicted by equivalent circuit, transport lattice, and MD modeling. They invoked mechanisms of assisted membrane discharge, two-step membrane lipid oxidation, and reverse transmembrane ion transport. Going back to a previous studies, Andrei pithed frogs for their sciatic nerves and his group found that some of the characteristics of bipolar pulses were different than those observed with other cell types. These studies are changing the paradigm of nanosecond pulse mechanisms on cell membranes and cell functions. You'll need to read about it (Schoenbach et al., 2015; Merla et al., 2017; Casciola et al., 2019).

In 2007, Rich Nuccitelli left ODU and moved BioElectroMed to Burlingame CA, where he ultimately licensed the ODU/EVMS intellectual property and continued to work on the effects of nsPEFs on skin and pancreatic cancer. He developed an endoscopic, ultrasound guided system called EndoPulse and continued nanosecond pulse research with SBIRs for pancreatic cancer. Rich and his collaborators continued to develop uses for nanosecond pulses and continued to publish papers that optimized nanoelectroablation, as he called the treatment, for treating melanoma (Nuccitelli et al., 2010, 2012a), basal cell carcinoma (Nuccitelli et al., 2012b), including the first clinical trial (Nuccitelli et al., 2014) and human pancreatic cancer (Nuccitelli et al., 2013). Rich's work took nanoelectroablation closer to clinical applications.

While I had been part of the concept of the Center for Bioelectrics and was considered a member of the Center, I was on the EVMS faculty until 2007. However, the concept of the Center was based on Karl Schoenbach's ODU pulsed power technology. Around 2006–2007, ODU wanted to be in complete control of the technology. Dr. Karim recruited Mike Stacey and me from EVMS, and we both joined the ODU faculty, first Mike and then me. Advanced changes were in the air.

In 2008, Karl was ready for a change and wanted to step down as the Director of the FRRCBE. In a national search, Dr. Karim recruited Dr. Richard Heller to be the Director of the Center. He also recruited Dr. Loree Heller into the faculty. While Richard and Loree worked closely together, each of them brought their own specific expertise to the Center. Karl continued to work in the Center but no longer wanted the demands put upon the Director. While Karl established the Center and built a strong foundation for its success, with the support of Dr. Karim and President Runte, Richard Heller significantly expanded the Center and made it the stalwart in Bioelectrics it is today. First, he and Loree brought the classical electric field technology of electroporation to ODU. The FRRCBE now included a full range of electric field investigations from the nanosecond, and even picosecond, to millisecond time domains. This also expanded the possible applications to include new approaches to tumor treatment and the delivery of drugs and genes to cells for therapeutic purposes. Richard also negotiated to improve the infrastructure of the Center, significantly enhancing the technological potential of the research there. In 2010, FRRCBE moved into a new building on the eastern side of the ODU campus in the new ODU village. Most importantly, this new building housed a state-of-the-art animal facility, now making it much more convenient to conduct animal studies that were needed to develop the nsPEF technology for cancer treatment as well as to support gene delivery studies in animals for Richard's studies. Richard had a distinctive bioelectric history that he brought and added to the FRRCBE.

While Karl, Frank, and I with the rest of the Bioelectrics nucleus in Norfolk were establishing uses of nanosecond pulses in the late 1990s, Richard and his colleagues at the University of South Florida were continuing to investigate and improve uses of micro- and millisecond pulses in electroporation. We were using pulses with durations between 10 and 300 ns with electric fields as high as 300 kV/cm. Richard and his colleagues were using pulses with durations between 99 μs and 20 ms with electric fields between 0.1 and 1.3 kV/cm. While we were just beginning to address the physical phenomena of nanosecond pulse effects on cells *in vitro*, electroporation technology had begun addressing these *in vitro* issues over two decades ago. While we were treating bacteria, brine shrimp, and mammalian cells and just beginning to treat animals, Richard's group had been treating animals and had begun clinical trials for human cancers. The Norfolk and South Florida groups were on opposite sides of an "electrical coin". Richard's arrival in Norfolk significantly enhanced the value of that electrical currency.

Application of both long electroporation pulses and short nanosecond pulses superimposes an induced transmembrane potential on a resting membrane potential that causes dielectric breakdown of the plasma membrane to cause permeabilization. However, effects on the plasma membrane can be different depending on the electric field strength. Furthermore, nanosecond pulses affect intracellular structures and their functions. One of the primary applications of nanosecond pulses is to eliminate tumor cells by regulated cell death. For gene or drug delivery by electroporation, the dielectric breakdown of the plasma membrane is meant to be reversible. The group of Richard Heller, Mark Jaroszeski, Richard Gilbert, and Loree Heller significantly advanced reversible electroporation for drug and gene delivery.

The first study demonstrating effects of electroporation in humans, called bleomycin-mediated electrochemotherapy (ECT), was carried out by this group (Glass et al., 1996a). Electroporation was applied after systemic administration of the drug in two patients with basal cell carcinoma. Partial and complete responses were seen in lesions without significant side effects. Heller's group also showed efficacy with ECT after intralesional delivery of bleomycin in five patients with metastatic melanoma. There were complete responses in 78% of the tumors and 17% partial responses. Although ECT was not a cure for melanoma, it was shown to be an effective alternative to palliative surgery or radiation (Glass et al., 1996b). That same year, a phase I/II clinical trial demonstrated ECT safety and efficacy for several cutaneous and subcutaneous tumors including melanoma, basal cell carcinoma, and metastatic adenocarcinoma. There were complete responses, decreases in tumor size, and objective responses showing that ECT was an effective treatment for most of the tumor nodules (Heller et al., 1996a).

Another strategy pioneered by Richard and his colleagues was the delivery of genes by electroporation, which he later called GET or gene electrotransfer. The major strategy was to avoid the possible hazards of viral gene delivery. Gene delivery by electroporation has a number of advantages over viral gene transfer. Unlike electroporation delivery, viral gene transfer can cause significant immunogenicity that induces inflammation, causing deterioration of the affected tissue, possible mutation due to DNA insertions, toxin production, and possible mortality (Gardlík et al., 2005; Katare and Aeri, 2010). However, gene delivery by electroporation is less efficient compared to viral delivery.

In one of the earliest studies, Heller and colleagues electroporated rat liver and delivered luciferase or beta-galactosidase (β-gal), which resulted in β-gal in 30%–40% of cells after 48 h and 5% of that level after 21 days (Heller et al., 1996b). Later studies optimized delivery of IL-12 showing that low-voltage, millisecond pulses were superior to high-voltage microsecond pulses. This established electroporation plasmid DNA delivery as an effective immunotherapy protocol (Lucas and Heller, 2003). Going a step further, he and his colleagues combined ECT with GET. They found that plasmid delivery of IL-2 and/or granulocyte macrophage – colony stimulating factor (GM-CSF) alone did not significantly reduce tumor growth and that ECT resulted in short-term, complete B16 melanoma tumor regression; however, there was no resistance to challenge with live cells. However, the combination resulted in long-term immunity and resistance to challenge in 25% of mice (Heller et al., 2000).

Richard's focus on IL-12 would pay off with a clinical trial by 2008. He and his collaborators delivered IL-12 to mouse skeletal muscle, with systemic expression of the molecule it induces, IFNγ (Lucas and Heller, 2001); into mouse dermis (skin), where IFNγ expression was also increased (Heller et al., 2001); and continued delivery into B16-f10 melanoma tumors. They demonstrated that intratumoral (i.t.) IL-12 was superior to intramuscular (i.m.) IL-12 with half of mice "cured" and 70% of those resistant to challenge injections after i.t. IL-12. IFNγ levels were elevated, and T-cells infiltrated tumors that were less vascular (Lucas et al., 2002). Treatment of a primary tumor resulted in significantly reduced growth of secondary tumors and i.m. IL-12 significantly decreased lung nodules from IV injected B16f10 cells (Lucas et al., 2003). Since IL-12 has been shown to induce severe toxicity, it was important to show that systemic toxicity with IL-12 delivery caused no increases in serum IL-12 levels or systemic side effects (Heller et al., 2006). Near the time Richard and Loree joined the FRRCBE, the first phase 1 dose escalation studies were published (Duad et al., 2008). The study was the first report of gene transfer utilizing *in vivo* DNA electroporation. Posttreatment biopsies showed dose-dependent increases in IL-12 protein, lymphocyte infiltration, and significant tumor necrosis. Ten percent of nineteen patients, or two patients showed complete regression of all lesions, while 42% showed stabilization or partial responses. Importantly there were no significant side effects other than transient pain after electroporation.

Since arriving at the FRRCBE, Richard's group has continued using electroporation for gene transfer for several other applications besides tumor treatment. He and his collaborators have applied gene transfer of vascular endothelial growth factor (VEGF) to promote wound healing and increase circulation to prevent necrosis in skip flaps and hind limb ischemia (Ferraro et al., 2009, 2010; Basu et al., 2014). Not fearing to live on the edge, Richard electroporated the heart to enhance blood flow after ischemia. They demonstrated electroporation-mediated gene transfer to the porcine heart with a plasmid expressing VEGF showing enhanced VEGF protein expression and enhanced perfusion in the ischemic heart (Marshall et al., 2010; Hargrave et al., 2013, 2014).

Dr. Loree Heller and her collaborators in Ljubljana, Slovenia, discovered that B16f10 melanoma tumors in mice completely regressed by the electrotransfer of single- and double-stranded DNA. Involvement of signaling pathways of multiple nucleic acid sensors in the innate immune system was suggested since this occurred in immunocompetent and immunodeficient mice (Heller et al., 2013). This was confirmed by significant increases in IFNβ mRNA and protein levels in tumors as well as upregulation of mRNAs of several DNA sensors (Znidar et al., 2016, 2018).

The FRRCBE received an unexpected bonus when Dr. Christian Zemlin took a position in the Department of Electrical and Computing Engineering. He was given an office and lab in the FRRCBE. This was an excellent fit for him, and it brought new skills and knowledge to the Center. Christian received his BS, MS, and PhD degrees in Berlin before coming to the SUNY Upstate Medical University, Syracuse, as a postdoctoral fellow. His knowledge in physics, mathematics, and biomedical imaging brought yet greater diversity to the FRRCBE. With his studies in cardiovascular imaging, he and his colleagues are investigating the possibility that nsPEFs may be used in the treatment of arrhythmias, especially atrial fibrillation, atrial flutter, and ventricular tachycardia (Xie et al., 2015). Since nsPEFs are low in energy, if possible, they could be used for cardiac defibrillation, where reduction in energy is needed. Nanosecond defibrillation uniformly depolarizes a tissue with an energy threshold nearly an order of magnitude lower than that required for defibrillation with standard conditions. There was no apparent damage due to electroporation, little or no change in duration of action potential or diastolic interval, and no apparent tissue damage. Thus, this technology could also be used for defibrillation with significant reduction in energy (Varghese et al., 2017).

One of the staples in Bioelectrics at ODU was Dr. Shu Xiao, associate professor of Electrical and Computing Engineering. Shu's early work focused on high-powered switching mechanisms (Xiao et al., 2003) before building pulse power devices for biological studies, which significantly enhanced studies of Pakhomov's group and mine. Shu's most recent interest is the effects of picosecond pulses on biological cells. In collaboration with Andrei Pakhomov's group, picosecond electric pulses (psEPs) have opened this time domain to a new bioelectric biological paradigm. Because the pulses are so short, in some experiments, many pulses must be delivered. However, in rat hippocampal neurons, even a single pulse (500 ps, 190 kV/cm) caused an increase in membrane conductance and depolarized cells. There were large inward currents but only at negative membrane potentials. This did not occur in CHO cells that lack VGCCs, did not occur in the absence of extracellular calcium, and was blocked by VGCC channel blockers. These findings indicated that psEP did not cause conventional lipid-phase electroporation or membrane depolarization by opening VG Na⁺ channels. The pulse duration being several orders of magnitude shorter than the channel opening time suggests either a nonconventional membrane electroporation where pore detection fails or nonconventional mechanism of channel opening, such as bypassing the shift of the voltage sensor in the VGCC (Semenov et al., 2015, 2016).

Shu's extended interest into the picosecond domain is because psEPs can be delivered noninvasively by antennae. With pulses of 100s of picoseconds, it is possible to noninvasively expose a $1\,cm^2$ spot on a tissue at a depth of about $2\,cm$ (Petrella et al., 2016). Under similar conditions, using a 3D bioprinter, they showed that neuronal stem cells (NSCs) and mesenchymal stem cells (MSCs) remained viable but exhibited increased metabolism. The picosecond pulses decreased proliferation in NSCs and increased expression of glial fibrillary acidic protein, indicative of astrocyte cell differentiation (Petrella et al., 2018).

Several new positions were added to the FRRCBE. In 2012, Dr. Michael Kong took the Battened Endowed Chair of Bioelectrics that was previously filled by Dr. Schoenbach. Kong's wife, Dr. Hai-Lan Chen, also joined the FRRCBE as a research associate professor. She too investigates cold plasmas for possible medical applications. Michael is perhaps the world's leading scientist in biomedical application of cold atmospheric plasmas. This added a new application of pulsed power to the Center in the emerging field of plasma medicine. Applications include decontamination of surgical instruments, disinfecting skin and living tissues, stimulating healing of chronic wounds, and suppression of tumor growth. For example, Michael's group demonstrated that cold plasma induces apoptosis in multiple myeloma and may provide an effective therapy (Xu et al., 2018). Cold plasma also increased the sensitivity of multiple myeloma to bortezomib and may be used in combination with plasma treatment to enhance current chemotherapy (Xu et al., 2016a). His group has shown that cold plasma might provide an efficient technique for the delivery of siRNA and miRNA in 2D and 3D culture models (Xu et al., 2016b).

While pulsed power technology was alive and flourishing on the East coast and before the formation of Norfolk's FRRCBE, pulsed power was also applied by Martin Gundersen's group at the University of Southern California (USC) in Los Angles. Martin and Karl had been friends since they had met at Texas Tech. Martin too had received support from Dr. Bob Barker at AFOSR. Two of his former students joined the FRRCBE. One was Dr. Chunqi Jiang, who joined the Center in July 2013. She was a student with Karl before going to the Gundersen lab as a postdoc. Her group's work enhanced the Center's strength in cold plasma or ionized gases for environmental and biomedical applications.

In January 2013, Dr. P. Thomas Vernier moved to the FRRCBE from the Gundersen lab. Tom was also one of the pioneers in using pulsed power on biological cells; he had already made a significant impact using nanosecond pulses before coming to Norfolk. In 2003, he and his collaborators in the Gundersen lab imaged real-time calcium bursts and PS externalization in Jurkat cells in response to nsPEFs (Vernier et al., 2003, 2004). Tom and his colleagues were also the first to show that nsPEFs did in fact have effects on the plasma membrane; this was an effect missed by the reagents that were used for membrane permeabilization. Using both *in vitro* and *in silico* strategies, electric field charging of the plasma membrane caused the formation of nanometer-diameter pores through which the negatively charged PS head group electrophoretic migrated (Vernier et al., 2006). Tom brought yet new perspectives to the FRRCBE with his expertise in engineering, modeling/simulation, and biology. For example, while at USC, he and Zack Levine resolved the evolution of electropores in heterogeneous lipid bilayers and characterized roles for Ca^{2+} and phosphatidylserine in pore creation and annihilation (Levine and Vernier, 2012). In addition, he and Dr. Gale Craviso, the University of Nevada, Reno, characterized effects of nanosecond pulses in bovine chromaffin cells, demonstrating that Ca^{2+} mobilization by a single 5 ns, 50 kV/cm EP required the opening of L-type VGCCs which was dependent on the tetrodotoxin-insensitive Na^+ uptake, possibly also due to nanoporation (Craviso et al., 2010).

Richard provided another addition to the faculty of the FRRCBE by recruiting Dr. Siqi Guo to the position of research assistant professor. Dr. Guo received his MD in Clinical Medicine from Zhejiang University Hangzhou, an MS in Oncology from the Academy of Military Medical Sciences in Beijing before coming to the U.S. He did postdoctoral work in immunology at the University of Alabama at Birmingham and Virginia Commonwealth University, Richmond, before coming to ODU. With his background in oncology and immunology, he was an excellent fit for studies on gene delivery in cancer and immune effects after nsPEF treatment. In collaboration with Richard and me, he established the mouse breast cancer model that led to immune responses after nsPEF treatment (Guo et al., 2018). Siqi and I continue to work together to better understand and enhance immune mechanisms induced by nsPEFs.

In 2013, Dr. John Catravas joined the FRRCBE from Medical College of Georgia, Georgia Health Sciences University, where he was a Regents Professor of Vascular Biology and Pharmacology and Toxicology. Appointed as Sentara Chair in Bioelectrics, Dr. Catravas is one of the world's leading scientists in vascular cell research. He also works closely with his spouse, Christiana Dimitropoulou, who is also an expert in airway and vascular function in pulmonary disease. While not specifically an expert in Bioelectrics, John's research brought expertise in the area of vascular endothelium, inflammation, and functions for heat shock proteins as well as leadership skills across pharmacological and physiological disciplines.

By 2015, a new Vice President for research, Dr. Morris Foster, was appointed by a new President, Dr. John Broderick. While there appeared to be support at the top, Dr. Foster's attitudes toward the FRRCBE were not as sanguine as those of Dr. Karim. Dr. Foster was more supportive of a departmental strategy and was less impressed with the FRRCBE. From 2008–2015 under Richard's directorship, the number of faculty members increased from 8 to 18, and the number of researchers (faculty, postdocs, students, and technicians) increased from 20 to over 70. Funding from external sources more than tripled during this time, with funds predominately coming from the U.S. federal government National Institutes of Health and Department of Defense (NIH and DOD). In addition, during this time, the Center moved into a new space and increased the space to ~30,000 ft². The FRRCBE's reputation around the Commonwealth, country, and world has grown tremendously, and the Center is recognized as the leading research institution in the area of research. The FRRCBE was recognized by Virginia Bio in 2019 by being selected to receive the Outstanding Contributions to Life Sciences award. In 2015, after the growth and establishment of the Center, Dr. Heller thought the Center was in excellent shape and it was time to step down. The hope was to recruit a new Director to enhance the Center further.

Dr. Catravas's leadership skills were tapped when Dr. Richard Heller stepped down as the FRRCBE's Director in 2015. Dr. Catravas took the position as Director and maintained and guided the Center for the next 2 years. Perhaps one of his strongest roles as Director was presenting the strengths of the FRRCBE as the strongest research component of ODU. The university had asked outside reviewers to evaluate the research environment at ODU and determine the best approaches for continued development of graduate programs. Comparing research in the FRRCBE with those of other departments, the strengths of the Center with its interdisciplinary environment, its strong contribution through funding, and its record of success even beyond the research environment were things the reviewers agreed should remain intact. Thanks to the strength of John's clear overview of the FRRCBE, there is now a tract or series of courses in the graduate program for Bioelectrics. John resigned from his position as the Director in 2017 when he too did not appreciate the lack of support and apparent efforts to undermine the value of the research conducted in the Center. Dr. Andrei Pakhomov took charge as interim Director until Dr. Gymama Slaughter took the Director's position in 2018.

It is particularly interesting and disappointing that this discord took place after the FRRCBE had developed intellectual property that formed the basis of a company, that is, taking ODU's pulsed power technology closer to the clinic. It's no secret that research costs money. However, the stature of a university is most often based on its research strengths, which repays the institution by enhanced student recruitment, alumni donations, and other ways that are not directly evident. However, in the case of the FRRCBE, the investment in research paid off directly. Nevertheless, the university somehow found ways to look in other directions, perhaps political and/or self-serving but certainly not logical.

And yet the Center continues to produce new intellectual property to better understand how nsPEFs affect cell structures and functions and to find new potentials for using nsPEFs, cold plasma, and gene delivery for medical applications. Perhaps one of the most interesting findings about pulsed power and nanosecond pulses was that they could induce an immune response against the cancer that was treated. This means that NPS essentially vaccinated animals against the treated cancer. This was first observed in ectopic liver tumors in mice (Beebe et al., 2011). Around 2010–2012, I was working with Xinhua (Sing) Chen treating melanoma and liver cancer. Sing was one of the students Frank Reidy and Karl recruited with scholarship money for President Runte. She got her PhD in Biology at ODU and was doing postdoctoral studies in my lab. Since she was from China, where liver cancer is a major problem, she worked on that disease and showed that after elimination, liver cancer did not recur (Chen et al., 2012). In six mice, after liver tumors were treated and eliminated with nanosecond pulses, the same live tumor cells were injected under the skin as before; however, these inoculations did not grow tumors, while untreated naïve animals grew tumors (Beebe et al., 2011). This strongly suggested an immune response must be present. In a later study supported by Ethicon Endo-Surgery, a division of Johnson & Johnson, examining liver cancer in rats, where the tumors were implanted in the liver (orthotopic), the same results were observed; rats that had tumor eliminated by nsPEFs did not grow tumors when implanted with the same live cells after treatment. This was called a vaccine-like effect (Chen et al., 2014).

In yet another study, the direct identification of innate and adaptive immune response was shown in the same rat model (Lassiter et al., 2018). After treatment, activated natural killer cells and effector memory and central memory T-cells were present in the tumor microenvironment (TME). Immunosuppressive cells were decreased, and dendritic cells were increased in the TME after treatment. Similar immune phenotypes were present in blood and spleen. After challenging rats with live cells, innate and adaptive immune responses were also observed, but they were different than the responses after treatment. Very similar results were observed in an orthotopic mouse breast model. Again, there was a vaccine-like effect that eliminated immunosuppressive phenotypes and activated adaptive immune phenotypes (Guo et al., 2018). These findings demonstrate that NPS induces an *in situ* vaccination against the treated cancer.

Other evidences for immune responses were also observed in Nuccitelli's group. For example, nsPEF treatment was superior to tumor excision at accelerating secondary tumor rejection in immune-competent mice, and there were CD^{4+} T-cells in treated as well as untreated tumors (Nuccitelli et al., 2012b). In another study with orthotopic liver tumors (Nuccitelli et al., 2015), the growth of secondary tumors was severely inhibited compared to tumor growth in CD^8-depleted rats. CD^{8+} T-cells were highly enriched in the secondary tumors exhibiting slow growth. This study also demonstrated that vaccinating mice with nsPEF-treated tumor cells inhibited growth of secondary tumors in a CD^{8+}-dependent manner. All these observations suggested immune responses.

It had been more than a decade since pulsed power and nanosecond pulse technology had remodeled our thinking about using electric fields in biology and medicine. In 2012, a company called Theliopulse was formed by the Alfred Mann Institute (AMI) based on pulsed power developed by Drs. Martin Gundersen and P. Thomas Vernier and the USC. Martin and Karl Schoenbach were both working with pulsed power at Texas Tech before going to the west and east coasts, respectively. AMI had licensed ODU/EVMS patents and intended to commercialize it for skin treatments. They did clinical trials showing that pulses as short as 20 ns could eliminate warts with high repetition rate.

In 2014, after nearly 2 years of negotiations, MDB Capital Group pulled together patents from BioElectroMed, Theliopulse, and ODU/EVMS and started a company called Electroblate, later changed to Pulse Biosciences, Inc. Pulse is now a publicly traded company in California (ticker symbol, PLSE). Several patents were obtained from Theliopulse and BioElectroMed while 21 patents came from 19 ODU researchers. This was the highest commercialization success in ODU's history, generating more than $41 million. Based on their business model, MDB recognized that the nanosecond pulse technology had a potential to have an impact on a large commercial market and provide a meaningful benefit to humanity through cancer treatment, among other technologies. Pulse Biosciences, Inc. is now a medical therapy company developing this technology for oncology, dermatology and esthetics, minimally invasive treatments, and veterinary applications. The technology is referred to now as Nano-Pulse Stimulation™ (NPS™) or nanopulse electrostimulation (NPES). They are looking at commercializing their proprietary CellFX™ System utilizing NPS™ technology. Based on the work accomplished at ODU/EVMS and BioElectroMed, treatment of solid tumors is one of the more promising applications of this technology. As already indicated, in preclinical studies, NPS effectively eliminates tumors and initiates an immune-mediated vaccine effect by inducing immunogenic-regulated cell death in treated cells.

In the context of our present knowledge, there are several footnotes in the data Bioelectrics hypotheses that can be reiterated. The general hypothesis that shorter pulses with shorter rise–fall times have greater probability for intracellular effects continues to stand the test of time. While we have generally considered this concept in the time domain, it may be more practical to give it serious consideration in the frequency domain. Considering nanosecond pulse effects *in vitro* works in both the time domain of about 75 ns and the corresponding frequency domain. However, the membrane charging time constant is much longer *in vivo*, perhaps ~1 μs. Yet the unique effects of nsPEFs *in vivo* still occur at these shorter durations and rise–fall times, where the frequency domain, but not the time domain, is still applicable. It remains to be determined if this hypothesis is correct.

Two reconsiderations of these hypotheses appear at the plasma membrane. The models that predicted the occurrence of these intracellular effects in the absence of plasma membrane effects did not sufficiently consider sizes of ions or molecules that could be transported across plasma membranes. Many of the early studies proposed unique intracellular effects without plasma membrane effects; however, they showed plasma membranes were permeabilized at higher electric field intensities. Yet caspase activation and cytochrome *c* release could be seen before cells are permeabilized using ethidium homodimer or PI. As indicated earlier, using two different strategies, it was revealed by Dr. P. Thomas Vernier et al. (2006) and by Dr. Andrei Pakhomov et al. (2007), both of whom are now research professors in the FRRCBE, that membrane pores are actually formed, but they are too small to allow classical permeabilization markers cross the cell membrane; the markers were larger than the "nanopores" formed by such short pulses (Pakhomov et al., 2007).

The other caveat at the plasma membrane was the use of annexin-V binding to PS as a marker for apoptosis. It turned out that the pulses themselves could pull the PS molecules through nanopores formed in plasma membranes. Thus, using PS externalization measured by annexin-V can occur in the absence of apoptosis. We actually demonstrated this by showing annexin-V binding occurred immediately after pulses, but not 5 and 10 min after pulses, but didn't draw the conclusion that this occurs through nanopores (Beebe et al., 2003). This was conclusively demonstrated by P. Thomas Vernier in collaboration with Dr. Martin Gundersen at the USC.

The FRRCBE had several valuable characteristics that gave strength to its successes. First and foremost, it had support from individuals with foresight and boldness. These individuals include Mr. Frank Reidy and Dr. Bob Barker; Dr. Mohammed Karim and President Roseann Runte at ODU; and Dr. Don Combs and Mr. David Theil at EVMS. Importantly, the FRRCBE is capitalized with an interdisciplinary group of scientists that are housed in the same building, which promotes collegial and intellectual interactions. The atmosphere is not only interdisciplinary but is strengthened by global contributions. Scientists in the Center are African American,

American, Chinese, German, Italian, and Russian. Scientists from around the world have visited the FRRCBE, and scientists from the FRRCBE have visited and worked abroad. Also, the love and appreciation of bioelectricity runs in the family. There are four "pairs" of scientists that built their careers as scientific duos researching on Bioelectrics – Richard and Loree Heller, Andrei Pakhomov and Olga Pakhomova, Michael Kong and Hai-Lan Chen, and John Catravas and Christiana Dimitropoulou. Each of these eight scientists have their own expertise and individual contributions, yet working together synergizes their efforts and makes their contributions exponentially stronger. It turns out that these recruitments have brought greater depth and breadth to the FRRCBE.

There is an intangible aspect to research in the FRRCBE that infiltrates the experimental environment. Generally, there are collaborations in the plain sight of competition. Almost everyone in the FRRCBE is interested in how nanosecond pulses affect cells, so it is reasonable that there would be overlaps in investigation strategies. For example, given that it was predicted that nanosecond pulses could have intracellular effects, several of us through the years have investigated calcium mobilization including Steve Buescher (Buescher et al., 2004); Jody White, Peter Blackmore, and me (White et al., 2004); Iurii Semenov and Andrei Pakhomov (Semenov et al., 2013a,b); Thomas (Vernier et al., 2003); and Thomas with Dr. Gail Craviso from Reno (Vernier et al., 2008; Craviso et al., 2010, 2012). Each of these works was done often knowing of the others' studies and with an attitude of healthy competition. In fact, the work on calcium mobilization improved with time such that the preceding studies provided a foundation for improvements in studies that followed. The same could be said about the regulation of cell death. Studies from my group (Beebe et al., 2002, 2003; Ren et al., 2012) and studies from Olga's group (Pakhomova et al., 2013) clarified that nanosecond pulses could induce apoptosis and that nsPEF-induced regulated cell death was dependent on cell type. This was supported by other studies outside of the Center (Morotomi-Yano et al., 2014).

As might be expected, there are presently multiple investigations concerning immune mechanisms among all the other commonalities within the Center. Since each of us is generally aware of the other's work, there is no encroachment. This FRRCBE professional comradery has developed over the years and is essentially taken for granted. It involves respecting our colleagues and our common goals. I have been to other institutions were work in a common area has caused competition problems between laboratory groups. Perhaps an important and lasting aspect of a research center is based on respect for the scientific method and respect for your colleagues that seek a common goal.

In August 2018, Dr. Gymama Slaughter became the Director of the FRRCBE. She came from the University of Maryland where she was the director of the Bioelectronics Laboratory, where her successes have been in the area of medical devices and biological sensors. In addition to her research expertise, she has placed emphasis on education and community outreach and engagement for the Center. While bioelectrics remains a major focus of the FRRCBE, the Center is now expanding to include bioelectronics, where biology and electronics converge to develop hardware as sensors and actuators (Walker et al., 2009).

REFERENCES

Basu G, Downey H, Guo S, Israel A, Asmar A, Hargrave B, Heller R. Prevention of distal flap necrosis in a rat random skin flap model by gene electro transfer delivering VEGF(165) plasmid. *J Gene Med*. 2014;16(3–4):55–65.

Beebe SJ, Reimann EM, Schlender KK. Purification and characterization of a cAMP- and Ca^{2+}-calmodulin-independent glycogen synthase kinase from porcine renal cortex. *J Biol Chem*. 1984;259:1415–1422.

Beebe SJ, Redmon JB, Blackmore PF, Corbin JD. Discriminative insulin antagonism of stimulatory effects of various cAMP analogs on adipocyte lipolysis and hepatocyte glycogenolysis. *J Biol Chem*. 1985;260:15781–15788.

Beebe SJ, Oyen O, Sandberg M, Frøysa A, Hansson V, Jahnsen T. Molecular cloning of a tissue-specific protein kinase (C gamma) from human testis--representing a third isoform for the catalytic subunit of cAMP-dependent protein kinase. *Mol Endocrinol*. 1990;4:465–475.

Beebe SJ, Fox PM, Rec LH, Buescher ES, Somers K, Schoenbach KH. Nanosecond pulsed electric field (nsPEF) effects on cells and tissues: Apoptosis induction and tumor growth inhibition. *IEEE Trans Plasma Sci*. 2002;30:286–292.

Beebe SJ, Fox PM, Rec LJ, Willis LK, Schoenbach KH. Nanosecond, high intensity pulsed electric fields induce apoptosis in human cells. *FASEB J*. 2003;17:1493–1495.

Beebe SJ, Ford WE, Ren W, Chen X. Pulse power ablation of melanoma with nanosecond pulsed electric fields. In: R. Morton (Ed.) *Treatment of Metastatic Melanoma*, In Tech Croatia, 2011, pp. 231–268.

Beebe SJ, Chen YJ, Sain NM, Schoenbach KH, Xiao S. Transient features in nanosecond pulsed electric fields differentially modulate mitochondria and viability. *PLoS One*. 2012;7:e51349.

Beebe SJ, Lassiter BP, Guo S. Nanopulse stimulation (NPS) induces tumor ablation and immunity in orthotopic 4T1 mouse breast cancer: A review. *Cancers*. 2018;10(4):E97.

Blackmore PF, Beebe SJ, Danforth DR, Alexander N. Progesterone and 17 alpha-hydroxyprogesterone. Novel stimulators of calcium influx in human sperm. *J Biol Chem*. 1990;265:1376–1380.

Bowman AM, Nesin OM, Pakhomova ON, Pakhomov AG. Analysis of plasma membrane integrity by fluorescent detection of Tl(+) uptake. *J Membr Biol*. 2010;236:15–26.

Buescher ES, Smith RR, Schoenbach KH. Submicrosecond intense pulsed electric field effects on intracellular free calcium: Mechanisms and effects. *IEEE Trans Plasma Sci*. 2004;32:1563–1572.

Casciola M, Xiao S, Apollonio F, Paffi A, Liberti M, Muratori C, Pakhomov AG. Cancellation of nerve excitation by the reversal of nanosecond stimulus polarity and its relevance to the gating time of sodium channels. *Cell Mol Life Sci*. 2019. doi: 10.1007/s00018-019-03126-0.

Chen R, Sain NM, Harlow KT, Chen YJ, Shires PK, Heller R, Beebe SJ. A protective effect after clearance of orthotopic rat hepatocellular carcinoma by nanosecond pulsed electric fields. *Eur J Cancer*. 2014;50:2705–2713.

Chen X, Kolb JF, Swanson RJ, Schoenbach KH, Beebe SJ. Apoptosis initiation and angiogenesis inhibition: Melanoma targets for nanosecond pulsed electric fields. *Pigment Cell Melanoma Res*. 2010;23:554–563.

Chen X, Zhuang J, Kolb JF, Schoenbach KH, Beebe SJ. Long term survival of mice with hepatocellular carcinoma after pulse power ablation with nanosecond pulsed electric fields. *Technol Cancer Res Treat*. 2012;11:83–93.

Craviso GL, Choe S, Chatterjee P, Chatterjee I, Vernier PT. Nanosecond electric pulses: A novel stimulus for triggering Ca^{2+} influx into chromaffin cells via voltage-gated Ca^{2+} channels. *Cell Mol Neurobiol*. 2010;30:1259–1265.

Craviso GL, Choe S, Chatterjee I, Vernier PT. Modulation of intracellular Ca^{2+} levels in chromaffin cells by nanoelectropulses. *Bioelectrochemistry.* 2012;87:244–252.

Daud AI, DeConti RC, Andrews S, Urbas P, Riker AI, Sondak VK, Munster PN, Sullivan DM, Ugen KE, Messina JL, Heller R. Phase I trial of interleukin-12 plasmid electroporation in patients with metastatic melanoma. *J Clin Oncol.* 2008;26:5896–5903.

Ferraro B, Cruz YL, Coppola D, Heller R. Intradermal delivery of plasmid VEGF(165) by electroporation promotes wound healing. *Mol Ther.* 2009;17:651–657.

Ferraro B, Cruz YL, Baldwin M, Coppola D, Heller R. Increased perfusion and angiogenesis in a hindlimb ischemia model with plasmid FGF-2 delivered by noninvasive electroporation. *Gene Ther.* 2010;17:763–769.

Ford WE, Ren W, Blackmore PF, Schoenbach KH, Beebe SJ. Nanosecond pulsed electric fields stimulate apoptosis without release of pro-apoptotic factors from mitochondria in B16f10 melanoma. *Arch Biochem Biophys.* 2010;497:82–89.

Gardlík R, Pálffy R, Hodosy J, Lukács J, Turna J, Celec P. Vectors and delivery systems in gene therapy. *Med Sci Monit.* 2005;11:RA110–21.

Glass LF, Fenske NA, Jaroszeski M, Perrott R, Harvey DT, Reintgen DS, Heller R. Bleomycin-mediated electrochemotherapy of basal cell carcinoma. *J Am Acad Dermatol.* 1996a;34:82–86.

Glass LF, Pepine ML, Fenske NA, Jaroszeski M, Reintgen DS, Heller R. Bleomycin-mediated electrochemotherapy of metastatic melanoma. *Arch Dermatol.* 1996b;132:1353–1357.

Green DR, Amarante-Mendes GP. The point of no return: Mitochondria, caspases, and the commitment to cell death. *Results Probl Cell Differ.* 1998;24:45–61.

Guo S, Jackson DL, Burcus NI, Chen YJ, Xiao S, Heller R. Gene electrotransfer enhanced by nanosecond pulsed electric fields. *Mol Ther Methods Clin Dev.* 2014;1:14043.

Guo S, Jing Y, Burcus NI, Lassiter BP, Tanaz R, Heller R, Beebe SJ. Nano-pulse stimulation induces potent immune responses, eradicating local breast cancer while reducing distant metastases. *Int J Cancer.* 2018;142:629–640.

Hanahan D, Weinberg RA. The hallmarks of cancer. *Cell.* 2000;100:57–70.

Hanahan D, Weinberg RA Hallmarks of cancer: The next generation. *Cell.* 2011;144:646–674.

Hargrave B, Downey H, Strange R Jr, Murray L, Cinnamond C, Lundberg C, Israel A, Chen YJ, Marshall W Jr, Heller R. Electroporation-mediated gene transfer directly to the swine heart. *Gene Ther.* 2013;20:151–157.

Hargrave B, Strange R Jr, Navare S, Stratton M, Burcus N, Murray L, Lundberg C, Bulysheva A, Li F, Heller R. Gene electro transfer of plasmid encoding vascular endothelial growth factor for enhanced expression and perfusion in the ischemic swine heart. *PLoS One.* 2014;9:e115235.

Heller L, Pottinger C, Jaroszeski MJ, Gilbert R, Heller R. In vivo electroporation of plasmids encoding GM-CSF or interleukin-2 into existing B16 melanomas combined with electrochemotherapy induces long-term antitumour immunity. *Melanoma Res.* 2000;10:577–583.

Heller L, Merkler K, Westover J, Cruz Y, Coppola D, Benson K, Daud A, Heller R. Evaluation of toxicity following electrically mediated interleukin-12 gene delivery in a B16 mouse melanoma model. *Clin Cancer Res.* 2006;12:3177–3183.

Heller L, Todorovic V, Cemazar M. Electrotransfer of single-stranded or double-stranded DNA induces complete regression of palpable B16.F10 mouse melanomas. *Cancer Gene Ther.* 2013;20:695–700.

Heller R, Jaroszeski MJ, Glass LF, Messina JL, Rapaport DP, DeConti RC, Fenske NA, Gilbert RA, Mir LM, Reintgen DS. Phase I/II trial for the treatment of cutaneous and subcutaneous tumors using electrochemotherapy. *Cancer.* 1996a;77:964–971.

Heller R, Jaroszeski M, Atkin A, Moradpour D, Gilbert R, Wands J, Nicolau C. In vivo gene electroinjection and expression in rat liver. *FEBS Lett.* 1996b;389:225–228.

Heller R, Schultz J, Lucas ML, Jaroszeski MJ, Heller LC, Gilbert RA, Moelling K, Nicolau C. Intradermal delivery of interleukin-12 plasmid DNA by in vivo electroporation. *DNA Cell Biol.* 2001;20:21–26.

Hristov K, Mangalanathan U, Casciola M, Pakhomova ON, Pakhomov AG. Expression of voltage-gated calcium channels augments cell susceptibility to membrane disruption by nanosecond pulsed electric field. *Biochim Biophys Acta Biomembr.* 2018;1860:2175–2183.

Ibey BL, Ullery JC, Pakhomova ON, Roth CC, Semenov I, Beier HT, Tarango M, Xiao S, Schoenbach KH, Pakhomov AG. Bipolar nanosecond electric pulses are less efficient at electropermeabilization and killing cells than monopolar pulses. *Biochem Biophys Res Commun.* 2014;443:568–573.

Katare DP, Aeri V. Progress in gene therapy: A Review. *IJTPR.* 2010;1:33–41.

Kerr JFR, Wyllie AH, Currie AR. Apoptosis: A basic biological phenomenon with wide-ranging implications in tissue kinetics. *Br J Cancer.* 1972;26:239–257.

Kroemer G, El-Deiry WS, Golstein P, Peter ME, Vaux D, Vandenabeele P, Zhivotovsky B, Blagosklonny MV, Malorni W, Knight RA, Piacentini M, Nagata S, Melino G. Nomenclature committee on cell death. Classification of cell death: Recommendations of the nomenclature committee on cell death. *Cell Death Differ.* 2005;2:1463–1467.

Lassiter BP, Guo S, Beebe SJ. Nano-pulse stimulation ablates orthotopic rat hepatocellular carcinoma and induces innate and adaptive memory immune mechanisms that prevent recurrence. *Cancers.* 2018;10(3):E69.

Levine ZA, Vernier PT. Calcium and phosphatidylserine inhibit lipid electropore formation and reduce pore lifetime. *J Membr Biol.* 2012;245:599–610.

Lucas ML, Heller L, Coppola D, Heller R. IL-12 plasmid delivery by in vivo electroporation for the successful treatment of established subcutaneous B16.F10 melanoma. *Mol Ther.* 2002;5:668–675.

Lucas ML, Heller R. IL-12 gene therapy using an electrically mediated nonviral approach reduces metastatic growth of melanoma. *DNA Cell Biol.* 2003;22:755–763.

Marshall WG Jr, Boone BA, Burgos JD, Gografe SI, Baldwin MK, Danielson ML, Larson MJ, Caretto DR, Cruz Y, Ferraro B, Heller LC, Ugen KE, Jaroszeski MJ, Heller R. Electroporation-mediated delivery of a naked DNA plasmid expressing VEGF to the porcine heart enhances protein expression. *Gene Ther.* 2010;17:419–423.

Merla C, Pakhomov AG, Semenov I, Vernier PT. Frequency spectrum of induced trans-membrane potential and permeabilization efficacy of bipolar electric pulses. *Biochim Biophys Acta Biomembr.* 2017;1859:1282–1290.

Morotomi-Yano K, Akiyama H, Yano K. Different involvement of extracellular calcium in two modes of cell death induced by nanosecond pulsed electric fields. *Arch Biochem Biophys.* 2014;555–556:47–54.

Muratori C, Pakhomov AG, Xiao S, Pakhomova ON. Electrosensitization assists cell ablation by nanosecond pulsed electric field in 3D cultures. *Sci Rep.* 2016;6:23225.

Muratori C, Pakhomov AG, Heller L, Casciola M, Gianulis E, Grigoryev S, Xiao S, Pakhomova ON. Electrosensitization increases antitumor effectiveness of nanosecond pulsed electric fields in vivo. *Technol Cancer Res Treat.* 2017;1:1533034617712397.

Nuccitelli R. Nano-Pulse Stimulation for the treatment of skin lesions. *Bioelectricity* 2019;1:235–239.

Nuccitelli R, Pliquett U, Chen X, Ford W, James Swanson R, Beebe SJ, Kolb JF, Schoenbach KH. Nanosecond pulsed electric fields cause melanomas to self-destruct. *Biochem Biophys Res Commun.* 2006;343:351–360.

Nuccitelli R, Chen X, Pakhomov AG, Baldwin WH, Sheikh S, Pomicter JL, Ren W, Osgood C, Swanson RJ, Kolb JF, Beebe SJ, Schoenbach KH. A new pulsed electric field therapy for melanoma disrupts the tumor's blood supply and causes complete remission without recurrence. *Int J Cancer.* 2009;125:438–445.

Nuccitelli R, Tran K, Sheikh S, Athos B, Kreis M, Nuccitelli P. Optimized nanosecond pulsed electric field therapy can cause murine malignant melanomas to self-destruct with a single treatment. *Int J Cancer.* 2010;127:1727–1736.

Nuccitelli R, Sheikh S, Tran K, Athos B, Kreis M, Nuccitelli P, Chang KS, Epstein EH Jr, Tang JY. Nanoelectroablation therapy for murine basal cell carcinoma. *Biochem Biophys Res Commun.* 2012a;424:446–450. Erratum in: *Biochem Biophys Res Commun.* 2016;480:288.

Nuccitelli R, Tran K, Lui K, Huynh J, Athos B, Kreis M, Nuccitelli P, De Fabo EC. Non-thermal nanoelectroablation of UV-induced murine melanomas stimulates an immune response. *Pigment Cell Melanoma Res.* 2012b;25:618–629.

Nuccitelli R, Huynh J, Lui K, Wood R, Kreis M, Athos B, Nuccitelli P. Nanoelectroablation of human pancreatic carcinoma in a murine xenograft model without recurrence. *Int J Cancer.* 2013;132:1933–1939.

Nuccitelli R, Wood R, Kreis M, Athos B, Huynh J, Lui K, Nuccitelli P, Epstein EH Jr. First-in-human trial of nanoelectroablation therapy for basal cell carcinoma: Proof of method. *Exp Dermatol.* 2014;23:135–137.

Nuccitelli R, Berridge JC, Mallon Z, Kreis M, Athos B, Nuccitelli P. Nanoelectroablation of murine tumors triggers a CD8-dependent inhibition of secondary tumor growth. *PLoS One.* 2015;10:e0134364.

Nesin V, Bowman AM, Xiao S, Pakhomov AG. Cell permeabilization and inhibition of voltage-gated Ca(2+) and Na(+) channel currents by nanosecond pulsed electric field. *Bioelectromagnetics.* 2012;33:394–404.

Nesin V, Pakhomov AG. Inhibition of voltage-gated Na(+) current by nanosecond pulsed electric field (nsPEF) is not mediated by Na(+) influx or Ca(2+) signaling. *Bioelectromagnetics.* 2012;33:443–451.

Pakhomov AG, Akyel Y, Pakhomova ON, Stuck BE, Murphy MR. Current state and implications of research on biological effects of millimeter waves: A review of the literature. *Bioelectromagnetics.* 1998;19:393–413.

Pakhomov AG, Mathur SP, Doyle J, Stuck BE, Kiel JL, Murphy MR. Comparative effects of extremely high power microwave pulses and a brief CW irradiation on pacemaker function in isolated frog heart slices. *Bioelectromagnetics.* 2000;21:245–254.

Pakhomov AG, Kolb JF, White JA, Joshi RP, Xiao S, Schoenbach KH. Long-lasting plasma membrane permeabilization in mammalian cells by nanosecond pulsed electric field (nsPEF). *Bioelectromagnetics.* 2007;28:655–663.

Pakhomov AG, Bowman AM, Ibey BL, Andre FM, Pakhomova ON, Schoenbach KH. Lipid nanopores can form a stable, ion channel-like conduction pathway in cell membrane. *Biochem Biophys Res Commun.* 2009;385:181–186.

Pakhomov AG, Grigoryev S, Semenov I, Casciola M, Jiang C, Xiao S. The second phase of bipolar, nanosecond-range electric pulses determines the electroporation efficiency. *Bioelectrochemistry.* 2018;122:123–133.

Pakhomova ON, Gregory BW, Khorokhorina VA, Bowman AM, Xiao S, Pakhomov AG. Electroporation-induced electrosensitization. *PLoS One.* 2011;6:e17100.

Pakhomova ON, Gregory BW, Semenov I, Pakhomov AG. Two modes of cell death caused by exposure to nanosecond pulsed electric field. *PLoS One.* 2013;8:e70278.

Pakhomova ON, Gregory B, Semenov I, Pakhomov AG. Calcium-mediated pore expansion and cell death following nanoelectroporation. *Biochim Biophys Acta.* 2014;1838:2547–2554.

Parvathenani LK, Buescher ES, Chacon-Cruz E, Beebe SJ. Type I cAMP-dependent protein kinase delays apoptosis in human neutrophils at a site upstream of caspase-3. *J Biol Chem.* 1998;273:6736–6743.

Petrella RA, Schoenbach KH, Xiao S. A dielectric rod antenna for picosecond pulse stimulation of neurological tissue. *IEEE Trans Plasma Sci IEEE Nucl Plasma Sci Soc.* 2016;44:708–714.

Petrella RA, Mollica PA, Zamponi M, Reid JA, Xiao S, Bruno RD, Sachs PC. 3D bioprinter applied picosecond pulsed electric fields for targeted manipulation of proliferation and lineage specific gene expression in nural stem cells. *J Neural Eng.* 2018;15:056021.

Ren W, Sain NM, Beebe SJ. Nanosecond pulsed electric fields (nsPEFs) activate intrinsic caspase-dependent and caspase-independent cell death in Jurkat cells. *Biochem Biophys Res Commun.* 2012;421:808–812.

Rossi A, Pakhomova ON, Pakhomov AG, Weygandt S, Bulysheva AA, Murray LE, Mollica PA, Muratori C. Mechanisms and immunogenicity of nsPEF-induced cell death in B16F10 melanoma tumors. *Sci Rep.* 2019;9:431.

Semenov I, Xiao S, Pakhomova ON, Pakhomov AG. Recruitment of the intracellular Ca^{2+} by ultrashort electric stimuli: The impact of pulse duration. *Cell Calcium.* 2013a;54:145–150.

Semenov I, Xiao S, Pakhomov AG. Primary pathways of intracellular Ca(2+) mobilization by nanosecond pulsed electric field. *Biochim Biophys Acta.* 2013b;1828:981–989.

Semenov I, Xiao S, Kang D, Schoenbach KH, Pakhomov AG. Cell stimulation and calcium mobilization by picosecond electric pulses. *Bioelectrochemistry.* 2015;105:65–71.

Semenov I, Xiao S, Pakhomov AG. Electroporation by subnanosecond pulses. *Biochem Biophys Rep.* 2016;6:253–259.

Schoenbach KH, Kristiansen M, Schaefer G. A review of opening switch technology for inductive energy storage. *Proc IEEE.* 1984:72:1019–1040.

Schoenbach KH, Peterkin FE, Alden RW, Beebe SJ. The effect of pulsed electric fields on biological cells: Experiments and applications. *IEEE Trans Plasma Sci* 1997;25:284–292.

Schoenbach KH, Joshi RP, Stark RH, Dobbs F, Beebe SJ, Bacterial decontamination of liquids with pulsed electric fields, invited review paper. *IEEE Trans Dielectr Electr Insul* 2000;7:637.

Schoenbach KH, Beebe SJ, Buescher ES. Intracellular effect of ultrashort electrical pulses. *Bioelectromagn Sept* 2001;22(6):440–448.

Schoenbach KH, Katsuki S, Stark RH, Buescher ES, Beebe SJ. Bioelectrics: New applications for pulsed power technology. *IEEE Trans Plasma Sci.* 2002;30:293–300.

Schoenbach KH, Joshi RP, Beebe SJ, Baum CE. A scaling law for membrane permeabilization with nanopulses. *IEEE Trans Dielectr Electr Insul.* 2009;16:1224–1235.

Schoenbach KH, Pakhomov AG, Semenov I, Xiao S, Pakhomova ON, Ibey BL. Ion transport into cells exposed to monopolar and bipolar nanosecond pulses. *Bioelectrochemistry.* 2015;103:44–51.

Schwan HP. Dielectric properties of cells and tissues. In: A Chiabrera, C Nicolini, HP Schwan (Eds). *Interactions between Electromagnetic Fields and Cells*, New York and London: Pergamon, 1985, pp. 75–97.

Stacey M, Stickley J, Fox P, Statler V, Schoenbach K, Beebe SJ, Buescher S. Differential effects in cells exposed to ultra-short, high intensity electric fields: cell survival, DNA damage, and cell cycle analysis. *Mutat Res.* 2003;542:65–75.

Tolstykh GP, Beier HT, Roth CC, Thompson GL, Payne JA, Kuipers MA, Ibey BL. Activation of intracellular phosphoinositide signaling after a single 600 nanosecond electric pulse. *Bioelectrochemistry.* 2013;94:23–29.

Varghese F, Neuber JU, Xie F, Philpott JM, Pakhomov AG, Zemlin CW. Low-energy defibrillation with nanosecond electric shocks. *Cardiovasc Res.* 2017;113:1789–1797.

Vernier PT, Sun Y, Marcu L, Salemi S, Craft CM, Gundersen MA. Calcium bursts induced by nanosecond electric pulses. *Biochem Biophys Res Commun.* 2003;310:286–295.

Vernier PT, Sun Y, Marcu L, Craft CM, Gundersen MA. Nanoelectropulse-induced phosphatidylserine translocation. *Biophys J.* 2004;86:4040–4048.

Vernier PT, Ziegler MJ, Sun Y, Gundersen MA, Tieleman DP. Nanopore-facilitated, voltage-driven phosphatidylserine translocation in lipid bilayers--in cells and in silico. *Phys Biol.* 2006;3:233–247.

Vernier PT, Sun Y, Chen MT, Gundersen MA, Craviso GL. Nanosecond electric pulse-induced calcium entry into chromaffin cells. *Bioelectrochemistry.* 2008;73:1–4.

Walker GM, Ramsey JM, Cavin III RK, Herr DJC, Merzbacher CI, Zhirnov V. A framework for bioelectronics: discovery and innovation (PDF). National Institute of Standards and Technology. February 2009. p. 42.

White JA, Blackmore PF, Schoenbach KH, Beebe SJ. Stimulation of capacitative calcium entry in HL-60 cells by nanosecond pulsed electric fields. *J Biol Chem.* 2004;279:22964–22972.

Xiao S, Kolb J, Kono S, Katsuki S, Joshi RP, Laroussi M, Schoenbach KH. High power, high recovery rate water switch. *Digest of Technical Papers. PPC-2003. 14th IEEE International Pulsed Power Conference*, Dallas, TX, 2003;1:649–652.

Xie F, Varghese F, Pakhomov AG, Semenov I, Xiao S, Philpott J, Zemlin C. Ablation of myocardial tissue with nanosecond pulsed electric fields. *PLoS One.* 2015;10:e0144833.

Xu D, Luo X, Xu Y, Cui Q, Yang Y, Liu D, Chen H, Kong MG. The effects of cold atmospheric plasma on cell adhesion, differentiation, migration, apoptosis and drug sensitivity of multiple myeloma. *Biochem Biophys Res Commun.* 2016b;473:1125–1132.

Xu D, Wang B, Xu Y, Chen Z, Cui Q, Yang Y, Chen H, Kong MG. Intracellular ROS mediates gas plasma-facilitated cellular transfection in 2D and 3D cultures. *Sci Rep.* 2016a;6:27872.

Xu D, Xu Y, Cui Q, Liu D, Liu Z, Wang X, Yang Y, Feng M, Liang R, Chen H, Ye K, Kong MG. Cold atmospheric plasma as a potential tool for multiple myeloma treatment. *Oncotarget.* 2018;9:18002–18017.

Znidar K, Bosnjak M, Cemazar M, Heller LC. Cytosolic DNA sensor upregulation accompanies DNA electrotransfer in B16.F10 melanoma cells. *Mol Ther Nucleic Acids.* 2016;5:e322.

Znidar K, Bosnjak M, Semenova N, Pakhomova O, Heller L, Cemazar M. Tumor cell death after electrotransfer of plasmid DNA is associated with cytosolic DNA sensor upregulation. *Oncotarget.* 2018;9:18665–18681.

Printed and bound by CPI Group (UK) Ltd, Croydon, CR0 4YY

17/10/2024

01775681-0007